AN INTRODUCTION TO ORGANIC CHEMISTRY
Aliphatic and Alicyclic Compounds

AN INTRODUCTION TO ORGANIC CHEMISTRY

Aliphatic and Alicyclic Compounds

E. D. Morgan
and
Sir Robert Robinson, O.M.

HUTCHINSON OF LONDON

Hutchinson & Co (Publishers) Ltd
3 Fitzroy Square, London W1

London Melbourne Sydney Auckland
Wellington Johannesburg Cape Town
and agencies throughout the world

First published 1975
© E. D. Morgan and
Sir Robert Robinson 1975
© Illustrations: Hutchinson & Co
(Publishers) Ltd 1975

Printed in Great Britain by
The Camelot Press Ltd, Southampton
and bound by Wm Brendon & Son Ltd
of Tiptree, Essex

ISBN 0 09 112840 4 cased
        0 09 112841 2 paperback

Contents

Authors' Preface

In this volume two related groups of compounds are considered. These are the open chain hydrocarbons and certain of their mono-substituted derivatives, that is to say, open chain compounds with a single functional group. Thus, we find the opportunity to study, in the simplest possible cases, the characteristic properties conferred on organic compounds by some of the more common substituents. The material has been selected so that general principles may be illustrated. It is not intended as a work of reference, but as a text which the student should read from cover to cover.

It is believed that a study of this material will serve to illustrate some of the basic principles and methods of organic chemistry. Incidentally, the matter discussed is of considerable industrial interest. The characters conferred by single substituents are, to a large extent, noted also in substances whose molecules contain more than one such atom or group, designated as poly-functional. We hope to develop this aspect, and variations from the normal, in a further volume.

E.D.M. would like to acknowledge the help of colleagues who have read some or all of the manuscript, and particularly Professor I. T. Millar and Dr G. Jones for their helpful criticisms.

E.D.M.   R.R.

Introduction

Organic chemistry is the study of carbon compounds. There are already roughly two and a half million such compounds known and they are being discovered or created at the rate of several hundred thousand new compounds a year. This should not dismay the new student, for he will never be expected to meet more than a fraction of them, but it gives an indication of the tremendous variety of combinations possible for this simple element and he will learn in due course how these compounds fall into systematic groups and how these groups differ in their physical, chemical and biological properties.

Life, at least as we know it on this earth, is conducted through the reactions of organic compounds. Early chemists, playing the game of animal, vegetable and mineral realized that there was a difference between *mineral* and *animal* and *vegetable* chemistry.

BERZELIUS in 1808 introduced the term 'Organic' chemistry. Organic compounds were those made by living organisms. Indeed, chemists at that time thought that some unknown 'vital force' was necessary to make them, and that the crude forces available in the laboratory were adequate only to make the simpler mineral or inorganic compounds. In 1828, one of BERZELIUS' pupils, WÖHLER, made ammonium cyanate and heated it, and it was converted to urea, which had previously been isolated from living materials and was considered to be available only via the vital force. One by one, more organic compounds were made in the laboratory, until the vital force theory had to be abandoned. As knowledge of the organic substances accumulated, it was realized that they were all compounds of carbon and hydrogen, with lesser amounts of oxygen, nitrogen, sulphur and phosphorus, and occasionally other elements.

By 1831, the method of combustion of organic compounds had been perfected by LIEBIG, to give an accurate value of the proportion of carbon and hydrogen present, and this method remains essentially unchanged to this day. Chemists had by this time obtained many pure compounds of carbon with widely differing properties and appearance. For example the formula for indigo, $C_{16}H_{10}N_2O_2$, gave no indication why it should be a dye, or why morphine $C_{17}H_{19}NO_3$ should be a colourless pain-killing drug. At this stage there was no clear idea how the atoms in molecules were grouped in space. An enormous step forward came in 1857 when Archibald COUPER of Scotland and August KEKULÉ of Germany both proposed the constant tetravalency of carbon. COUPER went further and represented for the first time a unit of valency as a line joining two atoms in a formula.

It still was not known whether all the bonds of carbon were equal or whether some were stronger or different from others, or how they were arranged in space. The equivalence of the bonds and their tetrahedral arrangement in space was put forward in the theory due to VAN'T HOFF in 1874. Upon these foundations was built up a vast amount of information about organic compounds and their reactions, which in our own time has been systematized in the electronic theory of structure and mechanism in organic chemistry.

The accumulation of facts and their assimilation into order and theory has been a great part of the history of Organic Chemistry. The process is far from complete; though we can make many predictions and explain many phenomena there is still a great deal that remains unexplained and unpredictable.

Valency

Fortunately, in organic chemistry we are nearly always dealing with simple valencies from one to four. The ideas of electrovalency and covalency should be familiar from general chemistry, so when we write H:Ö:H it should be remembered that a pair of dots between two atoms represents two electrons forming a bond, that the dots above and below the oxygen represent *unshared pairs* of electrons, and finally, that writing the formula like this should not suggest that it is a linear molecule, it is simply a convenience of printing or drawing, and the real shape of the molecule can best be shown with three-dimensional models.

Modern molecular orbital theory of valency helps to explain a great deal in organic chemistry that would otherwise be inexplicable or clumsily explained by simple valency theory. Unfortunately the mathematical derivation of molecular orbitals is usually far beyond the ability of a student beginning organic chemistry. Here we must make use of the conclusions of the theory only.

We have said nothing yet about why carbon should have a branch of chemistry all to itself and why it should be unique in its ability to form so many compounds. We can best understand this by looking at a table of bond energies, as in Table 1A.

Table 1A Bond energies of some single bonds (kjoules/mole)

Bond	Energy	Bond	Energy	Bond	Energy
C—H	404	Si—H	286	N—N	155
C—C	340	Si—Si	172	N—P	126
C—O	335	Si—O	360	O—O	135
C—Cl	322				

Notice that the energies of the C—H, C—C, C—O and C—Cl bonds are all roughly the same, and contrast that with its nearest neighbour in Group IV of the Periodic Table, silicon. We begin to understand why carbon forms quite stable C—C bonds, but Si—Si bonds are rather easily broken and Si—O bonds are more stable. Of course this is not the whole story, and other factors play a part. The position of carbon at the middle of the first row of elements, means it has little tendency to ionize either to a cation or an anion. This limitation to covalency and the formation of stable C—H and C—C bonds explains why this element can form stable bonds with itself almost *ad infinitum*.

We can divide organic compounds into four major groups: aliphatic, alicyclic, aromatic and heterocyclic. Aliphatic compounds are those that have a linear chain of carbon atoms, the name being derived from the Greek word *aleiphatos* meaning *oil*, because plant and animal oils were among the first substances recognized as belonging to this group. Alicyclic compounds are those where the linear chains of aliphatic type are joined up into rings of carbon atoms. Aromatic compounds contain rings of carbon atoms with special properties, typified by the compound benzene. The heterocyclic compounds have one or more of the carbon atoms in a ring of an alicyclic or aromatic compound replaced by an atom of a different element, such as nitrogen or oxygen. In this volume we shall consider simple compounds of the first two groups.

1

The Alkanes

Carbon is unique among the elements because of the ability of its atoms to form stable single covalent bonds with each other to an unlimited extent. The *alkanes* or paraffin hydrocarbons are formed by the combination of carbon and hydrogen in the simplest possible way, and illustrate this chain-forming property in a striking manner. The carbon atoms can be linked together in a straight chain or in a structure that is branched to any extent. The skeleton or backbone of carbon atoms thus formed is clothed with hydrogen atoms to the maximum combining power of the carbon atoms. The alkanes form a uniform series of compounds, starting with one carbon atom per molecule, and increasing by one carbon atom at a time. There is theoretically no limit to their number; many alkanes occur naturally and many have been synthesized.

The first member of the group is methane, CH_4, a colourless, odourless gas. The second is ethane, C_2H_6, also a colourless, odourless gas, but more easily condensed than methane. The next member of the series is propane, C_3H_8, an easily condensed gas. By examination of the formulae, we see that each higher member of the series differs by the unit CH_2, and the next higher member can always be obtained by replacing one hydrogen atom by CH_3, termed a *methyl* group. All members of the series are included by the type formula C_nH_{2n+2}, where *n* is 1 for methane, 2 for ethane, 3 for propane and so on. Such a series is called *homologous series* and each of the members a *homologue*.

We know that carbon has a valency of 4 and hydrogen 1, and so using Couper's graphic formulae we can write the structure of methane, ethane and propane as indicated below. Though the formulae are written flat and the chain of carbon atoms, as in propane, as a straight line, it must be remembered that the valencies of carbon are distributed towards the corners of a regular tetrahedron as suggested in Fig. 1 for methane and in Fig. 2 for propane, and so 'straight' chains are, in fact, zig-zag.

4

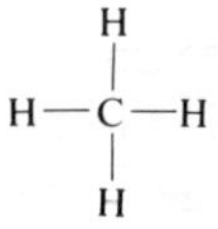

H—C—H
methane

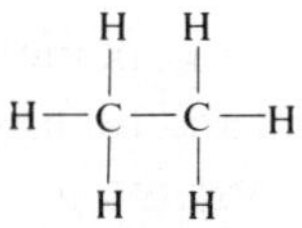

H—C—C—H
ethane

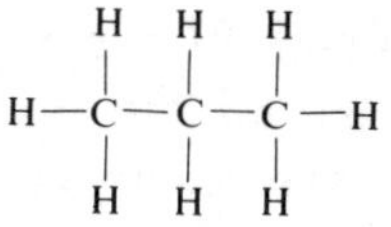

H—C—C—C—H
propane

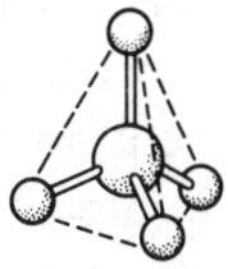

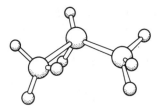

Fig. 1   Methane, showing the tetrahedral arrangement of bonds

Fig. 2   Propane; the carbon valencies are arranged tetrahedrally and the chain is not straight.

To form the next higher homologue of the series (C_4H_{10}), a CH_2 unit can be added to propane in two ways, and corresponding to these, two compounds are known to exist with the formula C_4H_{10}. They have different physical and chemical properties. The one with the straight chain is called normal butane or n-butane and the one with the branched chain isobutane.

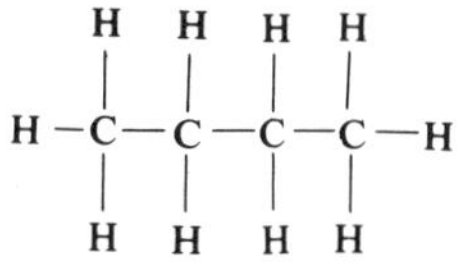

H—C—C—C—C—H

n-butane b.p. −0·5°

H—C—H
H—C—C—C—H

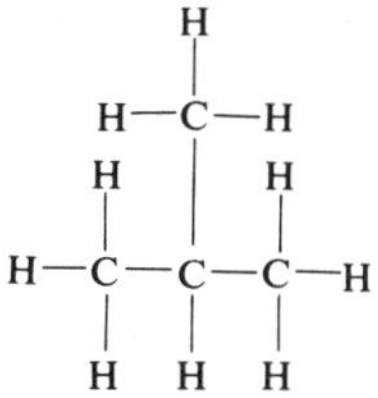

isobutane b.p. −12°

When more than one structural formula can be written for a given molecular formula of an alkane, the compounds represented by the structural formulae are called *structural isomers*. Thus for butane, C_4H_{10}, there are two isomers. In each case the isomer with the unbranched chain is called the *normal* isomer (symbol n).* Further

---
* The symbol n is usually omitted, unless we wish to emphasize it, e.g. if we write *pentane*, it is understood to mean the normal isomer.

members of the normal series of alkanes are given in Table 1, together with their melting points and boiling points. The last column gives the calculated number of isomers possible for the given molecular formula in column 2. For the higher members of the series very few of the possible isomers have been made or isolated. Note how the physical state of the alkanes changes smoothly from gases to liquids to low-melting solids with increasing molecular weight. Notice too how all the names end in *-ane*, the characteristic ending for this series and that above butane, the names are all derived from Latin and Greek numerals corresponding to the number of carbon atoms. The alkane names between C_{12} and C_{31} are given in Table 2.

Table 1.   Physical constants of n-alkanes

Name	Molecular formula	m.p. °C	b.p. °C	Number of isomers
Methane	CH_4	− 183	− 161	1
Ethane	C_2H_6	− 173	− 88	1
Propane	C_3H_8	− 187	− 42	1
Butane	C_4H_{10}	− 138	− 0·5	2
Pentane	C_5H_{12}	− 130	36	3
Hexane	C_6H_{14}	− 94	69	5
Heptane	C_7H_{16}	− 90	98	9
Octane	C_8H_{18}	− 56	125	18
Nonane	C_9H_{20}	− 53	150	35
Decane	$C_{10}H_{22}$	− 29	174	75
Undecane	$C_{11}H_{24}$	− 26	196	159
Eicosane	$C_{20}H_{42}$	36	345	366, 319
Triacontane	$C_{30}H_{62}$	66	305^{15}*	$4·1 \times 10^9$

It is not necessary to write formulae out in full as is done for the butanes above. These can be contracted

$$CH_3-CH_2-CH_2-CH_3 \qquad CH_3-\overset{\overset{\displaystyle CH_3}{|}}{CH}-CH_3$$

butane   isobutane

These can be contracted still further:

$$CH_3(CH_2)_2CH_3 \qquad (CH_3)_2CHCH_3$$

butane   isobutane

By inserting a $-CH_2-$ or replacing an $-H$ by $-CH_3$ in the two butanes, we arrive at three possible isomers of pentane.

* Meaning b.p. 305°C at a reduced pressure of 15 mm mercury.

$$CH_3-CH_2-CH_2-CH_2-CH_3 \qquad CH_3-\overset{\displaystyle CH_3}{\underset{|}{CH}}-CH_2-CH_3 \qquad CH_3-\overset{\displaystyle CH_3}{\underset{\underset{\displaystyle CH_3}{|}}{\overset{|}{C}}}-CH_3$$

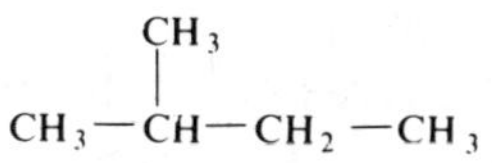

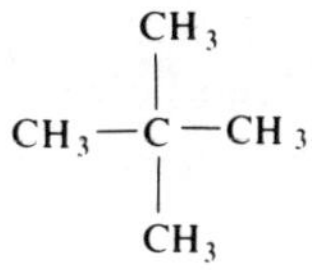

pentane	isopentane	neopentane
b.p. 36°	b.p. 28°	b.p. 9·5°

Table 2. Names of Higher Alkanes

Name	Number of carbons	Name	Number of carbons
Dodecane	12	Docosane	22
Tridecane	13	Tricosane	23
Tetradecane	14	Tetracosane	24
Pentadecane	15	Pentacosane	25
Hexadecane	16	Hexacosane	26
Heptadecane	17	Heptacosane	27
Octadecane	18	Octacosane	28
Nonadecane	19	Nonocosane	29
Eicosane	20	Triacontane	30
Heneicosane	21	Hentriacontane, etc.	31

A simple method of considering the number of possible isomers, is to write down the C-skeleton of the next lower homologue, and note how many different kinds of C-atoms are present. A new C-atom can be introduced at each point. Thus the n-pentane skeleton has identically situated C-atoms at 1 and 5, also at 2 and 4; C-atom 3 is unique. Hence 3 isomeric hexanes can be made by addition of a further C-atom to n-pentane.

$$\underset{1}{C}-\underset{2}{C}-\underset{3}{C}-\underset{4}{C}-\underset{5}{C}$$

With the increasing number of isomers possible with higher members of the series, a system was necessary so that each compound would have one name that distinguished it from all others, and given either the name or the formula, the other could be known from it. Such a system was started by chemists meeting at Geneva in 1892 and continued by the International Union of Pure and Applied Chemistry (IUPAC). The system is therefore known either as the Geneva or IUPAC system. It covers all types of organic compounds and is understood and accepted (but not always used) by chemists all over

7

the world. Many compounds in constant use have names dating from before the Geneva system; for others, the systematic names are cumbersome, and they are given shorter names; these are called *common* or *trivial* names and are frequently encountered.

To name an alkane by the Geneva system we select the longest unbranched chain in the molecule and name branches from this with a system of radical names themselves derived from the alkanes (Table 3). The position of the branches on the main chain is given by numbering the carbon atoms of the main chain and numbering from the

Table 3.   Radical names

Parent hydrocarbon	Radical	Abbreviation*
Methane	methyl CH_3-	Me
Ethane	ethyl CH_3CH_2-	Et
Propane	propyl $CH_3CH_2CH_2-$	Pr
	isopropyl $CH_3\overset{|}{C}HCH_3$	Pr^i
Butane	butyl $CH_3CH_2CH_2CH_2-$	Bu
	s-butyl $CH_3CH_2\overset{|}{C}HCH_3$	Bu^s
Isobutane	isobutyl $(CH_3)_2CHCH_2-$	Bu^i
	t-butyl $(CH_3)_3C-$	Bu^t
Pentane	pentyl $CH_3(CH_2)_3CH_2-$	—
Any alkane†	alkyl $R-$	R

end which will give the lowest possible numbers to the branches. When there is more than one branch, they are taken in alphabetical order. When there are two branches with the same name, this is prefixed by *di-*, or if three, *tri-*, four, *tetra-*, and so on (*di-*, *tri-*, etc., do not count in the alphabetical order).

$$CH_3-CH_2-\overset{\overset{\displaystyle CH_3}{|}}{C}H-CH_2-CH_2-CH_2-CH_3$$
$$1 \quad 2 \quad 3 \quad 4 \quad 5 \quad 6 \quad 7$$

3-methylheptane

$$CH_3-CH_2-\overset{\overset{\displaystyle CH_3}{|}}{C}H-CH_2-\overset{\overset{\displaystyle CH_3}{|}}{C}H-CH_3$$
$$6 \quad 5 \quad 4 \quad 3 \quad 2 \quad 1$$

2,4-dimethylheptane

* These abbreviations are frequently used as a kind of shorthand in writing formulae and equations. The meaning of secondary (*s*) and tertiary (*t*) groups is explained later in the chapter. Iso as a prefix refers to a chain with a methyl branch on the last but one carbon atom.

† $R-$ is used as a general symbol for an alkyl radical and therefore RH represents any alkane.

8

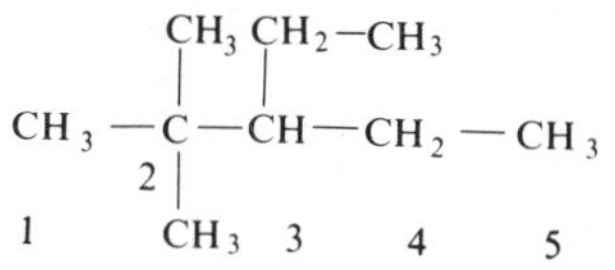

$$CH_3 - \underset{\underset{\underset{1}{CH_3}}{\overset{\overset{CH_3}{|}}{\underset{|}{C}}}}{\overset{2}{C}} - \underset{3}{CH} - \underset{4}{CH_2} - \underset{5}{CH_3}$$

where the ethyl branch is CH_2-CH_3

3-ethyl-2,2-dimethylpentane

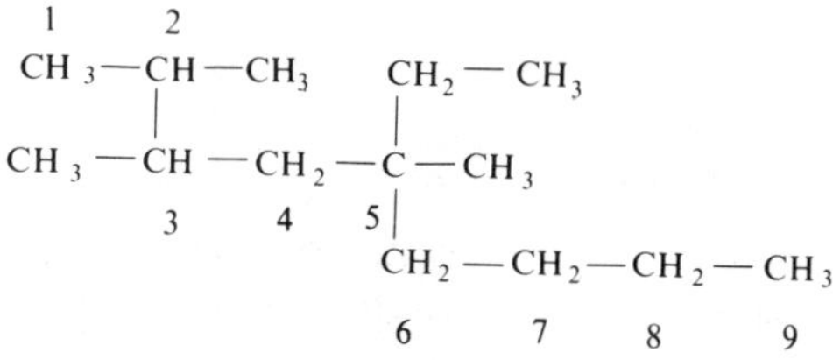

5-ethyl-2,3,4-trimethylnonane

For a few common alkanes, the trivial names are accepted. These are:

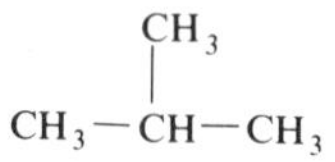

$$CH_3 - \overset{\overset{CH_3}{|}}{CH} - CH_3$$

isobutane

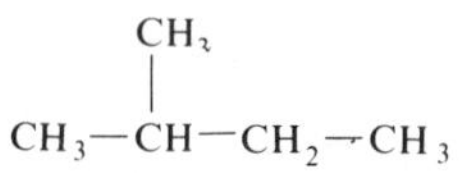

$$CH_3 - \overset{\overset{CH_3}{|}}{CH} - CH_2 - CH_3$$

isopentane

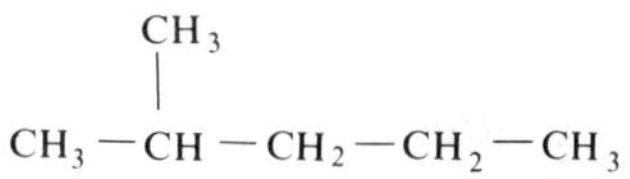

$$CH_3 - \overset{\overset{CH_3}{|}}{CH} - CH_2 - CH_2 - CH_3$$

isohexane

$$CH_3 - \overset{\overset{CH_3}{|}}{\underset{\underset{CH_3}{|}}{C}} - CH_3$$

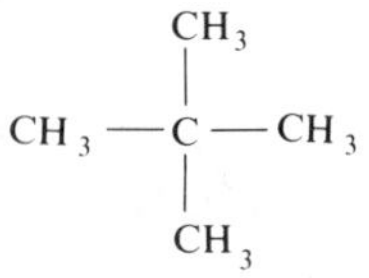

neopentane

For more complicated molecules further rules are provided, but these will not be necessary for the present work. All the rules can be found in the IUPAC handbook *Nomenclature of Organic Compounds* 3rd edition (1971).

The first compound of the alkane series, *methane*, occurs widely in nature as the chief constituent of natural gas, as marsh gas given off by decaying vegetation, as fire-damp in coal mines and sometimes in petroleum when the petroleum contains appreciable amounts of volatile compounds. It is also the principal constituent of coal gas and is produced in the bacterial decomposition of sewage. It is non-poisonous and odourless. It burns with a quiet flame, but explodes when mixed with air and ignited (hence the danger of fire-damp in mines). It is used extensively as a fuel (as natural gas and coal gas), but has many other uses, as a starting material for the production of other chemicals on an industrial scale.

It is used as a source of hydrogen for making ammonia and other chemicals by mixing it with steam and passing it over a nickel catalyst:

$$CH_4 + H_2O \underset{750°}{\rightleftharpoons} CO + 3H_2$$

Methane is decomposed to the elements by heating to 1000°, the carbon being left as a very fine powder, called carbon black. Large quantities of this are used in printing ink, paint and as a 'filler', e.g. it is mixed with rubber in vehicle tyres.

$$CH_4 \underset{1000°}{\rightleftharpoons} C + 2H_2$$

Both these reactions are reversible and carbon or carbon monoxide can be used to synthesize methane, with a nickel catalyst and at high pressure.

Ethane also occurs in natural gas and petroleum. It is non-poisonous and odourless. It burns with a quiet flame. It is also the starting point for many chemicals. This will be described in the section on petroleum.

Propane is similar to ethane both in its occurrence, physical properties, reactions and industrial uses.

For methane, ethane and propane, given their molecular formulae, there is only one possible structural formula one can write, i.e. those given on page 5. For the higher hydrocarbons, where isomers are possible, the structure must be determined in some way, such as by spectroscopic methods, to be discussed later, or by chemical reactions or synthesizing the compound from known fragments in a way that will give an unambiguous product.

The alkanes with up to about twelve carbon atoms occur in natural gas and up to about seventy carbon atoms in petroleum. The higher members also occur in animal and vegetable waxes to a small extent; hentriacontane ($C_{31}H_{64}$) is found in beeswax and candelilla wax, and nonacosane ($C_{29}H_{60}$) occurs in insect cuticle and some species of *Brassica* (cabbage).

Preparation of Alkanes

Alkanes are usually the starting materials for the preparation of other compounds. However, a few methods by which they can be prepared are worth examining here.

1. Historically, the first preparation of methane, by Berthelot in

1856, was by heating carbon disulphide and hydrogen sulphide with copper:

$$CS_2 + H_2S + 8Cu \longrightarrow CH_4 + 4Cu_2S$$

2. *From Alkyl Halides—Reduction.* Alkyl halides can be reduced either catalytically, or chemically to alkanes. Catalytic hydrogenation is carried out with hydrogen and a nickel catalyst under pressure.

$$CH_3Cl + H_2 \xrightarrow{Ni} CH_4 + HCl$$

Chemical reduction is achieved by dissolving any of a number of metals in acid or water, for example zinc dissolving in hydrochloric acid or magnesium amalgam in water.

$$CH_3CH_2Br + metal + H^{\oplus} \longrightarrow CH_3CH_3 + metal^{\oplus} Br^{\ominus}$$

3. *From Alkyl Halides—Wurtz Reaction.* Sodium reacts with alkyl halides to give an alkane of higher molecular weight. The sodium probably forms an alkyl sodium with one molecule of alkyl halide, which then reacts with a second halide molecule:

$$CH_3CH_2Br + 2Na \longrightarrow [CH_3CH_2Na] + NaBr$$

$$\downarrow CH_3CH_2Br$$

$$CH_3CH_2—CH_2CH_3 + NaBr$$

Square brackets around a formula indicate an intermediate that is probably formed but is not isolated. The overall result of the Wurtz reaction is usually written thus (X represents any halogen, so R—X is any alkyl halide):

$$R-X + 2Na + X-R \longrightarrow R—R + 2NaX$$

If two different alkyl halides (call them RX and R'X) are used in this reaction, a mixture of all possible products is formed, R—R, R—R' and R'—R', which can be difficult to separate unless, as in the example below, R and R' are sufficiently different from each other. The three compounds are produced approximately in the proportions 1:2:1.

$$CH_3Br + CH_3(CH_2)_4Br \longrightarrow CH_3CH_3 + CH_3(CH_2)_4CH_3 + CH_3(CH_2)_8CH_3$$

$$\text{b.p. } -88° \qquad \text{b.p. } 69° \qquad \text{b.p. } 174°$$

The yield in this reaction is usually poor, but it is useful for making straight chain alkanes.

4. *From Acids—Decarboxylation.* If the sodium salt of an organic

acid is heated with soda lime (a mixture of sodium hydroxide and calcium oxide), the acid loses carbon dioxide and an alkane is formed. The organic acid group —COOH is known as the carboxyl group and therefore removal of this group is called decarboxylation. The reaction is good for methane, but poor yields are obtained when trying to make higher alkanes by this method.

$$CH_3COONa + NaOH \longrightarrow CH_4 + Na_2CO_3$$

sodium acetate

5. *From Acids—Kolbe Electrolysis.* If an electric current is passed through a solution of a potassium or sodium salt of an organic acid, the salt is decarboxylated and an alkane of higher molecular weight is obtained.

With sodium acetate, the acetate ions migrate to the anode and give up an electron, forming acetoxy *free radicals* with one unpaired electron. These radicals are unstable and first decompose giving carbon dioxide and methyl radicals. The methyl radicals too are unstable and react together to give ethane:

$$CH_3COO^{\ominus} \longrightarrow e + CH_3COO^{\cdot} \longrightarrow \dot{C}H_3 + CO_2$$

$$\dot{C}H_3 + \dot{C}H_3 \longrightarrow CH_3{-}CH_3$$

At the cathode the following reactions can be considered as occurring:

$$Na^{\oplus} + e \longrightarrow Na$$

$$Na + H_2O \longrightarrow NaOH + \tfrac{1}{2}H_2$$

The overall reaction is to produce hydrogen at the cathode and carbon dioxide and the alkane at the anode.

$$2CH_3COONa + 2H_2O \longrightarrow CH_3{-}CH_3 + 2CO_2 + 2NaOH + H_2$$

This method, like method 3 (above) gives alkanes with 2, 4, 6, 8, etc. carbon atoms. If two different salts are mixed and electrolysed, then all three possible alkanes are formed, so the yield of each one will be low.

Other methods of preparing alkanes will be encountered in later chapters, particularly the reduction of Grignard reagents (page 36),

12

reduction of alkenes (page 55) the reduction of ketones (page 154), and the decarboxylation of acids (p. 185).

Yields
In organic chemistry reactions are seldom quantitative; commonly reactions are incomplete, or give rise to more than one product, or other reactions compete to give by-products, so it is important to know how efficient a reaction is to give a certain product. We express this efficiency as a percentage given by the ratio of the actual amount of material produced to the amount that would be produced if the reaction were quantitative, and call this the *yield*. A reaction that gives only 1 % yield of a compound will not be very useful; we normally try to achieve a maximum yield. Actual yields vary a little from one experiment to another, and depend upon the purity of the reagents, the conditions of the reaction, and the techniques of the experimenter, as well as many other variables. In industry the increase in yield of a few percent can mean a large difference in the profitability of a process.

Physical Properties
All the alkanes are colourless, and tasteless. Methane, ethane, propane and butane are gases at room temperature. Pentane (b.p. 36°) and isopentane (b.p. 28°) are low boiling liquids; higher alkanes are liquids with increasing boiling points, up to hexadecane which is a soft waxy solid (m.p. 18°). Above that the normal alkanes are all white, waxy solids. The difference in boiling point between the lower homologues makes them easy to separate by fractional distillation but as molecular weight increases the small difference in boiling point between two homologues makes them very difficult to separate (see Table 1).

Normal alkanes have higher boiling points and melting points than branched isomers, and this is true throughout aliphatic compounds of all types. Generally, one can say the more branched a molecule, the lower the boiling and melting points.

Alkanes are insoluble in water and are less dense than water, therefore on mixing with water, a liquid alkane floats to the top and forms a separate layer. Alkanes are soluble in all organic solvents, e.g. alcohol, ether, benzene, acetone, but the solubility decreases as the molecular weight increases.

Chemical Properties
The alkanes are also called paraffins (Latin, *parum affinis*=small

affinity) because of their low reactivity towards the common reagents. In the technical processes of petroleum chemistry they can be made to undergo a great variety of reactions, with the aid of relatively high temperature, pressure and catalysts.

They are inert towards cold concentrated nitric, sulphuric or hydrochloric acid and also to strong alkalies. They are unaffected by oxygen at ordinary temperatures, but ignited in air they burn to carbon dioxide and water:

$$CH_4 + O_2 \longrightarrow CO_2 + H_2O$$

When the supply of oxygen is limited, carbon monoxide or other partly oxidized products are formed, or carbon is deposited. Carbon black for paints, inks and rubber can be produced in this way.

1. *Halogenation.* Halogens react with alkanes by substituting X for H, to give mono- or poly-halogenated compounds. The great French chemist Dumas discovered this reaction in a curious way. He was called in to find out why the candles used at a *soirée* in the Tuileries Palace in Paris one evening were giving off sharp and acrid fumes. He found that the merchant had bleached the wax for the candles with chlorine. Some of the chlorine was incorporated into the wax and was evolved as hydrogen chloride on burning. This was the first evidence that halogens could *replace* hydrogen in organic molecules. Dumas' experiments following up this discovery were a great stimulus to the advancement of chemistry.

The reaction of chlorine is the most convenient to consider. Light, or heat, is required as a catalyst. In direct sunlight chlorine and methane react explosively, in diffuse light they react smoothly. The effect of the light is to dissociate the chlorine molecules into atoms; using the symbol $h\upsilon$ to represent light energy, we write:

$$Cl_2 \xrightarrow{h\upsilon} Cl\cdot + Cl\cdot \qquad \text{(Eq. 1)}$$

The dot represents an unpaired electron, and we call such atoms (or groups of atoms) with an unpaired electron, *free radicals*. The chlorine free radical is the reactive agent, and attacks the methane molecule, removes an atom of hydrogen, with formation of a molecule of hydrogen chloride, leaving a methyl radical:

$$Cl\cdot + H - \overset{\displaystyle H}{\underset{\displaystyle H}{C}} - H \longrightarrow H - \overset{\displaystyle H}{\underset{\displaystyle H}{C}}\cdot + HCl \qquad \text{(Eq. 2)}$$

The methyl radical, also reactive, attacks a chlorine molecule, forming methyl chloride and releasing another chlorine radical:

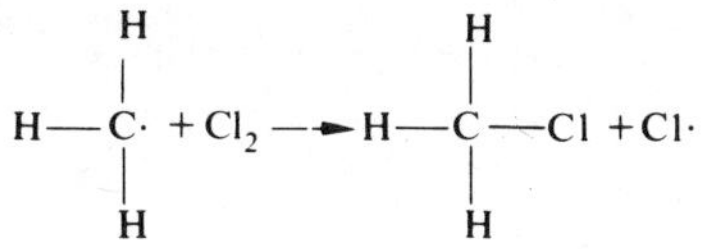

$$\underset{H}{\overset{H}{H-C\cdot}} + Cl_2 \longrightarrow \underset{H}{\overset{H}{H-C-Cl}} + Cl\cdot \qquad \text{(Eq. 3)}$$

The new chlorine radical replaces one formed by dissociation in equation 1, so once begun the reaction is self-perpetuating, it is called a *chain reaction*. Equation 1 is the *chain initiating* reaction, equations 2 and 3 are *chain propagating* reactions, and the chains are terminated usually by combination of radicals, as in equations 4, 5 and 6:

$$Cl\cdot + Cl\cdot \longrightarrow Cl_2 \qquad \text{(Eq. 4)}$$

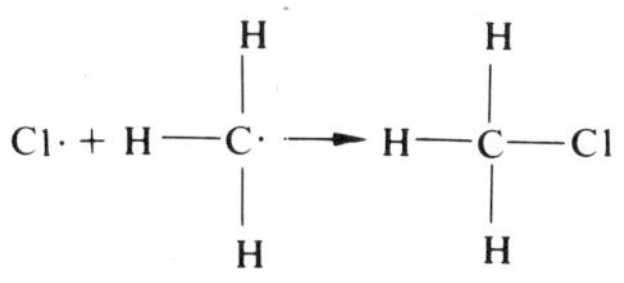

$$Cl\cdot + \underset{H}{\overset{H}{H-C\cdot}} \longrightarrow \underset{H}{\overset{H}{H-C-Cl}} \qquad \text{(Eq. 5)}$$

$$CH_3\cdot + CH_3\cdot \longrightarrow C_2H_6 \qquad \text{(Eq. 6)}$$

Equation 6 accounts for a small proportion of the methyl radicals. The chief overall reaction is then:

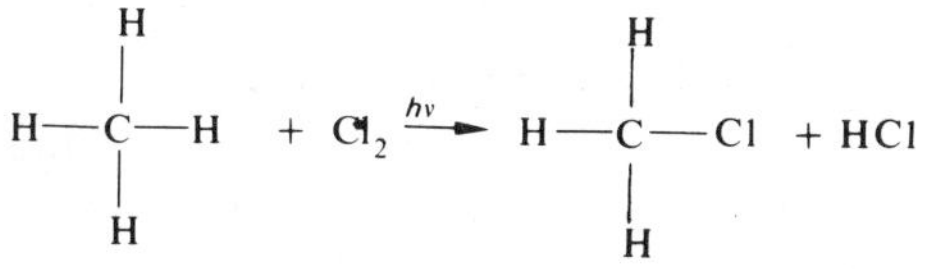

$$\underset{H}{\overset{H}{H-C-H}} + Cl_2 \xrightarrow{h\nu} \underset{H}{\overset{H}{H-C-Cl}} + HCl$$

The reaction can go further, depending upon concentration and temperature conditions, until all the hydrogens of methane have been replaced by chlorine:

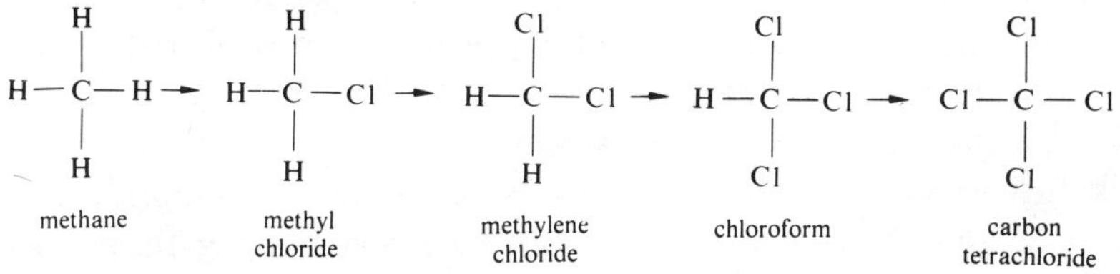

methane → methyl chloride → methylene chloride → chloroform → carbon tetrachloride

Bromine reacts similarly, but iodine reacts only sluggishly and the reaction is further impeded by the accumulation of hydrogen iodide.

15

Indeed, an alkyl iodide can be reduced to an alkane by heating with HI at 150°.

$$CH_4 + Br_2 \longrightarrow CH_3Br + HBr$$

$$CH_4 + I_2 \rightleftharpoons CH_3I + HI$$

Propane and higher alkanes can react to give more than one mono-chloro-derivative:

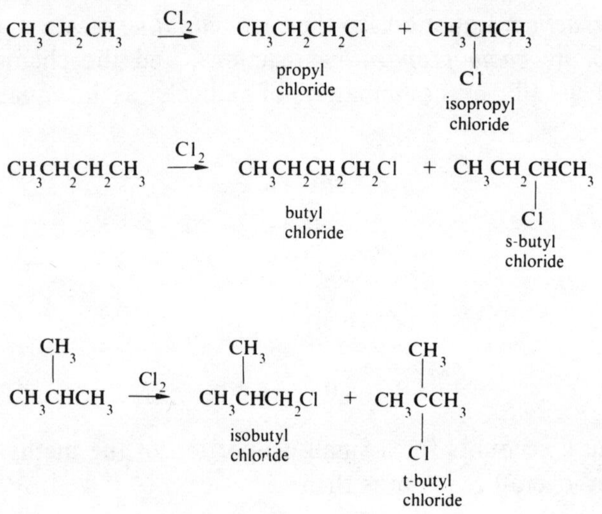

$$CH_3CH_2CH_3 \xrightarrow{Cl_2} CH_3CH_2CH_2Cl + CH_3CHCH_3$$

propyl
chloride

Cl

isopropyl
chloride

$$CH_3CH_2CH_2CH_3 \xrightarrow{Cl_2} CH_3CH_2CH_2CH_2Cl + CH_3CH_2CHCH_3$$

butyl
chloride

Cl

s-butyl
chloride

$$CH_3CHCH_3 \xrightarrow{Cl_2} CH_3CHCH_2Cl + CH_3CCH_3$$

CH_3

CH_3

CH_3

isobutyl
chloride

Cl

t-butyl
chloride

We encounter here a new homologous series, the alkyl chlorides, $C_nH_{2n+1}Cl$ or RCl; and further examples of isomerism. Propane gives two propyl chlorides and there are four butyl chlorides and correspondingly greater numbers for higher homologues. That these isomers are distinct compounds is easily discovered from their different physical and chemical properties. From the reactions of these chlorides we find that they fall into three types; propyl, butyl and isobutyl chlorides are all *primary* chlorides. In each case the chlorine atom is attached to a CH_2 group. Isopropyl and s-butyl chlorides are similar in reactivity. These are called *secondary* chlorides because the chlorine is attached to a carbon atom having two alkyl groups; and t-butyl chloride is of the *tertiary* chloride class because the chlorine is attached to a carbon bearing three alkyl groups. These three types are encountered throughout organic chemistry and for clarity they are shown diagrammatically below, Z representing any group or atom, such as hydrogen, chlorine,

16

hydroxyl etc. (but *not* carbon). The carbon atom ringed is called a primary, secondary and tertiary carbon atom respectively:

$$
\begin{array}{ccc}
\text{H} & \text{C} & \text{C} \\
| & | & | \\
\text{C}-\textcircled{C}-\text{z} & \text{C}-\textcircled{C}-\text{z} & \text{C}-\textcircled{C}-\text{z} \\
| & | & | \\
\text{H} & \text{H} & \text{C} \\
\text{primary} & \text{secondary} & \text{tertiary}
\end{array}
$$

Propane has two primary and one secondary carbon atoms and isobutane has three primary and one tertiary carbon atoms. These atoms all have different reactivity towards chlorine, in the order tertiary > secondary > primary. Note, however, that there are six primary hydrogen atoms in propane and only two secondary hydrogens that can be replaced, so that isopropyl chloride is formed only slightly faster than propyl chloride. Isobutane has only one very reactive tertiary hydrogen atom to nine primary ones, so more isobutyl chloride is formed.

2. *Isomerization.* When heated above 350°, alkanes slowly rearrange into a mixture of isomers. Hexane, for example, can give a mixture of methylpentanes and dimethylbutane. The reaction can be catalysed by a number of substances, chiefly Lewis acids such as aluminium chloride or boron trifluoride. A very effective isomerizing agent which operates at relatively low temperature is hexafluoroantimonic acid ($HSbF_6$). With this agent any hexane or heptane will quickly yield the equilibrium mixture of all the isomerides. The rearrangement takes place through the intermediate formation of carbonium ions, to be described later.

$$
\begin{array}{c}
 CH_3 CH_3 \\
 | | \\
CH_3CH_2CH_2CH_2CH_2CH_3 \longrightarrow CH_3CHCH_2CH_2CH_3 + CH_3CH_2CHCH_2CH_3
\end{array}
$$

$$
\begin{array}{c}
 CH_3 CH_3 \\
 | | \\
+ CH_3CHCHCH_3 + CH_3CCH_2CH_3 \\
 | | \\
 CH_3 CH_3
\end{array}
$$

3. *Pyrolysis.* Above 400°, the carbon–carbon bonds begin to break through thermal vibration. Alkanes with straight chains and those of higher molecular weight begin to decompose first. The products are mixtures of alkanes and alkenes of lower molecular weight. This pyrolysis (or *cracking* in the petroleum world) can be catalysed by a number of substances such as carbon, silica, clays and various metal

compounds. These substances catalyse the breaking of large molecules and reduce the amount of isomerization. Isomerization and pyrolysis are of little value in the laboratory, but industrially they are of immense importance.

4. *Oxidation*. The stability of alkanes towards oxidizing agents has been mentioned. At high temperature (400°) and pressure (100–200 atmospheres) alcohols can be formed from lower alkanes using air as oxidizing agent. Higher alkanes can be oxidized in this way to carboxylic acids. At the moment, the chief industrial use of this reaction is the preparation of formaldehyde, CH_2O, from natural gas, in areas where this is cheap and plentiful.

Identification

One soon learns in organic chemistry certain reactions for identifying classes of compounds. How then can one recognize and identify an alkane, when they are so unreactive? The answer is that alkanes are recognizable by their absence of reaction, in all the common tests. Very few classes of compounds will not dissolve in concentrated sulphuric acid. The alkanes are one of them. On combustion they give only carbon dioxide and water.

The modern chemist uses spectroscopy a great deal to help solve his problems of identifying compounds. Ultraviolet, infrared, nuclear magnetic resonance and mass spectra are all used to help solve organic chemical problems.

Very few of the compounds in this volume have ultraviolet absorption in the region we can easily measure (below 200 nm) and nuclear magnetic resonance and mass spectra, are better introduced at a later stage.

The infrared spectrum is the one easiest to measure and the one that gives the simplest information. All bonds vibrate, by bending and stretching. A change in the amplitude of vibration can be caused by the quantized absorption of light. The light energy associated with the changes occurs in the infrared region, and the spectrum between wavelengths of 2 and 15 μm (micrometres, 10^{-4} cm) can be measured by most instruments. This is equivalent, on a frequency scale to 4000 to 700 waves per centimetre (usually written cm^{-1}). The frequency scale is the more usual one. Most functional groups met with in organic chemistry have characteristic absorption in this region.

The alkanes have very simple spectra; a strong absorption at 2960–2850 cm^{-1} characteristic of the stretching of the simple C—H bond, and at about 1450 cm^{-1} for the bending of the C—H bond.

Bands due to other groups (e.g. C=C) are obviously absent.

In addition to the specific absorptions due to groups, there is usually a complex pattern of bands between about 1300 cm^{-1} and 1000 cm^{-1} which cannot be attributed to any particular vibrations but are specific for a particular compound, this is therefore called the fingerprint region. No two compounds can ever have identical infrared spectra in this region. A typical alkane spectrum is shown in Fig. 3.

Petroleum

Petroleum is a complex mixture of organic compounds in which the paraffin hydrocarbons predominate. Besides these compounds of carbon and hydrogen, it contains smaller, and widely varying amounts of compounds containing nitrogen, oxygen and sulphur. Next in importance to the open-chain hydrocarbons come cyclic hydrocarbons, called 'naphthenes', consisting chiefly of derivatives of cyclopentane and cyclohexane (see Chapter 11); and the aromatic hydrocarbons (e.g. benzene, toluene, xylene, naphthalene). The relative proportions of these types in crude oil vary widely from one region to another. For example Russian oil from the Caucasus is very rich in naphthenes, that from Borneo is rich in aromatic hydrocarbons, and Pennsylvanian oil is richest in aliphatic hydrocarbons. The nitrogen is chiefly present in the form of amines (e.g. pyridine, and its derivatives), the oxygen as carboxylic acid derivatives of the cyclic hydrocarbons (e.g. cyclopentane carboxylic acid and cyclohexane carboxylic acid) and the sulphur as thiols, sulphides (Chapter 10) and thiophen. In appearance again, the oil varies with the oilfield from which it comes. It may be a thick, black and tarry oil through all gradation to a mobile yellow fluid.

The origin of deposits of petroleum and natural gas is unknown. They may have been formed in the primaeval earth when there was a great abundance of hydrogen in the atmosphere, or they may be formed, like coal, by the accumulated decay of vegetable or animal material.

There is evidence for the biological origin of part, at least, of all petroleum. The chief pointers are optical activity, occurrence of porphyrins (a group of complex organic compounds containing nitrogen produced by living organisms) and the predominance of alkanes with an odd number of carbon atoms, all of which are associated with living materials and are found in petroleum other than that in rocks of very great age (circa 1000 million years old and

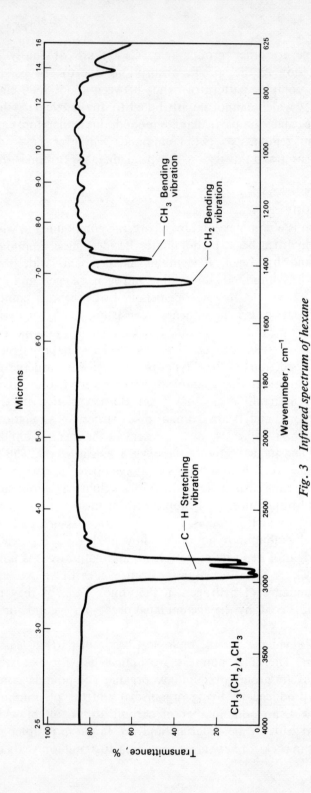

Fig. 3   Infrared spectrum of hexane

older). On the other hand, the non-biological origin of part of petroleum is conjectural, though cogent arguments favouring this view have been advanced.

Noah probably pitched his Ark with bitumen which escapes through rock faults in places in the Near East and rock oil was drunk as a medicine by North American Indians, but the first oil-well drilled was in Pennsylvania in 1859: oil was found at a depth of 69 feet! At first oil was distilled to provide lighting oil for lamps, and the lighter fractions were discarded as useless. Slowly, all the constituents found their uses, until today hardly anything in crude oil is wasted.

Refining of petroleum consisted essentially of a fractional distillation, giving first 'light petroleum' or ligroin (b.p. 35–100°), then gasoline or petrol (b.p. 50–220°), paraffin oil or kerosene (b.p. 180–325°) and then various heavier oil fractions, such as diesel engine fuel, furnace oil, lubricating oils, greases and wax, all boiling above 300°, and leaving behind a black residue of bitumen or asphalt. These fractions overlap a great deal; in practice the separation into fractions is governed by what the market demands.

The age of the family motor car created an enormous demand for the gasoline fraction of petroleum which could not be met from distillation alone, and so the process of catalytic 'cracking' or pyrolysis was introduced in 1935. In this process, the higher boiling fractions are heated (at 400–600°) with a variety of catalysts such as aluminium oxide or silica. The reaction proceeds through the formation of free radicals which can decompose and recombine in a great variety of ways. Choice of catalyst and conditions can control the kind of products formed to a large extent. Up to about 60% of the material can be converted to useful petrol by this process; appreciable quantities of methane, ethane and other hydrocarbon gases are produced as well, which at one time were burned as waste.

In an internal combustion engine, increasing compression leads to increased efficiency, but one of the limiting factors of increasing compression is 'knocking'. A smooth and rapid burning of the petrol-air mixture is required to produce a smooth increase in pressure against the piston. If the burning of the vapour mixture is too rapid, as can happen if the combustion mixture is highly compressed, there is a too sudden and explosive increase of pressure on the piston, accompanied by a rattling or knocking noise, and the efficiency of the engine falls sharply. Many factors seem to affect the rate of burning, among them the nature of the fuel. Branched chain hydrocarbons and aromatic hydrocarbons (compounds of the benzene type) burn smoothly, and normal alkanes are more liable to cause knocking. As a standard of quality for petrol, iso-octane or 2,2,4-trimethylpentane was chosen as a compound with low knocking tendency and given an arbitrary 'octane number' of 100, and heptane, which knocks badly, given an octane number of 0, so all petrols can be compared with mixtures of these two to determine octane number.

$$CH_3-\underset{\underset{CH_3}{|}}{\overset{\overset{CH_3}{|}}{C}}-CH_2-\underset{\overset{CH_3}{|}}{CH}-CH_3 \qquad\qquad CH_3CH_2CH_2CH_2CH_2CH_2CH_3$$

2,2,4-trimethylpentane, octane number 100   heptane, octane number 0

The petrol from the catalytic cracking process has a higher octane rating than the 'straight run' petrol (that obtained from crude oil by fractional distillation). To improve the quality of petrol further by introducing more branched hydrocarbons, the process of *isomerization* is used (see page 17) with aluminium chloride or other acidic catalyst. Alkenes formed in the cracking process can be *alkylated* or *polymerized* to branched hydrocarbons. (These processes are explained in Chapter 3.) Naphthenes or cyclic hydrocarbons can be dehydrogenated over platinum by the *reforming* process; for example, cyclohexane gives benzene, an aromatic hydrocarbon, in this way. The aromatic hydrocarbons have high octane numbers and are also valuable as starting materials for many chemical processes. The aromatic hydrocarbons can, if needed, be removed by selective extraction with solvents. By the suitable combination of straight distillation, cracking, polymerization, alkylating, isomerization and reforming, a refinery may vary the nature of its products over a wide range to meet market demands.

One of the most objectionable impurities in petroleum and natural gas is sulphur, because the sulphur compounds have an unpleasant odour, and in combustion they produce corrosive acids. It is not practical to treat crude oil to remove sulphur, this must be done to the separated fractions. The sulphur is removed chiefly by the process of hydrodesulphurizing, that is, treating the petroleum fraction with hydrogen under pressure with a hydrogenation catalyst (cobalt and molybdenum are used), the carbon–sulphur bonds are broken, and the sulphur converted to hydrogen sulphide, which can then be removed in a variety of ways. Other processes, such as passing the oil through sulphuric acid (which can be expensive for the quantities used) or passing the vaporized oil through an absorbant such as clay at 450°, or washing with sodium hydroxide solution can be used to reduce the amount of sulphur present. In some wells sulphur is present in sufficient quantities to make it worthwhile to extract it, and sell it as a by-product as is done at the natural gas field of Lacq in France.

The low molecular weight hydrocarbon gases produced during cracking were at first wasted, but in a comparatively short time they have become the basis of the enormous petroleum chemicals industry. In Great Britain in 1972, of 102 475 000 tons of petroleum con-sumed, 6 250 000 tons were for chemicals, still quite a small fraction, but this represented an increase of 200% in ten years. In the same year, 10 200 million therms of natural gas were consumed of which only 35 million (0·35%) were for the petrochemical industry. The chief constituents of refinery gas are hydrogen, methane, ethane, propane, butanes, ethylene, propylene and butylenes. These gases are converted by a variety of reactions, which will be described under the individual compounds, into further compounds and finished products. Table 4 gives, as an example, the principal use of ethy-lene in Great Britain in 1972.

Table 4.   Consumption of Ethylene

Produce	Millions of pounds
Polyethylene	1163
Ethylene oxide	579
Ethylene dichloride	552
Polystyrene	325
Ethyl alcohol	311
Others	70
Total	3000

In general, petrochemical processes are carried out at high temperature and pressures above atmospheric. Catalysts are fre-quently used. A student is often confused by the lack of order or system in the choice of catalysts. However, most catalysts can be placed into one of four groups. First, transition metals such as iron, nickel, platinum and palladium are used in reactions involving hydrogen, where hydrogen is either absorbed or given off (hydro-genation and dehydrogenation). Second, metal oxides and sulphides, chiefly those of copper, zinc and nickel for oxygen reactions (oxidation). Third, a group of oxides, including silica, alumina and magnesia, which are used in water reactions (hydration and dehydration). Fourth, acid catalysts such as sulphuric acid, phosphoric acid or aluminium chloride. Their uses are too varied to generalize. These four groups, though far from covering all cases, can help to reduce a long list of catalysts to some order.

Synthetic Petroleum

For countries possessing coal but no oil, a means of producing synthetic fuel from coal is attractive. During World War II, the Bergius process was developed in Germany, for the hydrogenation of coal at 200–700 atmospheres pressure and 400–500°. The product, an oil resembling petroleum, is separated by distillation into petrol, gas oil and a tarry residue. During the war, Bergius process plants supplied a large proportion of Germany's motor fuel needs.

Another method of producing oil from coal is the Fischer–Tropsch Process, by which carbon monoxide and hydrogen are combined to give hydrocarbons. Coke and steam together at high temperature give a mixture of carbon monoxide and hydrogen. This is enriched with more hydrogen and passed over a cobalt or iron catalyst at 200° and 5 to 35 atmospheres, giving a liquid-like petroleum, which is separated into similar fractions by distillation.

$$C + H_2O \rightarrow CO + H_2$$

$$nCO + (2n + 1)H_2 \rightarrow C_nH_{2n+2} + nH_2O$$

The Fischer–Tropsch process gives chiefly straight chain molecules, and could be used as a source of straight chain alkanes of higher molecular weight. Under present economic conditions, hydrocarbons produced from coal cannot compete with natural gas and petroleum, but the world reserves of coal are much greater than those of petroleum, and if the demand for hydrocarbons continues to grow as at present, the Fischer–Tropsch kind of process could become very important as a source of organic chemicals.

2
Halogenated Alkanes

The existence of homologous series of compounds simplifies the dividing up of the subject of organic chemistry. The alkanes form a homologous series, and certain of their reactions lead to the introduction of a *functional* group, such as halogen or hydroxyl. The properties of a particular compound are a combination of the properties dependent on its carbon skeleton and its functional group. Our attention focuses on the functional group because that undergoes some change during reaction; the carbon chain may or may not be affected. This chapter deals with the halo-alkanes, in which the functional group is a halogen atom.

Nomenclature

The simplest type of halo-alkane is a monohalide, which can be represented by the general formula R—X or $C_nH_{2n+1}X$. The systematic way to name these compounds is to prefix the name of the hydrocarbon chain with a name derived from the halogen (e.g. chloro-) and if necessary, a number to indicate the carbon atom at which the halogen is attached. Trivial names for these compounds are in common use and are equally acceptable. They are derived from the name of the alkyl group and halogen. Some examples illustrate both systems:

Formula	CH_3F	CH_3CH_2Cl	$CH_3CH_2CH_2Br$	$CH_3\overset{\displaystyle I}{\underset{\displaystyle |}{C}}HCH_3$
Systematic name	fluoromethane	chloroethane	1-bromopropane	2-iodopropane
Trivial name	methyl fluoride	ethyl chloride	propyl bromide	isopropyl iodide

Formula	$CH_3\overset{\displaystyle CH_3}{\underset{\displaystyle \underset{Cl}{|}}{-C-}}CH_3$	$CH_3\overset{\displaystyle CH_3}{\underset{\displaystyle |}{C}}HCH_2Cl$	$CH_3CH_2\underset{\displaystyle \underset{Cl}{|}}{C}HCH_3$
Systematic name	2–chloro–2–methylpropane	1–chloro–2–methylpropane	2–chlorobutane
Trivial name	t–butyl chloride	isobutyl chloride	s-butyl chloride

These monohalo-alkanes can be subdivided into three types, primary, secondary and tertiary, each with its own characteristic reactions and behaviour.

When an alkyl chain contains two halogen atoms, still more arrangements are possible. The two halogens may be attached to the *same* carbon atom; it is then called a geminal (=twin) dihalide; this is abbreviated to gem. If the halogens are on adjoining carbon atoms, it is called a vicinal dihalide (*vicinus*=neighbouring). The IUPAC system of naming all dihalogen compounds is simple and can easily be demonstrated with a few examples. The trivial naming is more complex. Geminal dihalides are named as derivatives of alkylidene radicals, ethyl gives ethylidene, propyl gives propylidene, etc.; methyl becoming methylene is an exception.

$$\underset{\underset{\displaystyle Cl}{|}}{\overset{\overset{\displaystyle H}{|}}{HCCl}} \qquad \underset{\underset{\displaystyle Br}{|}}{\overset{\overset{\displaystyle H}{|}}{CH_3CBr}} \qquad \underset{\underset{\displaystyle Cl}{|}}{\overset{\overset{\displaystyle Cl}{|}}{CH_3CCH_3}}$$

Systematic name	dichloromethane	1,1-dibromoethane	2,2-dichloropropane
Trivial name	methylene chloride	ethylidene bromide	isopropylidene chloride

Vicinal dihalides are commonly named as dihalides of the alkenes from which they can be made.

$$ClCH_2CH_2Cl \qquad \underset{}{\overset{\overset{\displaystyle Br}{|}}{CH_3CHCH_2Br}} \qquad \underset{\underset{\displaystyle Cl}{|}}{\overset{\overset{\displaystyle CH_3}{|}}{CH_3CCH_2Cl}}$$

Systematic name	1,2-dichloroethene	1,2-dibromopropane	1,2-dichloro-2-methyl-propane
Trivial name	ethylene dichloride	propylene dibromide	isobutylene dichloride

Dihalides with the halogen at the ends of a simple alkyl chain have special trivial names too.

$$BrCH_2CH_2CH_2Br \qquad ClCH_2CH_2CH_2CH_2Cl \qquad I(CH_2)_6I$$

Systematic name	1,3-dibropropane	1,4-dichlorobutane	1,6-di-iodohexane
Trivial name	trimethylene dibromide	tetramethylene dichloride	hexamethylene-di-iodide

If two different halogens are in the same molecule or when three or more functional groups are present, the systematic names must be used. The prefixes are taken in alphabetical order (di- and tri- do not count).

26

$$\text{ClCH}_2\text{CH}_2\text{CH}_2\text{F}$$

$$\underset{\begin{array}{cc}\text{Br} & \text{Cl}\end{array}}{\text{CH}_3\text{CH}_2\,\text{CH}_2\,\overset{\displaystyle\text{CH}_3}{\text{CH}}\text{CHCHCH}_2\text{Br}}$$

1-chloro-3-fluoropropane   1,4-dibromo-2-chloro-3-methylheptane

Preparation of Alkyl Halides

1. Reaction of an alkane with free halogen gives an alkyl halide.
The method is simple but its practical value is limited because the reaction does not stop at the introduction of one halogen atom per molecule and a mixture of products is formed. This disadvantage can be minimized by limiting the supply of halogen. The reactive particle is a halogen free radical, formed by splitting a halogen molecule by means of either heat or ultraviolet light (page 14).

$$\text{CH}_4 + \text{Cl}_2 \xrightarrow{\text{h}\nu} \text{HCl} + \text{CH}_3\text{Cl} + \text{CH}_2\text{Cl}_2 + \text{CHCl}_3 + \text{CCl}_4$$

$$\text{CH}_3\text{CH}_3 + \text{Br}_2 \xrightarrow{\text{h}\nu} \text{HBr} + \text{CH}_3\text{CH}_2\text{Br}$$

Fluorine reacts very violently, iodine does not react appreciably and chlorine and bromine react at a controllable rate. The hydrogen atoms of the alkane react in the order tertiary > secondary > primary. That is, tertiary hydrogen atoms are most easily replaced. The effect is not very noticeable with chlorine, which is a comparatively unselective reagent, but bromine is much more selective, and tends to give mono-bromo-compounds only.

$$\underset{}{\overset{\displaystyle\text{CH}_3}{\text{CH}_3\text{CHCH}_3}} + \text{Br}_2 \longrightarrow \underset{\text{Br}}{\overset{\displaystyle\text{CH}_3}{\text{CH}_3\text{CCH}_3}} + \text{HBr}$$

2. Addition of hydrogen halide to an alkene gives an alkyl halide.
The order of reactivity here is HI > HBr > HCl > HF. In a simple case such as ethylene, only one product is possible.

$$\text{HBr} + \text{CH}_2 = \text{CH}_2 \longrightarrow \text{CH}_3 - \text{CH}_2\text{Br}$$

However, where the alkene is unsymmetrical, as in propylene, theoretically two substances could be formed, $\text{CH}_3\text{CH}_2\text{CH}_2\text{Br}$ and $\text{CH}_3\text{CHBrCH}_3$. In fact, the latter is the normal product. Markovnikov put forward a rule in 1870 to the effect that when addition occurs at low temperatures to the double bond of an unsymmetrical

27

alkene, a system **YZ** (here hydrogen halide) adds so that the more negative group (here halogen) becomes attached to the carbon atom bearing the smallest number of hydrogen atoms; put the other way round, the carbon of C=C with the more hydrogen gets still more hydrogen. At comparatively higher temperatures the direction is reversed.

$$CH_3-CH=CH-CH_3 \quad + \ HBr \longrightarrow CH_3-CH_2-\underset{\underset{Br}{|}}{CH}-CH_3$$

(symmetrical)   (only one possibility)

$$CH_3-\underset{\underset{CH_3}{|}}{C}=CH-CH_3 \ + \ HBr \longrightarrow CH_3-\underset{\underset{Br}{|}}{CH}-CH_2-CH_3$$

(unsymmetrical)   (preferred product)

To Markovnikov this was a rule of thumb, but today we can explain why the hydrogen halide adds in one preferred way.

Clearly elements differ considerably in their electron-attracting powers. For example, sodium and chlorine illustrate the extreme case where an ionic bond is formed, but in covalent bonds between unlike atoms, the electrons of the bond are never shared equally between the two atoms, they are held more tightly to one or the other. Again, chlorine forms a covalent bond with carbon, but because chlorine is more electron-attracting (or electronegative) than carbon, the bonding electrons are held more tightly to chlorine. We can represent this by drawing an arrow in the bond pointing towards the more electronegative atom, or by using the symbols $\delta+$ and $\delta-$ to represent small differences in electron density:

$$C \longrightarrow Cl \quad \text{or} \quad \overset{\delta\oplus}{C} - \overset{\delta\ominus}{Cl}$$

We can draw up a list of elements in order of electronegativity, starting with fluorine, the most electronegative, followed by oxygen, then chlorine, nitrogen, bromine and so on. Carbon and hydrogen come near the middle of the table. If we place CH_3 in this table, we find that it is more electron releasing than hydrogen.

Now looking again at an alkene with alkyl groups attached, the $H^\oplus$ of a hydrogen halide can be considered as adding to the double bond as the first step of the reaction; see opposite, top of page.

The intermediate product may be one of two possible ions (a or b), but the ion with the larger number of alkyl groups attached to it (b) will be formed more readily. This is because the alkyl groups donate

28

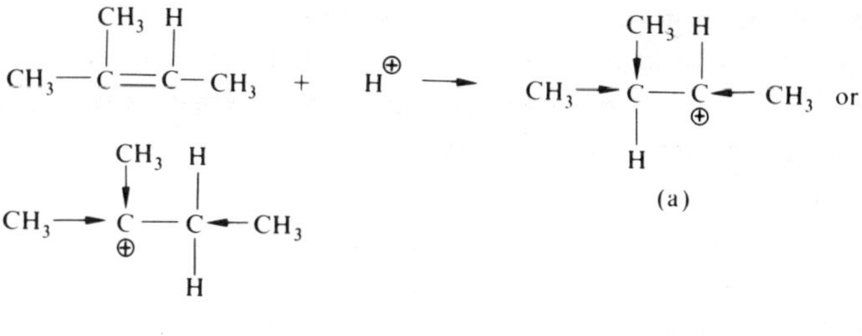

(a)

(b)

a little electronic charge to help neutralize the positive charge introduced into the molecule by the hydrogen ion. A carbon atom carrying a full positive or negative charge is a very unstable arrangement. The ion will react rapidly by the addition of $Br^{\ominus}$.

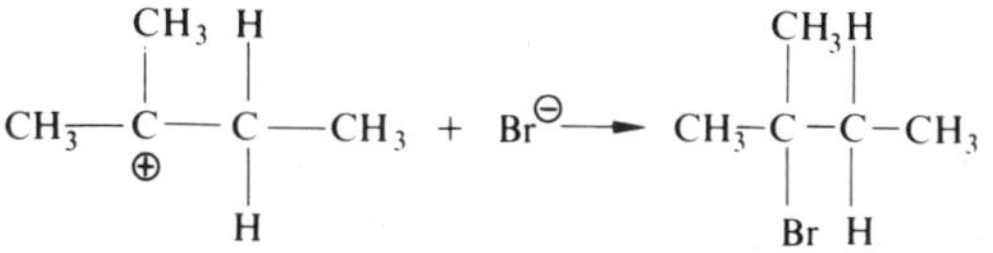

The existence of the intermediate ion must not be interpreted too literally, sometimes such ions exist, for a finite time, and can be detected, in other cases, the addition of $H^{\oplus}$ and $Br^{\ominus}$ may follow each other so closely that the intermediate ion may have no real existence at all. Nevertheless, this method of considering which will be the more stable intermediate, in the light of electron withdrawing or electron-donating groups enables us to predict the course of these additions and give a rational interpretation of Markovnikov's rule.

These intermediate positive carbon ions are called *carbonium* ions (compare ammonium ions $R_4N^{\oplus}$) and will be encountered frequently in later work. Considering the electron donating effect of alkyl groups, it follows that tertiary carbonium ions are the most stable and primary carbonium ions the least stable.

Stability of alkyl carbonium ions: tertiary > secondary > primary

For a time the addition of *HBr* to some double bonds appeared to be an exception to Markovnikov's rule and the ratio of the products formed varied from one preparation to another. In 1933 Kharasch showed that hydrogen bromide adds by a different mechanism when *light* or *peroxides* are present. In these circumstances the reaction proceeds *via* free radicals.

$$CH_3-CH{=}CH_2 + HBr \xrightarrow{\text{peroxides}} CH_3-CH_2-CH_2Br$$

Peroxides are formed in small quantities when alkenes have been left in contact with atmospheric oxygen. Therefore, an old sample of alkene, if not freed of peroxide can give an unexpected product. The peroxides decompose to give free radicals which then attack the alkene.

Just as the order of stability of carbonium ions is tertiary> secondary> primary, so for free radicals the same order applies, tertiary free radicals are the most stable and primary radicals the least, but here, bromine radicals are being added to the alkene first, and so a secondary or tertiary free radical is formed, which then abstracts a hydrogen atom from another HBr molecule to maintain the chain reaction:

$$R—O—O—R \longrightarrow 2R\cdot + O_2$$
(peroxide)

$$R\cdot + HBr \longrightarrow RH + Br\cdot$$

$$Br\cdot + CH_3—CH{=}CH_2 \longrightarrow CH_3—\dot{C}H—CH_2Br \xrightarrow{HBr} CH_3—CH_2—CH_2Br + \dot{B}r$$

Light also splits HBr into radicals, which in the absence of an alkene would rapidly recombine, but in the presence of an alkene start the radical chain reaction.

Further indication that the anti-Markovnikov addition of HBr is a radical reaction is found in the observation that inhibitors (e.g. hydroquinone) which inhibit radical reactions inhibit this reaction too.

Apparently only bromine radicals are formed in sufficient quantity and are of sufficient energy to maintain this reaction. Iodine radicals are very easily formed but are not sufficiently reactive to attack the alkene. At high temperatures too, hydrogen bromide adds by the free radical mechanism and this agrees with the second part of Markovnikov's statement.

3. *Alcohols can be converted to alkyl halides by hydrogen halide, often with the help of an acidic catalyst.* This method is important because alcohols are often readily available. There is a considerable variation in the ease with which different types of alcohol react; tertiary alcohols react very readily, secondary less, and primary least of all. Mixing a tertiary alcohol with aqueous hydrogen halide usually gives the tertiary halide without heat being applied though part of the alcohol may be dehydrated to an alkene (see page 47). Secondary alcohols react equally well with HBr and HI.

$$CH_3-\underset{\underset{CH_3}{|}}{\overset{\overset{CH_3}{|}}{C}}-OH + HCl \longrightarrow CH_3-\underset{\underset{CH_3}{|}}{\overset{\overset{CH_3}{|}}{C}}-Cl + H_2O$$

$$CH_3-\underset{\underset{OH}{|}}{CH}-CH_3 + HI \longrightarrow CH_3-\underset{\underset{I}{|}}{CH}-CH_3 + H_2O$$

Secondary alcohols usually need a catalyst (sulphuric acid or zinc chloride) with hydrochloric acid to give the chlorides, and primary alcohols require heating with concentrated sulphuric acid or zinc chloride:

$$CH_3CHCH_3 \xrightarrow[HCl]{ZnCl_2} CH_3CHCH_3 + H_2O$$
$$\quad\; | \qquad\qquad\qquad\qquad\; |$$
$$\quad OH \qquad\qquad\qquad\qquad Cl$$

$$CH_3CH_2CH_2OH \xrightarrow[H_2SO_4]{HCl} CH_3CH_2CH_2Cl + H_2O$$

Primary, secondary and tertiary iodides can be readily made by heating alcohols with potassium iodide and 95% phosphoric acid:

$$R-OH + KI + H_3PO_4 \longrightarrow R-I + KH_2PO_4 + H_2O$$

Alcohols by themselves are not reactive towards halide ions, i.e. the OH group cannot be displaced by halide, but in acid solution some of the alcohol molecules add hydrogen ion to form oxonium ions, $ROH_2^\oplus$ (compare $H_3O^\oplus$). In the case of tertiary alcohols, this ion can dissociate into water and a carbonium ion $R^\oplus$, which is stabilized by the electron-donating alkyl groups. The carbonium ion, once formed, rapidly reacts with halide ion to give the tertiary halide:

$$CH_3-\underset{\underset{CH_3}{|}}{\overset{\overset{CH_3}{|}}{C}}-\overset{\oplus}{\underset{H}{O}}H \rightleftharpoons H_2O + CH_3-\underset{\underset{CH_3}{|}}{\overset{\overset{CH_3}{|}}{C}}{}^\oplus \xrightarrow{Cl^\ominus} CH_3-\underset{\underset{CH_3}{|}}{\overset{\overset{CH_3}{|}}{C}}-Cl$$

The fewer the electron-donating groups attached to the C—OH, the less stable such an intermediate carbonium ion will be. A secondary alcohol reacts in this way more slowly, but it can be assisted by adding a reagent that will help to remove the OH groups. Sulphuric acid acts by donating protons, and increasing the concentration of oxonium ions, $ROH_2^\oplus$, which can slowly form carbonium ions:

$$CH_3 \underset{H}{\overset{CH_3}{-}} \!\!\!\overset{|}{\underset{|}{C}}\!\!\!-OH \quad \underset{}{\overset{H^\oplus}{\rightleftharpoons}} \quad CH_3\!\!-\!\!\underset{H}{\overset{CH_3\ H}{\underset{|}{C}}}\!\!-\!\!\underset{\oplus}{OH} \rightarrow CH_3\!\!-\!\!\underset{H}{\overset{CH_3}{\underset{|}{C}}}\!\!\overset{\oplus}{} \ + \ H_2O$$

$$Cl^\ominus \downarrow$$

$$CH_3\!\!-\!\!\underset{H}{\overset{CH_3}{\underset{|}{C}}}\!\!-\!\!Cl$$

Zinc chloride probably works in a similar way, forming a complex with the OH group, and splitting off a $Zn(OH)Cl_2^\ominus$ ion.

Primary alcohols are too reluctant to form carbonium ions, because there is little donation of electrons to stabilize them. It is found they react by a different route, also in the presence of a strong acid. A bimolecular collision is necessary to break the bond between carbon and oxygen and a new bond is simultaneously formed on the other side of the carbon atom.

$$CH_3 \!-\!OH \overset{H^\oplus}{\rightleftharpoons} CH_3\!-\!\underset{\oplus}{OH} \overset{Cl^\ominus}{\longrightarrow} \underset{H}{\overset{H}{Cl\cdots\overset{|}{\underset{|}{C}}\cdots\underset{\oplus}{OH}}} \longrightarrow CH_3Cl \ + \ H_2O$$

These two kinds of reaction, dissociation followed by combination of ions and attack on one side of a carbon atom followed by expulsion of a group or atom from the other side are fundamental kinds of reaction for organic molecules and will be encountered frequently.

4. *Phosphorus halides or thionyl halides react with alcohols to give the corresponding alkyl halides.* Phosphorus pentachloride, tribromide or tri-iodide or red phosphorus and bromine or iodine all can be used. Primary, secondary and tertiary alcohols can be used.

$$ROH \ + \ PCl_5 \longrightarrow RCl \ + \ POCl_3 \ + \ HCl$$

$$3ROH \ + \ PBr_3 \longrightarrow 3RBr \ + \ H_3PO_3$$

Thionyl chloride and thionyl bromide react readily with tertiary alcohols. Primary and secondary alcohols give better yields if pyridine* is added.

$$ROH \ + \ SOCl_2 \longrightarrow RCl \ + \ SO_2 \ + \ HCl$$

* Pyridine, C_5H_5N is a cyclic nitrogen compound with basic properties, often used as a reagent where a base is required.

5. *Halogen Exchange.* Alkyl chlorides and bromides can be converted to iodides by heating them in acetone with sodium iodide. The method is useful for converting an unreactive chloride into a more reactive iodide.

$$RCl + NaI \longrightarrow RI + NaCl$$

All the other three halides can be converted to fluorides by heating in ethylene glycol with anhydrous potassium fluoride.

$$RX + KF \longrightarrow RF + KX$$

6. When the silver salt of a carboxylic acid is treated with halogen (usually bromine) in the presence of a solvent, silver halide is formed, the acid is decarboxylated and an alkyl halide with one less carbon atom is formed. This is known as the Hunsdiecker Reaction.

$$RCOOAg + Br_2 \dashrightarrow RBr + AgX + CO_2$$

Properties of Alkyl Halides

The halogen derivatives of the alkanes do not occur in nature. The lower molecular weight compounds usually have a characteristic sweet odour and a sweet taste. Except for the lowest members of the series (see Table 5), which are gases, they are liquids with boiling points higher than the corresponding alkanes. As with the alkanes, the more branched the chains, the lower the boiling points. For the same carbon skeleton, boiling points increase in the order F, Cl, Br, I, as would be expected from the molecular weights. They are insoluble in water and soluble in organic solvents. Many compounds are toxic to animals, particularly certain compounds with one fluorine atom, which are extremely poisonous.

The direct fluorination of alkanes is too violent to be a useful preparative method, hence the fluoro-alkanes are made from alcohols or by other indirect processes; they are not common laboratory reagents. However, many fluorinated compounds have unusual and useful properties and in recent years fluoro-alkanes have received a great deal of study. Perfluoro-alkanes—those in which all hydrogen atoms have been replaced by fluorine—are surprisingly inert towards most chemical reagents.

Reactions of Alkyl Halides

The reactivity of monohalo-alkanes decreases in the order $I > Br > Cl > F$ and also in the order tertiary > secondary > primary.

There is also a tendency for halides with short alkyl chains to react faster than corresponding ones with longer chains.

Because a halide atom is strongly electron attracting, in an alkyl halide there is a net negative charge on the halogen atom and a net positive charge on the carbon atom to which it is attached, e.g. we may write:

$$CH_3 \overset{\delta+}{\underset{}{-}} CH_2 \overset{\delta-}{\underset{}{-}} Cl$$

Reagents that have available non-bonded electrons (e.g. $OH^\ominus$, $CN^\ominus$, $:NH_3$), originally known as *anionoid* reagents, now widely known as *nucleophilic* (or nucleus-seeking) reagents, react readily with a carbon atom that carries a partial positive charge, as the carbon atom attached to halogen does. *Cationoid* reagents, now known as *electrophilic* reagents, are those deficient in electrons, e.g. $H^\oplus$, $R^\oplus$. The reaction with nucleophilic reagents is one of the most important characteristics of the alkyl halides. Many diverse reactions are simply variations on the reaction:

$$Nu^\ominus + R \overset{\delta+}{\underset{}{-}} CH_2 \overset{\delta-}{\underset{}{-}} X \longrightarrow R \underset{}{-} CH_2 \underset{}{-} Nu + X^\ominus$$

where $Nu^\ominus$ represents any of a large number of nucleophiles.

A great deal of evidence has been collected to show that these reactions are bi-molecular at primary carbon atoms (abbreviated to S_N2, =substitution, nucleophilic, bimolecular) and unimolecular at secondary and tertiary carbon atoms (abbreviated to S_N1=substitution, nucleophilic, unimolecular), just as in the case of conversion of alcohols to halides.

For example, the exchange of chloride for iodide (page 33) is a typical case, the nucleophilic reagent being $I^\ominus$:

primary:

secondary and tertiary:

Both reactions are reversible, but a number of factors combine to favour formation of the alkyl iodide.

1. *Hydrolysis.* Reaction of a halide with an aqueous solution of a base or with water alone at 100°, gives an alcohol. The kinetics and

mechanism of the reaction have received a great deal of attention. In detail, the reaction, a nucleophilic displacement of X by $OH^\ominus$, varies with conditions and the particular compound, but essentially we may write the reaction thus:

$$CH_3CH_2Br \;+\; OH^\ominus \;\longrightarrow\; CH_3CH_2OH \;+\; Br^\ominus$$

With secondary and tertiary halides there is an increasing tendency to form an alkene and of a mixture of products is always obtained.

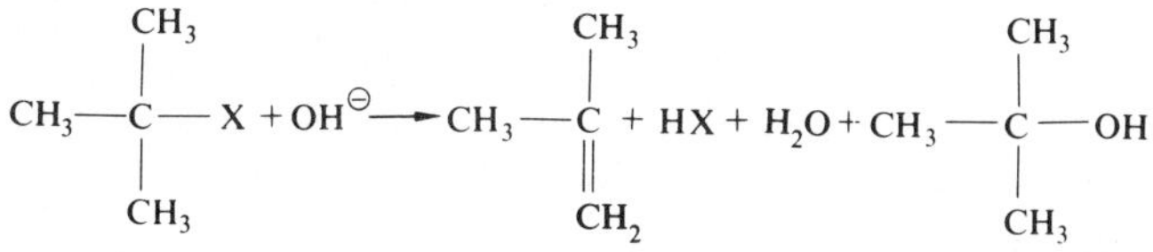

The carbonium ion formed as an intermediate can stabilize itself in three ways, by recombining with a halide ion (no net reaction) by combining with the hydroxyl ion (the desired substitution reaction) or by eliminating a proton from an adjoining carbon atom, leading to the formation of a double bond:

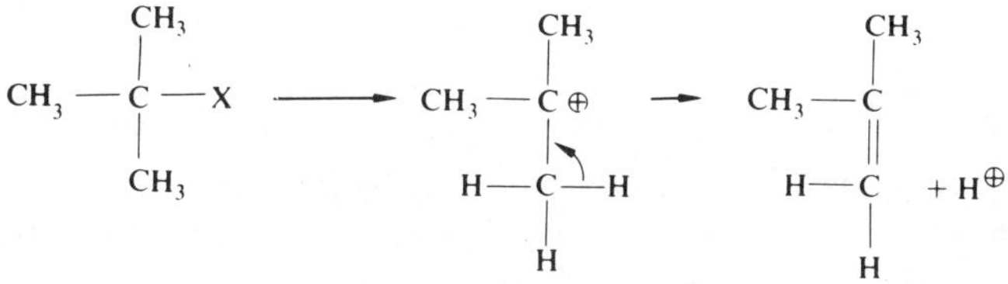

The bent arrow indicates the movement of a pair of electrons. The pair joining C to H are now used to join C to C, all valencies are satisfied and the positive charge is now on the hydrogen atom which has been eliminated, this is a more stable system than a positive charge on carbon. The factors affecting substitution (alcohol formation) versus elimination (double bond formation) will be discussed later.

2. *Reaction with Salts.* The reaction of alkyl halides with hydroxide ion is just one example of a very wide range of S_N2 and S_N1 reactions. The alkyl halides will react with a number of anions which are also good nucleophiles (e.g. acetate, cyanide, nitrite or alkoxide, $RO^\ominus$ derived from alcohols):

$$CH_3CH_2Br \;+\; CH_3\overset{O}{\overset{\|}{C}}O^\ominus \longrightarrow CH_3CH_2O\overset{O}{\overset{\|}{C}}CH_3 \;+\; Br^\ominus$$

$$CH_3I \;+\; CN^\ominus \longrightarrow CH_3CN \;+\; I^\ominus$$

$$RBr \;+\; CH_3O^\ominus \longrightarrow ROCH_3 \;+\; Br^\ominus$$

35

The formation of cyanides applies to primary and secondary halides only. Potassium cyanide is a basic reagent (due to hydrolysis to HCN and $Na^{\oplus}OH^{\ominus}$) and gives chiefly alkenes from tertiary halides by the elimination reaction described above.

3. *Reaction with Metals.* (a) With sodium, an alkyl sodium is probably first formed which immediately reacts with more alkyl halide to give an alkane (the Wurtz reaction, page 11):

$$CH_3I + 2Na \longrightarrow [CH_3Na] + NaI$$

The intermediate alkyl sodium can be thought of as containing another nucleophilic reagent (in this case $CH_3^{\ominus}$) which attacks another molecule of alkyl halide.

$$[CH_3Na] + CH_3I \longrightarrow CH_3-CH_3 + NaI$$

(b) With lithium, the alkyl lithium first formed is not quite so reactive, and may be used, in solution, for further reactions.

$$CH_3CH_2Br + 2Li \longrightarrow CH_3CH_2Li + LiBr$$

(c) With magnesium. Provided the reagents are very dry, magnesium reacts readily with simply alkyl halides to give complex alkylmagnesium halides, called *Grignard reagents* after their discoverer. The Grignard reagents are extremely useful synthetic tools and will be encountered again and again in later chapters. Alkyl iodides and bromides are usually employed for forming the Grignard reagent in the laboratory; chlorides react more sluggishly but find many such applications in industry.

$$CH_3I + Mg \xrightarrow{\text{in ether}} CH_3 - Mg - I$$
methyl magnesium iodide

Grignard reagents react with water, alcohols or ammonia to give an alkane. Compounds that react in this way with Grignard reagents are said to contain active hydrogen.

$$R - Mg - Br + \begin{cases} HOH & \longrightarrow & RH + MgBrOH \\ R'OH & \longrightarrow & RH + MgBrOR' \\ NH_3 & \longrightarrow & RH + MgBrNH_2 \end{cases}$$

If methylmagnesium iodide is used, methane gas is formed and can be collected and the volume estimated. If an excess of methylmagnesium halide is used, the amount of —OH or =NH in an unknown sample can be determined. The method is called the Zerevitinov determination.

Grignard reagents provide a means of reducing alcohols to alkanes:

$$R-OH \longrightarrow RBr \longrightarrow RMgBr \xrightarrow{H_2O} RH + MgBrOH$$

The basic magnesium salt formed is decomposed with dilute acid:

$$MgXOH + HX \longrightarrow MgX_2 + H_2O$$

The more important uses of Grignard reagents will be mentioned in later chapters.

(d) Other metals. Zinc and mercury react in a manner similar to magnesium, but the derivatives are more limited in usefulness than the Grignard reagents. A sodium-lead alloy reacts with ethyl chloride to give tetraethyl lead, a volatile organic compound of lead, which is added to petrol in 0·1% concentration to help reduce 'knocking' (see page 21):

$$4EtCl + 4Na-Pb \longrightarrow Et-\underset{\underset{Et}{|}}{\overset{\overset{Et}{|}}{Pb}}-Et + 4NaCl + 3Pb$$

4. *Reduction.* Reduction to an alkane can be carried out in four ways, but these are of limited preparative importance.

(a) With hydrogen and a catalyst, such as nickel and palladium, under pressure.

(b) With a dissolving metal:

$$CH_3Br \xrightarrow{Sn/HCl.} CH_4 + HBr$$

(c) Lithium aluminium hydride reduces chlorides, bromides and iodides under mild conditions in ether solution:

$$4RCl + LiAlH_4 \longrightarrow RH + LiCl + AlCl_3$$

(d) The Grignard reagents with water give alkanes, as described in section 3(c) above.

5. *Elimination of Hydrogen Halide.* Basic conditions (usually potassium hydroxide in ethanol) more drastic than those required to convert a halide to an alcohol, will eliminate hydrogen halide from

an alkyl halide to give an alkene. The details of this reaction are described in the next chapter.

$$CH_3CHBrCH_3 \xrightarrow[\text{EtOH}]{\text{KOH}} CH_3CH{=}CH_2$$

6. *Primary alkyl halides react with ammonia and amines to give substituted ammonium salts:*

$$CH_3CH_2Br + NH_3 \longrightarrow CH_3CH_2\overset{\oplus}{N}H_3\ \overset{\ominus}{Br}$$

ethyl ammonium bromide

The reaction proceeds by the same S_N2 mechanism as described earlier. The nucleophilic attacking reagent is ammonia or the amine. The reaction does not work as well for secondary or tertiary halides, which prefer S_N1 reactions.

$$H_3N: \curvearrowright \overset{\displaystyle CH_3}{\underset{|}{CH_2}} - Br \longrightarrow H_3\overset{\oplus}{N} - \overset{\displaystyle CH_3}{\underset{|}{CH_2}}\ \overset{\ominus}{Br}$$

The reaction does not stop there. Because the salt formed can exchange with ammonia, further reaction can take place, and a mixture of products is usually formed:

$$CH_3CH_2\overset{\oplus}{N}H_3\ \overset{\ominus}{Br} + NH_3 \rightleftharpoons CH_3CH_2NH_2 + \overset{\oplus}{N}H_4\ \overset{\ominus}{Br}$$

ethylamine

$$CH_3CH_2NH_2 + CH_3CH_2Br \longrightarrow (CH_3CH_2)_2\overset{\oplus}{N}H_2\ \overset{\ominus}{Br} \underset{}{\overset{NH_3}{\rightleftharpoons}} (CH_3CH_2)_2NH$$

diethylamine

$$(CH_3CH_2)_2NH + CH_3CH_2Br \longrightarrow (CH_3CH_2)_3\overset{\oplus}{N}H\ \overset{\ominus}{Br} \overset{NH_3}{\rightleftharpoons} (CH_3CH_2)_3N$$

triethylamine

$$(CH_3CH_2)_3N + CH_3CH_2Br \longrightarrow (CH_3CH_2)_4\overset{\oplus}{N}\ \overset{\ominus}{Br}$$

tetra-ethylammonium bromide

The mono-, di- and tri-ethylammonium salts are easily converted to the free amine with alkali, so the mixture of amines can be separated by distillation.

$$(CH_3CH_2)_2\overset{\oplus}{N}H_2\ \overset{\ominus}{Br} + NaOH \longrightarrow (CH_3CH_2)_2NH + NaBr + H_2O$$

The tetra-alkylammonium salts are more stable and are not affected by aqueous alkali.

Generally bromides and iodides are preferred as laboratory reagents in the reactions considered above because they react more easily, but bromine and iodine are extracted from sea water by a rather expensive process, so these elements and compounds containing them are expensive. Where cost becomes important, as in industrial processes, the cheaper chlorides are used and higher temperatures or pressures are used to obtain comparable rate of reaction.

Alkyl Polyhalides

Organic compounds containing more than one halogen atom per molecule are an important group of chemicals. Two or more halogen atoms can replace hydrogen in an alkane to give a number of isomers. Geminal and vicinal dihalides, defined earlier in this chapter, are two groups deserving special attention. Table 5 lists some of the familiar mono- and polyhalo-alkanes.

Table 5.   Halogen derivatives of alkanes

Common name	Systematic name	Formula	m.p. °C	b.p. °C
Methyl fluoride	fluoromethane	CH_3F	-115	-78
Methyl chloride	chloromethane	CH_3Cl	-98	-24
Methyl bromide	bromomethane	CH_3Br	-93	5
Methyl iodide	iodomethane	CH_3I	-66	42
Ethyl fluoride	fluoroethane	CH_3CH_2F	-143	-38
Ethyl chloride	chloroethane	CH_3CH_2Cl	-136	12
Ethyl bromide	bromoethane	CH_3CH_2Br	-118	38
Ethyl iodide	iodoethane	CH_3CH_2I	-111	72
Propyl bromide	1–bromopropane	$CH_3CH_2CH_2Br$	-110	71
Isopropyl bromide	2–bromopropane	$CH_3CHBrCH_3$	-89	60
Methylene chloride	dichloromethane	CH_2Cl_2	-96	40
Chloroform	trichloromethane	$CHCl_3$	-64	61
Carbon tetrachloride	tetrachloromethane	CCl_4	-22	77
Methylene bromide	dibromomethane	CH_2Br_2	-53	98
Bromoform	tribromomethane	$CHBr_3$	9	150
Carbon tetrabromide	tetrabromomethane	CBr_4	48	189
Methylene iodide	di-iodomethane	CH_2I_2	5	180
Iodoform	tri-iodomethane	CHI_3	119	—
Carbon tetra-iodide	tetra-iodomethane	CI_4	dec.	—
Ethylene chloride	1,2-dichloroethane	$ClCH_2CH_2Cl$	-35	83
Ethylidene chloride	1,1-dichloroethane	CH_3CHCl_2	-97	57

Preparation of Dihalo-alkanes

1. *Geminal dihalides are prepared by treating aldehydes or ketones (Chapter 8) with phosphorus halides.* Phosphorus pentabromide

cannot be used with ketones because replacement of H by Br occurs as well.

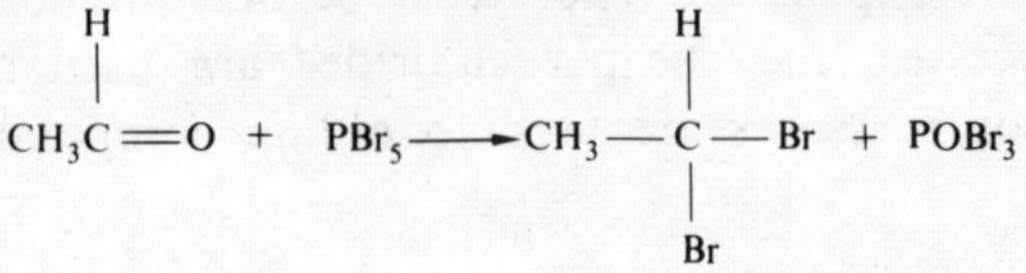

$$\underset{H}{\overset{H}{CH_3C}}{=}O \; + \; PBr_5 \longrightarrow CH_3 - \overset{\displaystyle H}{\underset{\displaystyle Br}{C}} - Br \; + \; POBr_3$$

ethylidene bromide

$$\underset{O}{\overset{\displaystyle \parallel}{CH_3CCH_3}} \; + \; PCl_5 \longrightarrow CH_3 - \overset{\displaystyle Cl}{\underset{\displaystyle Cl}{C}} - CH_3 \; + \; POCl_3$$

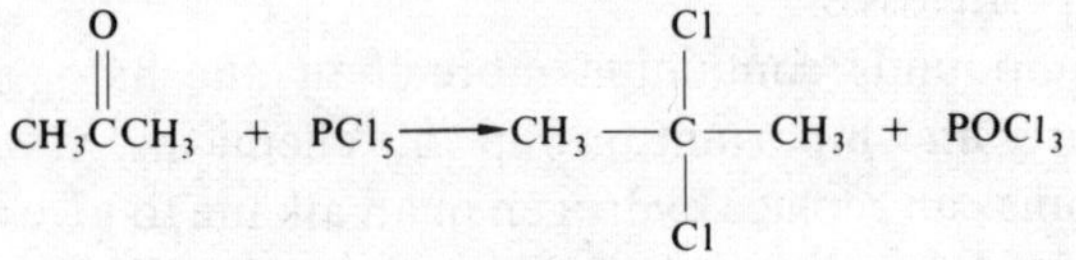

isopropylidene chloride

2. *Acetylenes and hydrogen halide give geminal dihalides.* The addition of the second molecule of HCl follows Markovnikov's rule:

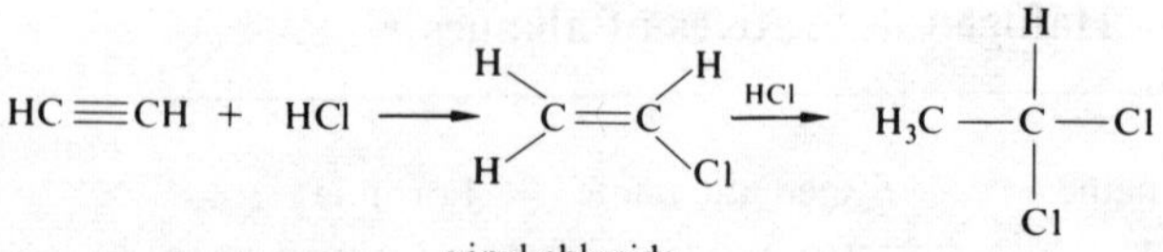

$$HC{\equiv}CH \; + \; HCl \longrightarrow \underset{H}{\overset{H}{C}}{=}\underset{Cl}{\overset{H}{C}} \overset{HCl}{\longrightarrow} H_3C - \overset{\displaystyle H}{\underset{\displaystyle Cl}{C}} - Cl$$

vinyl chloride

3. *Vicinal dihalides are prepared by the addition of halogen to alkenes:*

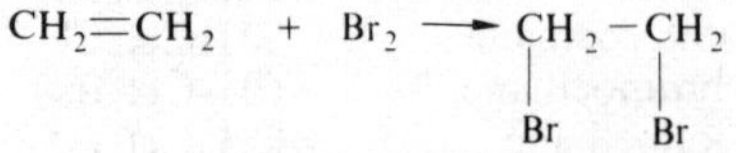

$$CH_2{=}CH_2 \; + \; Br_2 \longrightarrow \underset{Br}{\overset{}{CH_2}} - \underset{Br}{\overset{}{CH_2}}$$

ethylene dibromide

4. *Glycols (dihydroxy compounds) can be converted to dihalides with phosphorus halides as for monohalo-alkanes.* Some vicinal glycols (e.g. ethylene glycol, propylene glycol) are readily available.

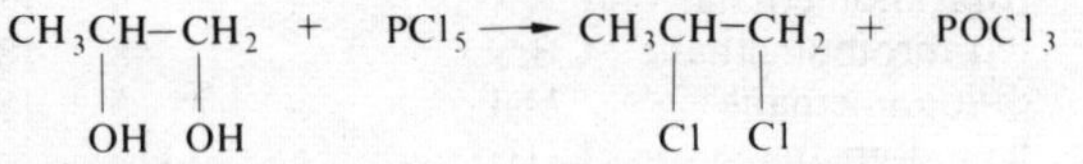

$$\underset{\substack{OH \; OH \\ \text{propylene glycol}}}{CH_3CH{-}CH_2} \; + \; PCl_5 \longrightarrow \underset{\substack{Cl \; Cl \\ \text{propylene dichloride}}}{CH_3CH{-}CH_2} \; + \; POCl_3$$

$$\underset{\text{trimethylene glycol}}{HO{-}CH_2CH_2CH_2{-}OH} \; + \; PCl_5 \longrightarrow \underset{\text{trimethylene dichloride}}{Cl{-}CH_2CH_2CH_2{-}Cl} \; + \; POCl_3$$

5. *The Hunsdiecker reaction, carried out on the silver salt of a dicarboxylic acid gives a polymethylene dihalide.*

$$\underset{\text{silver adipate}}{AgOOC(CH_2)_4COOAg} \; + \; Br_2 \longrightarrow \underset{\text{tetramethylene dibromide}}{Br(CH_2)_4Br} \; + \; CO_2 \; + \; AgBr$$

Reactions of Dihalo-alkanes

The dichloro- and dibromo-alkanes, in general, are stable compounds. Geminal dihalides are hydrolysed by dilute alkali to the corresponding aldehydes and ketones:

$$CH_3 - \underset{\underset{Br}{|}}{CH} - Br \ + \ H_2O \longrightarrow CH_3 - \underset{\underset{OH}{|}}{\overset{\overset{H}{|}}{C}} - OH \longrightarrow CH_3 - \overset{\overset{H}{|}}{C} = O \ + \ HBr$$

Vicinal dihalides more closely resemble the simple alkyl halides. For example, alkaline hydrolysis gives a glycol; these are simple examples of the bimolecular nucleophilic substitution reaction:

$$ClCH_2 CH_2 Cl \ + \ 2NaOH \ \xrightarrow{H_2O} \ HOCH_2 CH_2 OH \ + \ 2NaCl$$
$$\text{ethylene glycol}$$

In strongly alkaline conditions (potassium hydroxide in alcohol) elimination of two molecules of hydrogen halide also occurs to give some acetylene.

$$CH_2 BrCH_2 Br \ + \ 2 \ KOH \ \xrightarrow{EtOH} \ HC \equiv CH \ + \ 2KBr \ + \ 2 H_2O$$

A vicinal dihalide with zinc in alcohol gives an alkene. If the halogen atoms are separated by three or more carbon atoms, a cycloalkane is formed (Chapter 11).

$$CH_3CHBrCH_2 Br \ + \ Zn \longrightarrow CH_3CH = CH_2 + \ ZnBr_2$$

$$\underset{CH_2}{\overset{CH_2 - Br}{\diagup}}\overset{}{\underset{CH_2 - Br}{\diagdown}} \ + \ Zn \longrightarrow \underset{CH_2 -\!- CH_2}{\overset{CH_2}{\diagup \diagdown}} + \ ZnBr_2$$

$$\text{cyclopropane}$$

Reaction of a dihalide with ammonia gives a di-ammonium salt and with potassium cyanide, a dicyanide.

$$\underset{Br}{\overset{CH_2 -}{|}} \underset{Br}{\overset{CH_2}{|}} \ + \ 2KCN \longrightarrow \underset{CN}{\overset{CH_2 -}{|}} \underset{CN}{\overset{CH_2}{|}} \ + \ 2 \ KBr$$

Vicinal di-iodides lose iodide simply on heating to form an alkene. In this way a glycol can be converted to an alkene.

$$\underset{OH}{\overset{CH_2 -}{|}} \underset{OH}{\overset{CH_2}{|}} \xrightarrow{PBr_5} \underset{Br}{\overset{CH_2 -}{|}} \underset{Br}{\overset{CH_2}{|}} \xrightarrow{KI} \underset{I}{\overset{CH_2 -}{|}} \underset{I}{\overset{CH_2}{|}} \longrightarrow CH_2 = CH_2 + I_2$$

Important Halo-alkanes

Fluoroform (CHF_3) is an extremely stable, non-toxic gas (b.p. $-84°$). It can be prepared in many ways, for example from chloroform and silver fluoride.

Carbon tetrafluoride (CF_4) is similar to fluoroform in being an extremely inert gas (b.p. $-128°$). It can be prepared directly from carbon and fluorine (one of the very few organic compounds that can be prepared from the elements).

Mixed fluoro-chloromethanes (CCl_3F, CCl_2F_2 and $CClF_3$) are prepared by the action of HF on carbon tetrachloride or from antimony or cobalt trifluoride and carbon tetrachloride. They are all inert, non-toxic, odourless gases. Under the name of 'Freon' these gases are used as refrigerants and as the propellent gas in aerosols. Halothane (ICI 'Fluothane') $CF_3CHBrCl$ is a valuable anaesthetic.

Methyl chloride (CH_3Cl) is a sweet-smelling gas, soluble in alcohol and water. It is prepared industrially from methanol or methane.

Methylene chloride (CH_2Cl_2) is an important solvent. It has solvent properties and a boiling point (b.p. $40°$) similar to ether and has the advantage that it is less inflammable than ether. It is prepared by chlorinating methyl chloride or by reducing chloroform.

Chloroform ($CHCl_3$) is prepared industrially by reducing carbon tetrachloride with iron and dilute acid. It is a sickly, sweet-smelling liquid. It was once used as a general anaesthetic but is more toxic than other volatile anaesthetics, particularly to the heart and liver. It is an excellent solvent for many organic compounds, including natural rubber. Chloroform tends to oxidize slowly in light and air, producing the extremely poisonous gas, phosgene ($COCl_2$); it is therefore usually kept in brown bottles and about 1% of ethanol is added, which inhibits the decomposition. Alkaline hydrolysis, by a complex reaction, gives a salt of formic acid and some carbon monoxide. With potassium hydroxide and ammonia a vigorous reaction takes place at room temperature giving potassium cyanide, and other products.

$$CHCl_3 + NH_3 + 4KOH \longrightarrow KCN + 3KCl + 4H_2O$$

Chlorination of chloroform gives carbon tetrachloride and reduction with zinc gives methane.

Carbon tetrachloride (CCl_4) is made commercially by chlorinating carbon disulphide, or from carbon disulphide and sulphur monochloride.

$$CS_2 + 2S_2Cl_2 \longrightarrow 6S + CCl_4$$

It is used as a solvent, especially as a dry-cleaning agent, and as a fire extinguisher. At the temperature of the flames it can form phosgene, so that good ventilation is necessary after it has been used on a fire. Sodium should never be used to dry carbon tetrachloride in the laboratory, because together they form a mixture easily exploded by shock. Carbon tetrachloride is decomposed by boiling with alcoholic potassium hydroxide.

$$CCl_4 + 6KOH \longrightarrow K_2CO_3 + 3H_2O + 4KCl$$

Methyl bromide (CH_3Br) is a gas, used widely as a fumigant against insects, rats and mice (for example, in stored grain) and as an insecticide.

Ethylene bromide or dibromide (CH_2BrCH_2Br) is prepared commercially from ethylene and bromine. When tetraethyl lead is used as an anti-knock additive, ethylene bromide is added as well, to prevent the deposition of lead oxide on the inner surfaces of the engine and exhaust. The lead is carried away in the exhaust gases as volatile lead bromide. Partly because it is produced in large quantities for petrol, ethylene bromide is the cheapest organic compound of bromine. It is also used as a solvent. It is poisonous and the vapour is toxic; it is therefore used as a fumigant also.

Methyl iodide (CH_3I) is a liquid, and therefore more popular as a laboratory reagent than methyl chloride or bromide. It is prepared industrially from methanol, red phosphorus and iodine, or from potassium iodide and dimethyl sulphate.

$$CH_3SO_4CH_3 + KI \longrightarrow KSO_4CH_3 + CH_3I$$

With increasing molecular weight, the polyhalo-alkanes become less stable because the C—X bonds are weaker in $-CX_2-$ than in $-CHX-$ and because halogen atoms are very bulky (Fig. 4) and cannot fit compactly on the carbon skeleton. Fluorine is an exception, the atom is small and polyfluoro-compounds can be very stable. Polytetrafluoroethylene, a plastic made by polymerizing tetrafluoroethylene, is an extremely inert material and has valuable anti-stick properties.

$$n\ CF_2{=}CF_2 \longrightarrow \cdots -\underset{\underset{F}{|}}{\overset{\overset{F}{|}}{C}} - \underset{\underset{F}{|}}{\overset{\overset{F}{|}}{C}} - \underset{\underset{F}{|}}{\overset{\overset{F}{|}}{C}} - \underset{\underset{F}{|}}{\overset{\overset{F}{|}}{C}} - \cdots$$

43

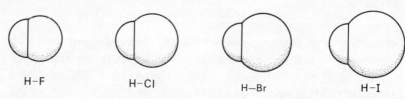

Fig. 4   The relative size of hydrogen and halogen atoms, covalently bound

Identification

Though in solubility properties the halogenated alkanes closely resemble the alkanes, they are easily distinguished by a chemical test for the halogen elements, known as the Lassaigne test or sodium fusion. A small sample of compound is melted in a dry test tube with a clean piece of sodium. The tube is then heated strongly, so that the organic material is completely destroyed, and the halogen is left as sodium halide. The halide ion is detected in the residue by dissolving it in water, neutralizing excess sodium carbonate with nitric acid, and testing with silver nitrate solution. Variations on this test can be used to determine whether the element is Cl, Br or I.

Nitrogen and sulphur can be detected by sodium fusion as well. Nitrogen, with some of the carbon of the organic compound, is converted to sodium cyanide, which is detected as the complex iron cyanide, Prussian blue. Sulphur is converted to sodium sulphide.

The C—Cl bond absorbs strongly around 700 cm^{-1} in the infrared spectrum, but so do many other groups, so evidence of absorption in this region is not reliable evidence on its own. C—Br and C—I bonds absorb at still lower frequency, out of the range of ordinary spectrometers.

3
Alkenes

The Double Bond

The removal of two hydrogen atoms from adjacent carbon atoms of an alkane leads to the formation of an *olefin* or *alkene* containing a *double bond*. This double bond can be conceived, drawn and derived in many ways, each having its usefulness and limitations. Essentially, it consists of four valence electrons instead of the usual two, each carbon atom contributing two electrons (Fig. 5).

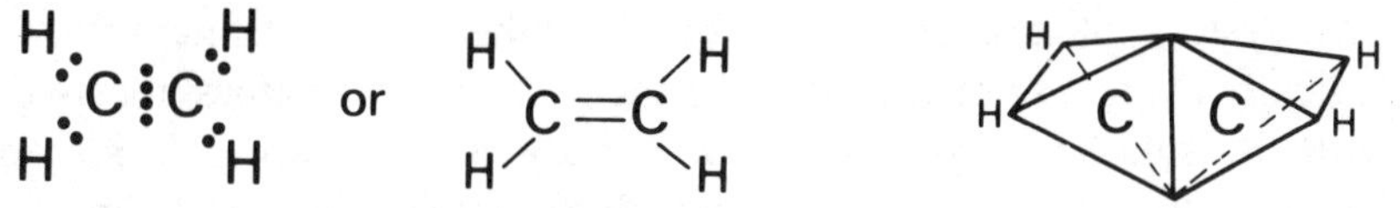

Fig. 5   Alternative pictures of the ethylene molecule

Two of the four valences of carbon are utilized in forming this bond, and we may then regard the two tetrahedral carbon atoms as connected along one edge of their tetrahedra (Fig. 5). Two consequences follow from this structure. First, that substituents attached to the double-bonded carbon atoms will all lie in the same plane; and secondly, that rotation about the double bond is impossible without rupturing one of the bonds. These concepts were developed to explain the properties of a group of compounds which we now know as the alkenes. For example, there are three alkenes with the formula C_4H_8 derived formally from n-butane by the loss of two hydrogen atoms:

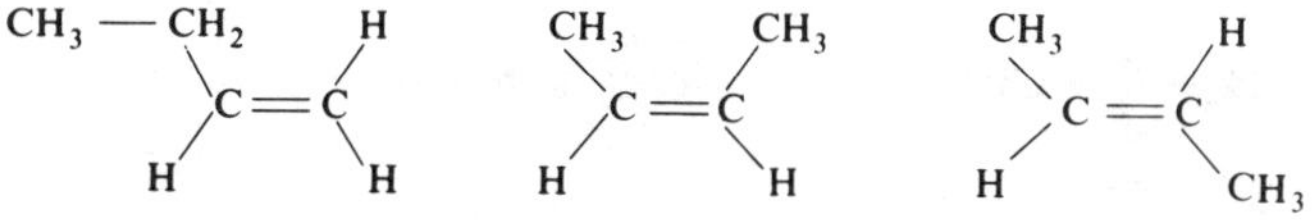

but–1–ene   *cis* – but–2– ene   *trans* –but–2– ene

The but-2-enes are physically and chemically distinct compounds. *Cis*-but-2-ene has a b.p. of 4° and *trans*-but-2-ene has a b.p. of 1°. The form with the same or similar groups on the same side of the

45

molecule is called the *cis*-form and the form with the similar groups on opposite sides of the molecule is called *trans*. This general pheno-menon of having the same structural elements with two possible arrangements of them around a double bond (or as we learn later, a ring of atoms) is called *geometrical isomerism* or *cis-trans* isomerism. Notice that in but-1-ene such isomerism cannot exist. For alkenes with three or four alkyl groups attached to the double bond, the isomer name is found by considering the longest alkyl chain passing through the double bond. For example, the carbon skeleton below would be named as a *trans*-hept-3-ene:

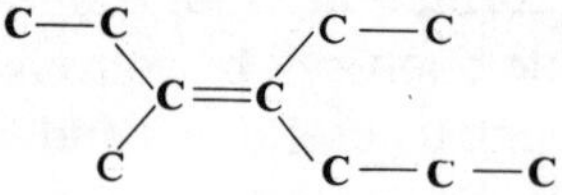

$$\begin{array}{ccc} C-C & & C-C \\ & C=C & \\ C & & C-C-C \end{array}$$

Nomenclature

The system of nomenclature for the alkenes or olefins follows logically from that of the alkanes; the same radical names are used, with the suffix *-ene* instead of *-ane* and a number which is placed between the stem and the suffix which gives the position of the first atom bearing the double bond. Some examples are given below. Some common alkenes have trivial names which are given in brackets.

$CH_2\!=\!CH_2$ ethene (ethylene)

$CH_3\!-\!CH\!=\!CH_2$ propene (propylene)

$CH_3\!-\!CH_2\!-\!CH\!=\!CH_2$ but-1-ene (1-butene or 1-butylene)

$CH_3\!-\!CH\!=\!CH\!-\!CH_3$ but-2-ene (2-butene or 2-butylene)

$$\begin{array}{c} CH_3 \\ | \\ CH_3\!-\!C\!=\!CH_2 \end{array}$$ 2-methylpropene (isobutylene)

$$\begin{array}{c} CH_3 \\ | \\ CH_3\!-\!CH_2\!-\!C\!-\!CH\!=\!CH_2 \\ | \\ H \end{array}$$ 3-methylpent-1-ene

When complex alkenes are to be named, the longest chain contain-ing the double bond is selected and numbered so as to give the double bond the lowest possible number. In the next example, the compound is 3-ethyl-4-methylhex-2-ene:

$$
\begin{array}{c}
CH_3 \\
| \\
CH_2 \\
\underset{6}{CH_3}-\underset{5}{CH_2}-\underset{4}{CH}-\underset{}{\underset{|}{C}}=\underset{2}{CH}-\underset{1}{CH_3} \\
\qquad\quad | \qquad 3 \\
\qquad\quad CH_3
\end{array}
$$

and not a hex-4-ene, or a diethylpentene. Alkenes are also named sometimes as alkyl derivatives of ethylene, so that propylene is methylethylene, 1-butylene is ethylethylene, and isobutylene can be called 1,1-dimethylethylene. Olefins with one double bond are grouped under the name *alkenes* and have the general formula C_nH_{2n}.

Preparation

The alkenes occur to a small extent in natural gas and petroleum, but large quantities are produced in the catalytic cracking of petroleum, and from this source, they are today important intermediates for the preparation of other chemicals. The chief laboratory methods, a few of which are also used industrially in special cases, are as follows:

1. *Dehydration of an Alcohol.* The removal of the elements of water from an alcohol gives an alkene. This is often carried out with sulphuric or phosphoric acid, in the liquid phase, or by passing the vapour of the alcohol over a catalyst such as phosphoric acid, pumice, or alumina at 350°.

$$
CH_3CH_2OH \xrightarrow{\;H^{\oplus}\;} CH_2{=}CH_2 + H_2O
$$

This method is used commercially when alcohol is available cheaply. In practice, some charring occurs using sulphuric acid, but if benzenesulphonic or toluenesulphonic acid is used, the charring is avoided. Secondary and tertiary alcohols react more readily, giving high yields of alkenes.

That secondary and tertiary alcohol dehydrate more easily than primary alcohols and why acids catalyse this reaction can be explained in terms of what we know about how these reactions proceed.

If we dissolve sulphuric acid in water, oxonium ions are formed by protonation of water by the strong acid:

$$
H{:}\ddot{O}{:}\,H + H_2SO_4 \rightleftharpoons H{:}\overset{\oplus}{\underset{H}{\ddot{O}}}{:}H + HSO_4^{\ominus}
$$

47

By dissolving sulphuric acid in alcohol, we get the analogous protonated alkyloxonium ion:

$$CH_3CH_2OH + H_2SO_4 \rightleftharpoons CH_3CH_2 \overset{\oplus}{\underset{H}{-OH}} + HSO_4^{\ominus}$$

Notice that by breaking the C—O bond now we would form a molecule of water and a primary carbonium ion:

$$CH_3CH_2\overset{\oplus}{\underset{H}{OH}} \not\leftarrow CH_3CH_2^{\oplus} + H_2O$$

This is an unfavourable reaction, because although the water molecule is very stable, we have pointed out that primary carbonium ions are very unstable and are also difficult to form. But if another molecule of alcohol is nearby, as it always is in the alcoholic solution, this can accept a proton from the adjacent carbon atom of alcohol, and if the removal of H and H₂O takes place simultaneously, the whole process becomes energetically favourable:

$$CH_3CH_2\overset{H}{O} \quad H_2C\overset{}{-}CH_2\overset{\oplus}{\underset{H}{-O-H}} \rightarrow CH_3CH_2OH + H_2C=CH_2 + H_2O$$

The arrows, as before, represent movement of electron pairs to form new bonds. The $HSO_4^{\ominus}$ ion does not enter into the reaction, and a new alcohol molecule is protonated as another is dehydrated.

Secondary carbonium ions are more easily formed: here protonation can lead directly to C—O bond breaking, without the intervention of another alcohol molecule:

$$CH_3-\overset{CH_3}{\underset{H}{C}}-OH + H_2SO_4 \rightleftharpoons CH_3-\overset{CH_3}{\underset{H}{C}}\overset{\oplus}{\underset{H}{-O-H}} \underset{\substack{(slow \\ stage)}}{\rightarrow} CH_3-\overset{CH_3}{\underset{H}{C^{\oplus}}} + H_2O$$

And this can happen still more readily with a tertiary alcohol:

$$CH_3-\overset{CH_3}{\underset{CH_3}{C}}-OH \underset{}{\overset{H^{\oplus}}{\rightleftharpoons}} CH_3-\overset{CH_3}{\underset{CH_3}{\overset{}{C}}}\overset{\oplus}{\underset{H}{-OH}} \underset{\substack{(slow \\ stage)}}{\rightarrow} CH_3-\overset{CH_3}{\underset{CH_3}{C^{\oplus}}} + H_2O$$

As was described in the last chapter, these carbonium ions can stabilize themselves by addition of a cation (here it would be $HSO_4^{\ominus}$

or $SO_3^=$, but these are stable ions and have no strong tendency to form covalent bonds) or by elimination of hydrogen from an adjacent carbon atom.

$$\underset{\underset{H}{|}}{\overset{\overset{H_2C\text{---}H}{|}}{CH_3\text{---}C^\oplus}} \longrightarrow \underset{\underset{H}{|}}{\overset{\overset{H_2C \;+\; H^\oplus}{\parallel}}{CH_3\text{---}C}}$$

The reaction described for primary alcohols requires two events to take place more or less simultaneously, that is, collision with a second alcohol molecule to remove a proton and also breaking of a C—O bond. Therefore this reaction proceeds more slowly than the reaction of secondary and tertiary alcohols, where the stages of the reaction can take place in sequence—protonation of the alcohol, loss of water to give the carbonium ion, and loss of $H^\oplus$ to give the alkene.

The dehydration of a primary alcohol is a sequence of reactions, but the slowest stage of the sequence, the one that governs the overall rate of the reaction, is a bimolecular process, so this type of reaction is symbolized by E2 (elimination, bimolecular) and the secondary and tertiary alcohol route is E1 (elimination, unimolecular) since here the slow, rate-determining stage is the splitting of the protonated alcohol to give the carbonium ion. E1 reactions go so readily for many tertiary alcohols that simply distilling them in the presence of acidic impurities will convert them into the alkene.

Sometimes an intermediate ester is formed with the acid which is then decomposed. In the case of sulphuric acid and ethyl alcohol, this is an S_N2 reaction followed by an E2 reaction.

$$H_2SO_4 \;+\; CH_3CH_2OH \;\rightleftharpoons\; HSO_4^\ominus \;+\; CH_3CH_2\overset{\oplus}{\underset{H}{O}}H$$

$$HSO_4^\ominus \;+\; \underset{H_2 \;\; H}{\overset{CH_3}{\underset{|}{C\text{---}\overset{\oplus}{O}H}}} \;\xrightarrow{S_N2}\; HO\overset{O}{\underset{\underset{O}{\parallel}}{\overset{\parallel}{S}}}O \;\cdots\; \underset{H_2 \;\; H}{\overset{CH_3}{\underset{|}{C}}} \;\cdots\; OH \longrightarrow HO\overset{O}{\underset{\underset{O}{\parallel}}{\overset{\parallel}{S}}}O\text{---}CH_2 \;+\; H_2O$$

$$CH_3CH_2\underset{..}{O}H$$
$$\searrow H$$
$$\underset{\underset{O}{\parallel}}{H_2C\overset{\frown}{\underset{\smallfrown}{\text{---}}}CH_2\text{---}O\overset{O}{\underset{\underset{O}{\parallel}}{\overset{\parallel}{S}}}OH} \;\xrightarrow{E_2}\; CH_3CH_2\overset{\oplus}{\underset{H}{O}}H \;+\; CH_2\!=\!CH_2 \;+\; HSO_4^\ominus$$

Aliphatic and Alicyclic Compounds

Notice that the protonation of the alcohol makes the S_N2 reaction go more easily because $HSO_4^{\ominus}$ has to expel H_2O, a stable molecule, which requires less energy than producing $OH^{\ominus}$. S_N2 reactions with alcohols in general go more easily in an acidic medium, because then one is dealing with a protonated alcohol.

2. *Some active metals, such as zinc (often used as the alloy, zinc-copper couple) remove the halogen from vicinal dihalo-alkanes giving alkenes.* This method gives very pure ethylene from 1,2-dibromo-ethane and is useful for preparing small quantities, but as the dihalides are usually made from the ethylene itself, it is not of general preparative usefulness.

$$RCH-CHR' + Zn \longrightarrow RCH=CHR' + ZnBr_2$$
$$\quad | \quad\ |$$
$$\quad Br \quad Br$$

The reaction is particularly useful for 'protecting' double bonds. For example, in a molecule containing several functional groups, we may wish to carry out a reaction on another group but leave the double bond intact. If the double bond is sensitive to the reagent, we can first convert it to the dibromo-derivative, carry out the reaction on the other functional group and then regenerate the double bond with zinc.

3. *An alkyl halide gives an alkene with alcoholic potassium hydroxide, by elimination of hydrogen halide:*

$$CH_3CH_2CH_2I + KOH \xrightarrow{EtOH} CH_2=CH-CH_3 + KI + H_2O$$

For ethyl halides the yield in this reaction is small, but for higher alkyl halides it can be very good. A primary alkyl halide gives a terminal alkene (i.e. where the double bond is at the end of an alkyl chain):

$$CH_3CHCH_2CH_2Br \xrightarrow[EtOH]{KOH} CH_3CHCH=CH_2$$

Secondary halides can usually give two possible products. According to Saytzev's rule, the alkene that will be preferred is the one having *least* hydrogen attached to the double bond carbons (the double bond prefers to be flanked by alkyl groups):

$$CH_3CH_2CHCH_3 \text{ (X)} \longrightarrow CH_3CH=CHCH_3, \text{ not } CH_3CH_2CH=CH_2$$

$$CH_3CHCHCH_2CH_3 \text{ (CH}_3\text{)(X)} \longrightarrow CH_3C=CHCH_2CH_3 \text{ (CH}_3\text{)}, \text{ not } CH_3CHCH=CHCH_3 \text{ (CH}_3\text{)}$$

This reaction warrants a closer examination because of its general usefulness and because it sometimes intrudes when we do not want it.

In Chapter 2 it was explained how a nucleophilic reagent such as hydroxyl ion can *replace* halogen in an alkyl halide by an S_N1 or S_N2 reaction. Now we are saying the same reagent can cause elimination. In fact both reactions, substitution and elimination, are competing reactions and we must choose our conditions to favour one reaction or the other.

The dehydrohalogenation of a primary alkyl halide is an E2 reaction very similar to the dehydration of an alcohol:

$$CH_2\text{—}CH_2\text{—}Cl \text{ (H)(HO}^\ominus\text{)} \longrightarrow H_2O + CH_2=CH_2 + Cl^\ominus$$

It requires a collision of hydroxyl ion with a β-hydrogen atom (the carbon atom next to the one bearing a group we are considering is called the a-carbon atom, the next one farther away is the β-carbon and so on). The rate of reaction is dependent upon the concentration of hydroxyl ion and alkyl halide but also the mechanism shows that the controlling step is a bimolecular reaction, it belongs to the E2 class.

Tertiary halides undergo what appears to be a spontaneous ionization to halide ion and carbonium ion, but the rate of this ionization is strongly dependent on the solvent. Just as polar liquids (particularly water) are good solvents for ionic substances by surrounding the ions and shielding their electronic charges from mutual attraction and repulsion (Fig. 6) so they appear to surround and shield the carbonium and halide ions as these are formed and help to separate them. Certainly the reaction is much faster in polar solvents. This shielding of ions by solvent molecules is called *solvation*.

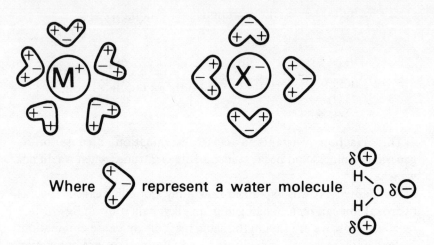

Where ⟩ represent a water molecule

Fig. 6   Diagrammatic representation of solvation of metal and halide ions in water

In S_N2 and E2 reactions, no carbonium ions are formed, so the efficiency of solvation does not affect the reaction rate. Indeed, the change in rate of reaction with change of solvent can be used to decide between S_N1 and S_N2 types.

Kinetic studies show that the initial ionization to a carbonium ion is a slow process, followed by a rapid elimination of a proton to give the double bond, the rate of reaction is therefore dependent only on the concentration of the tertiary halide, and belongs to the E1 or unimolecular class.

Secondary halides, intermediate in character between primary and tertiary halides, tend to resemble tertiary halides more closely in their reaction mechanisms.

We understand a great deal about the *stereochemistry*, that is, the arrangement of the reactants in space for this type of reaction. In order to illustrate it, a simplified diagrammatic convention is needed to take the place of perspective drawings as in Fig. 7a which represents an ethyl halide molecule. Two systems are commonly used, either the *Newman projection* (Fig. 7b), which shows the molecule viewed end-on with C_1 directly in front of C_2, or the 'saw-horse' simplified perspective shown in Fig. 7c, which is close to the perspective view of 7a. Notice that the groups are staggered, although they can rotate about the C—C bond; the mutual repulsion of the

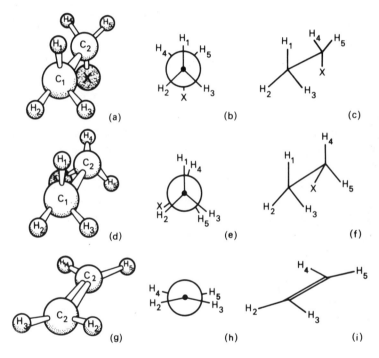

Fig. 7 *Elimination from an ethyl halide (a) Staggered arrangement, perspective view of model (b) Corresponding Newman projection and (c) Saw-horse diagram (d) Eclipsed arrangement of groups with (e) Newman projection and (f) Saw-horse model (g) Perspective view of ethylene model with (h) Newman projection and (i) Saw-horse model*

hydrogen and halogen atoms will cause the molecule on average to take up this position. The picture shown in 7e is much less favoured because H_1 is brought closer to H_4, H_3 to H_5 and H_2 to X.

When elimination takes place, it is H_1 that is lost from C_1 and X from C_2, that is the groups that are eliminated are on opposite sides of the molecule. This is called *trans* elimination. If instead of H_2, H_3, H_4 and H_5 we had different groups R_2, R_3, R_4 and R_5 we can see that one particular alkene will be formed, in place of ethylene, 7g, and from the knowledge we have of the way the elimination proceeds we can predict the arrangement of the groups R_2, R_3, R_4 and R_5 around the double bond. Further, if the carbon-to-carbon bond is prevented in any way from rotating freely so that a hydrogen atom cannot arrange itself on the opposite side from the halogen atom then elimination of hydrogen halide can become very difficult or completely inhibited. This can be caused if there are bulky groups attached to C_1 or C_2 or if the carbon atoms are joined up in a ring.

Alkyl halides with large bulky groups and tertiary halides prefer to undergo elimination rather than substitution. But given a particular compound, how can one influence it to give a substitution or elimination product? Using a solvent of high polarity (e.g. water) and a low concentration of the nucleophilic reagent, will favour substitution. For elimination it is better to use a solvent of low polarity (alcohol, ether or benzene) and a high concentration of the nucleophile to remove the proton on the β-carbon atom. Higher temperatures also favour elimination reactions. If we could choose a reagent which was a strong nucleophile (would undergo nucleophilic substitution readily) but a weak base (would not eliminate the β-hydrogen) that would be ideal for substitution, and the opposite conditions would be good for elimination, but in general basicity and nucleophilicity go hand in hand. Tertiary amines are sometimes better as bases for elimination than hydroxyl ion.

The more highly branched alkenes are the more stable compounds (we can check this from other data such as the heat of hydrogenation of isomeric compounds) and so these are preferentially formed. Saytzev's rule applies for all simple compounds but with very complex molecules other factors come into action.

4. *The Wittig Reaction*. This is a method introduced in 1954 whereby an alkene can be built up from two components, a halide and an aldehyde or ketone. The halide is first reacted with triphenylphosphine, a phosphorus compound similar in behaviour to a tertiary amine. With the halide, a quaternary phosphonium salt is obtained, which is treated with a strong base, giving a phosphorane. This reacts with a ketone or aldehyde, the phosphorus has a strong affinity for oxygen, triphenylphosphine oxide is formed and the two other fragments combine to give an alkene:

$$\begin{array}{c} \text{Ph} \\ | \\ \text{Ph}-\text{P} \\ | \\ \text{Ph} \end{array} + \text{CH}_3\text{CH}_2\text{Br} \longrightarrow \begin{array}{cc} \text{Ph} & \text{Br}^{\ominus} \\ | & \\ \text{Ph}-\overset{\oplus}{\text{P}}-\text{CH}_2\text{CH}_3 \\ | \\ \text{Ph} \end{array} \xrightarrow{\text{NaOEt}} \begin{array}{c} \text{Ph} \\ | \\ \text{Ph}-\text{P}=\text{CHCH}_3 \\ | \\ \text{Ph} \end{array}$$

triphenylphosphine   phosphonium salt   phosphorane

$$\text{Ph}_3\text{P}=\text{CHCH}_3 + \text{RC}=\text{O} \longrightarrow \overset{\text{H}}{\underset{|}{\text{O}}}=\text{CH}-\text{R} \longrightarrow \text{Ph}_3\text{P}=\text{O} + \text{CH}_3\text{CH}=\text{CHR}$$

$$\text{Ph}_3\text{P}=\text{CH}-\text{CH}_3$$

triphenylphosphine oxide

Ph = phenyl

5. As already mentioned, the catalytic cracking of petroleum fractions to give mixtures of alkene and hydrogen have been extensively studied, and by a suitable choice of petroleum fraction, temperature, pressure and catalyst, a great variety of olefins can be produced.

Reaction of Alkenes

The alkenes show a greater range of reaction types than the alkanes or alkyl halides. Therefore although their physical properties are close to those of the alkanes (see Table 6) the two groups can easily be distinguished chemically and hence separated. Alkenes burn in air with a yellow flame due to the production of incandescent carbon particles.

Table 6. Some normal alk-l-enes

Name	Formula	m.p. °C	b.p. °C
Ethylene	$CH_2{=}CH_2$	−169	−102
Propylene	$CH_3CH{=}CH_2$	−185	−48
But-1-ene	$CH_3CH_2CH{=}CH_2$	−185	−7
Pent-1-ene	$CH_3(CH_2)_2CH=CH_2$	−165	30
Hex-1-ene	$CH_3(CH_2)_3CH{=}CH_2$	−138	64
Hept-1-ene	$CH_3(CH_2)_4CH{=}CH_2$	−119	93
Oct-1-ene	$CH_3(CH_2)_5CH{=}CH_2$	−104	123
Non-1-ene	$CH_3(CH_2)_6CH{=}CH_2$		145
Dec-1-ene	$CH_3(CH_2)_7CH{=}CH_2$	−87	172
Undec-1-ene	$CH_3(CH_2)_8CH{=}CH_2$	−50	189
Dodec-1-ene	$CH_3(CH_2)_9CH{=}CH_2$	−35	213

1. *Reduction*. Hydrogen adds to the double bond, under the influence of a catalyst (usually platinum, palladium or nickel) giving

an alkane. With an active, finely divided platinum catalyst, the reduction can be carried out in the laboratory at normal temperature and pressure. Reduction with hydrogen gas is called hydrogenation.

$$CH_2 = CH_2 + H_2 \longrightarrow CH_3 - CH_3$$

This is an example of a heterogeneous reaction. Many of the vapour phase reactions described in the section on petroleum belong to this class. We do not know the exact details of this reaction, but we do know that both hydrogen and alkanes are adsorbed onto the surface of many metals, such as platinum or nickel and that addition of hydrogen is always *cis*, that is, to the same side of the molecule.

It would seem that the hydrogen and alkene are both adsorbed onto the metal surface near each other (Fig. 8) and that while they are in this adsorbed condition (represented by dotted lines in Fig. 9) there is a simultaneous rearrangement of bonds (represented by dashed lines), H—H bond and C—C bond breaking with H—C and H—C bond formation. The alkane is not so strongly adsorbed and leaves the metal free for further adsorption of hydrogen and alkene.

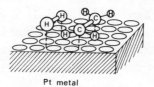

Pt metal

Fig. 8   Adsorption of ethylene and hydrogen on a platinum catalyst surface

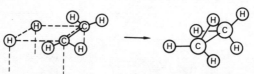

Fig. 9   The cis *addition of hydrogen to ethylene*

2. *Oxidation.* The product depends on the strength of the oxidizing agent. A comparatively mild reagent, such as potassium permanganate in neutral solution produces a dihydroxy compound or *glycol*:

$$3 CH_2 = CH_2 + 2 KMnO_4 + 4 H_2O \longrightarrow 2 MnO_2 + 2 KOH + 3 \underset{\underset{OH}{|}}{CH_2} - \underset{\underset{OH}{|}}{CH_2}$$

ethylene glycol

This reaction is used as a test for a double bond (the Baeyer test). Substances containing a double bond when shaken with cold dilute potassium permanganate solution immediately decolorize the permanganate. Some other easily oxidized substances interfere with the test. The hydroxyl groups are added *cis.*

Stronger oxidizing agents split the glycol first formed, and give two fragments which can be either an acid or a ketone, depending upon the number of alkyl groups attached to the carbon of the double bond:

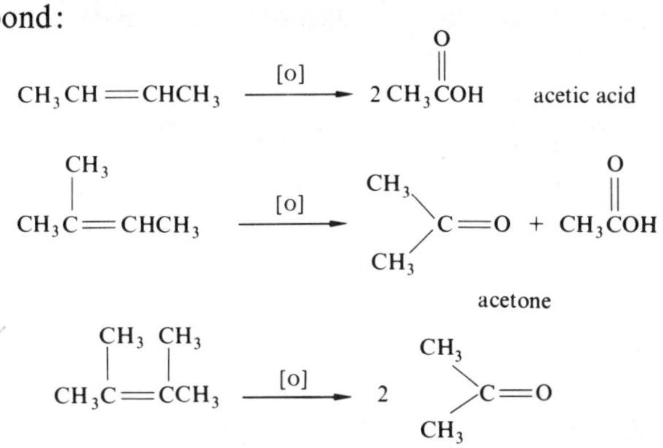

$$CH_3CH\!\!=\!\!CHCH_3 \xrightarrow{\text{[o]}} 2\,CH_3\overset{\displaystyle O}{\overset{\|}{C}}OH \quad \text{acetic acid}$$

$$\underset{CH_3}{\overset{\displaystyle CH_3}{\underset{|}{CH_3C}}}\!\!=\!\!CHCH_3 \xrightarrow{\text{[o]}} \underset{CH_3}{\overset{CH_3}{\diagdown}}C\!\!=\!\!O + CH_3\overset{\displaystyle O}{\overset{\|}{C}}OH$$

$$\text{acetone}$$

$$\underset{CH_3C}{\overset{CH_3\ CH_3}{\underset{|\quad|}{CH_3C}}}\!\!=\!\!CCH_3 \xrightarrow{\text{[o]}} 2\ \underset{CH_3}{\overset{CH_3}{\diagdown}}C\!\!=\!\!O$$

Osmium tetroxide (OsO_4) or concentrated hydrogen peroxide in formic acid oxidizes alkenes in a manner similar to permanganate to give glycols, except that the peroxidation reaction is *trans.*

A gentle method of oxidizing alkenes that splits them into two fragments is to treat them at room temperature with ozone; the reaction is rapid and complete. The ozone adds across the double bond, probably first through the formation of a molozonide as shown below, which is unstable and decomposes to an aldehyde or ketone (represented by $R_2C\!\!=\!\!O$) and another fragment. In a neutral solvent, such as chloroform, these recombine to the normal ozonide. In a reactive solvent such as ethanol they give the $R_2C\!\!=\!\!O$ compound and a hydroperoxide. The hydroperoxides are themselves oxidizing agents, giving off hydrogen peroxide with water and this can easily oxidize any aldehydes formed in the reaction; so the ozonide or hydroperoxide is reduced with a metal dissolving in acid, hydrogen and a catalyst or a reducing agent such as titanous dichloride ($TiCl_2$):

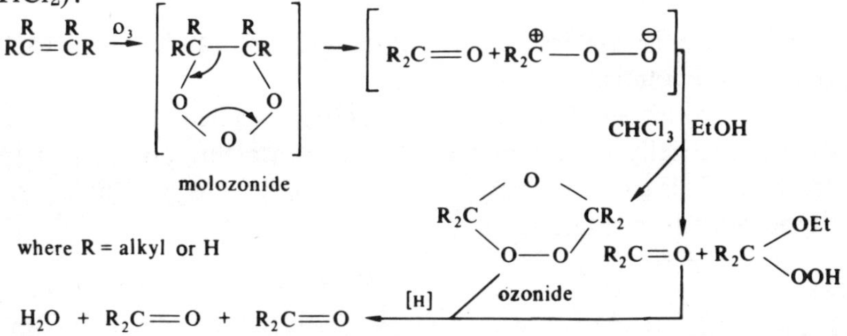

where R = alkyl or H

$$H_2O + R_2C\!\!=\!\!O + R_2C\!\!=\!\!O$$

The products are mixtures of aldehydes and ketones depending on the number of alkyl groups attached to the double bond. The aldehydes and ketones are easily identified, so that ozonolysis is a particularly useful method for finding just exactly where the double bond is in an unknown alkene. For example but-1-ene gives one molecule of propionaldehyde and one of formaldehyde (CH_2O) on ozonolysis, while but–2–ene gives two molecules of acetaldehyde:

$$CH_3CH_2CH{=}CH_2 \xrightarrow[CHCl_3]{O_3} CH_3CH_2CH \underset{O-O}{\overset{O}{\diagdown O \diagup}} CH_2 \xrightarrow[H_2O]{H_2} CH_3CH_2C{\overset{H}{=}}O + CH_2O + H_2O$$

$$CH_3CH{=}CHCH_3 \xrightarrow[CHCl_3]{O_3} CH_3CH \underset{O-O}{\overset{O}{\diagdown \diagup}} CHCH_3 \xrightarrow[H_2O]{H_2} 2\,CH_3C{\overset{H}{=}}O + H_2O$$

2-Methylbut-2-ene would give one molecule of acetone and one of acetaldehyde:

$$(CH_3)_2C{=}CHCH_3 \xrightarrow{O_3} CH_3C \underset{O-O}{\overset{CH_3 \quad O}{\diagdown \diagup}} CHCH_3 \xrightarrow{H_2} CH_3\overset{CH_3}{C}{=}O + CH_3\overset{H}{C}{=}O$$

Ozone is readily prepared in the laboratory. The method is also used for synthesis even on an industrial scale.

3. *Epoxidation.* When an alkene is treated with a peracid (e.g. peracetic acid) the double bond adds oxygen and an epoxide is formed. The peracetic acid can be prepared from acetic acid and hydrogen peroxide with a little sulphuric acid as catalyst:

$$CH_3\overset{O}{\overset{\|}{C}}OH + H_2O_2 \xrightarrow{H^{\oplus}} CH_3\overset{O}{\overset{\|}{C}}OOH + H_2O$$

peracetic acid

$$RCH{=}CHR' + CH_3\overset{O}{\overset{/\!/}{C}}OOH \longrightarrow RCH\overset{O}{\overset{\diagup\diagdown}{-}}CHR' + CH_3\overset{O}{\overset{\|}{C}}OH$$

Epoxides are themselves reactive compounds and can be converted to glycols or alcohols.

Direct oxidation of ethylene to its epoxide (called ethylene oxide) is an industrially important reaction as the starting point for many processes. It is carried out by direct reaction with oxygen (air) over a supported silver catalyst:

$$CH_2\!=\!CH_2 + O_2 \quad \xrightarrow[250°]{Ag} \quad CH_2\!\!-\!\!CH_2$$
$$\underset{O}{\diagdown\!\!\diagup}$$ ethylene oxide

Propylene gives a poor yield of epoxide by this method, and much CO_2 is formed, but in this case isobutane can be used as a carrier, that is, isobutane forms a peroxide which then oxidizes propylene. Propylene oxide also is an important industrial chemical.

$$\underset{\displaystyle CH_3CHCH_3}{\overset{\displaystyle \overset{CH_3}{|}}{}} + O_2 \longrightarrow (CH_3)_3C \longrightarrow (CH_3)_3COH + O\cdot$$
$$\underset{O-OH}{|}$$

$$\Big| CH_3CH\!=\!CH_2$$

$$\downarrow$$

$$CH_3CH\!\!-\!\!CH_2$$
$$\underset{O}{\diagdown\!\!\diagup}$$
propylene
oxide

4. *Addition of Halogens and Hydrogen Halides.* Chlorine and bromine add readily to double bonds, usually at or below room temperature, giving dihalides. Ethylene was originally called *olefiant gas* because with chlorine it gave an oily liquid (dichloroethane) and it was from this that the alkenes became known as olefins.

$$CH_2\!=\!CH_2 + Cl_2 \longrightarrow ClCH_2CH_2Cl$$

Hydrogen halides also add readily, hydrogen iodide most readily; hydrogen chloride usually requires a high temperature. The addition follows Markovnikov's Rule (page 27):

$$CH_3CH\!=\!CH_2 + HCl \longrightarrow CH_3CHCH_3$$
$$\underset{Cl}{|}$$

When an alkene is chlorinated or brominated in aqueous solution, a halo-alcohol (or halohydrin) is formed:

$$CH_3CH\!=\!CH_2 + Br_2 + H_2O \longrightarrow CH_3CHCH_2Br + HBr$$
$$\underset{OH}{|}$$

The addition of hydrogen halide according to Markovnikov's Rule was explained in Chapter 2 in terms of the relative stabilities of carbonium ions. Now we can extend the explanation given there to bring these three reactions into a uniform mechanism.

Some reservations were made in the earlier explanation about the real existence of carbonium ions and no explanation was offered as to how ethylene itself adds hydrogen halide. Is a primary carbonium ion the intermediate here?

59

Aliphatic and Alicyclic Compounds

The answer is that the addition of HX to ethylene seems to be a simultaneous process, with $H^\oplus$ adding at one end of the double bond and $Br^\ominus$ at the other. Not only that, but the addition is *trans*, $H^\oplus$ adding from one side and $Br^\ominus$ from the other (Fig. 10). The addition is simply the reverse of the E2 elimination reaction mechanism.

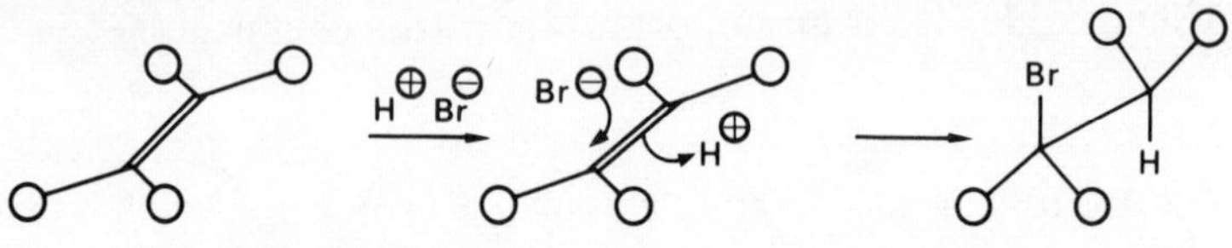

Fig. 10   The trans *addition of HBr to an alkene*

Carbon-to-carbon double bonds can be regarded as nucleophiles. Double bonds are rich in electrons and the electrons are less firmly held that in a single bond. We can regard the double bond itself as the attacking reagent, and the movement of electrons when it attacks an electron deficient species (e.g. $H^\oplus$) thus:

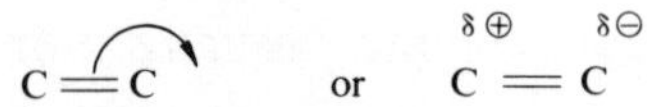

At the same time as negative charge is building up on one C atom, the other C atom is becoming negatively charged and this attracts the negative $Br^\ominus$ ion. In solution, the alkene molecule is surrounded by $H^\oplus$ and $Br^\ominus$ ions, so the appropriate ion to complete the concerted process should always be close by:

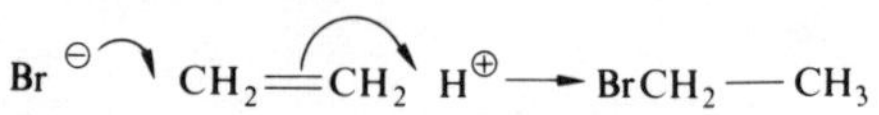

In the formation of the dihalide, the double bond is attacking molecules of halogen, but these molecules themselves are easily polarized, i.e. induced to accept a $\delta^\oplus$ or $\delta^\ominus$ charge:

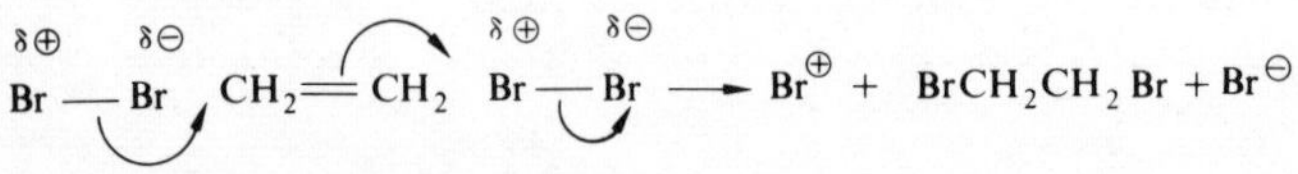

$Br^\oplus$ and $Br^\ominus$ would be expected to quickly recombine to Br_2 or to be absorbed in reaction with another molecule of ethylene. Studies with unsymmetrically substituted ethylenes have shown that the mechanism involves a *trans*-addition (as in Fig. 10) for bromine and all the other examples given in this section.

When the reaction is carried out in aqueous solution, a water molecule completes the reaction in place of the second halogen molecule and the halohydrin is formed:

60

$$H—O \quad CH_2 =\!\!=CH_2 \quad Br—Br \longrightarrow H^\oplus + HOCH_2CH_2Br + Br^\ominus$$
$$\underset{H}{|}$$

As the reaction proceeds, HBr accumulates in the solution and the following reaction takes place as well:

$$Br^\ominus \quad CH_2 =\!\!=CH_2 \quad Br—Br \longrightarrow BrCH_2CH_2Br + Br^\ominus$$

In fact any of a great variety of reagents can take part in the concerted process, e.g. an added halide ion, $CN^\ominus$, an alcohol or an acid:

$$Cl^\ominus + CH_2 =\!\!=CH_2 + Br_2 \longrightarrow ClCH_2CH_2Br + Br^\ominus$$

$$C\!\equiv\!N^\ominus + CH_2 =\!\!=CH_2 + Br_2 \longrightarrow N\!\equiv\!CCH_2CH_2Br + Br^\ominus$$

$$CH_3OH + CH_2 =\!\!=CH_2 + Br_2 \longrightarrow H^\oplus + CH_3OCH_2CH_2Br + Br^\ominus$$

$$\overset{O}{\underset{\|}{CH_3COH}} + CH_2 =\!\!=CH_2 + Br_2 \longrightarrow H^\oplus + \overset{O}{\underset{\|}{CH_3COCH_2CH_2Br}} + Br^\ominus$$

Propylene has an electron-donating methyl group, so it can be expected to polarize with negative charge away from the methyl group:

$$Z^\ominus \overset{\delta\oplus}{\underset{H}{\overset{CH_3}{C}}}\!\!=\!\!CH_2^{\delta\ominus} \quad X—X \longrightarrow Z\!-\!\overset{CH_3}{\underset{H}{C}}\!-\!CH_2X + X^\ominus$$

rather than:

$$Z^\ominus \quad CH_2 =\!\!=\!\!\overset{CH_3}{\underset{H}{C}} X—X \longrightarrow Z\!-\!CH_2\!-\!\overset{CH_3}{\underset{H}{C}}\!-\!X + X^\ominus$$

This mechanistic picture rationalizes not only Markovnikov's Rule for the addition of HX, but can predict the arrangement of halohydrins and other products of this group of reactions. We can explain this reaction without calling on carbonium ions at all. The two pictures of concerted reactions and carbonium ion formation are complementary, it may be a matter of a very short time interval between addition of $H^\oplus$ and $Br^\ominus$, so short that at present we are unable to distinguish it from an instantaneous process.

5. *Allylic Chlorination and Bromination.* Chlorine and bromine can be *substituted* for hydrogen as a carbon atom next to a double bond,

by a free radical reaction. The group CH_2=CH—CH_2— is called the *allyl* group, so this reaction is called allylic substitution. It is a very useful and important method for introducing a second functional group into an alkene.

If chlorine is dissociated by light or heat into chlorine free radicals (page 14) these can attack propylene in two ways. First, by addition to give a chloropropyl radicals which can then either add a second chlorine atom to give dichloropropane or decompose, giving back propylene:

$$Cl_2 + 2Cl \cdot \xrightarrow{\text{CH}_2\text{ CHCH}_3} ClCH_2\dot{C}HCH_3$$

$$ClCH_2\dot{C}HCH_3 \; + \; Cl_2 \longrightarrow ClCH_2CHClCH_3 \; + \; Cl.$$

$$ClCH_2\dot{C}HCH_3 \longrightarrow CH_2{=}CHCH_3 + Cl\cdot$$

Secondly, chlorine can remove a hydrogen atom from the methyl group, to give an allyl radical, which can only become a stable molecule by addition of a chlorine atom to give allyl chloride:

$$Cl\cdot + CH_2{=}CHCH_3 \longrightarrow CH_2{=}CH\dot{C}H_2 + HCl$$

$$CH_2{=}CH\dot{C}H_2 + Cl_2 \longrightarrow CH_2CHCH_2Cl + Cl\cdot$$

At higher temperatures the reaction to give allyl chloride is favoured. Indeed this reaction is used industrially to give allyl chloride, contaminated with only a little dichloropropane by chlorination of propylene above 400°. The allyl chloride is important for the synthetic production of glycerol.

$$CH_2{=}CHCH_2Cl \xrightarrow[\text{OH}^{\ominus}]{\text{H}_2\text{O}} CH_2{=}CHCH_2OH \xrightarrow[\text{H}_2\text{O}]{\text{Cl}_2}$$

$$\underset{\overset{|}{OH}\;\;\;\overset{|}{Cl}\;\;\;\overset{|}{OH}}{CH_2-CH-CH_2} \xrightarrow{\text{NaOH}} \underset{\overset{|}{OH}\;\;\overset{|}{OH}\;\;\overset{|}{OH}}{CH_2-CH-CH_2}$$

glycerol

Bromine reacts similarly at high temperatures. High temperature gas reactions of this kind are not easy to handle in the laboratory, but there is an alternative reagent, N–bromosuccinimide, which in solution gives a low concentration of bromine, which similarly gives bromoalkyl and allylic radicals. Because the concentration of bromine is low and the reaction to give allyl radicals is slightly favoured, only allyl bromides are formed. In general free radicals remove hydrogen on carbon *next* to a double bond (allylic hydrogen) more easily than normal alkyl hydrogen or hydrogen attached to the double bond, so again because of the low concentration of bromine radicals, only allylic substitution takes place and alkyl groups are not affected.

$$Br\cdot + RCH=CHCR_2H \longrightarrow RCH=CH\dot{C}R_2 + HBr$$

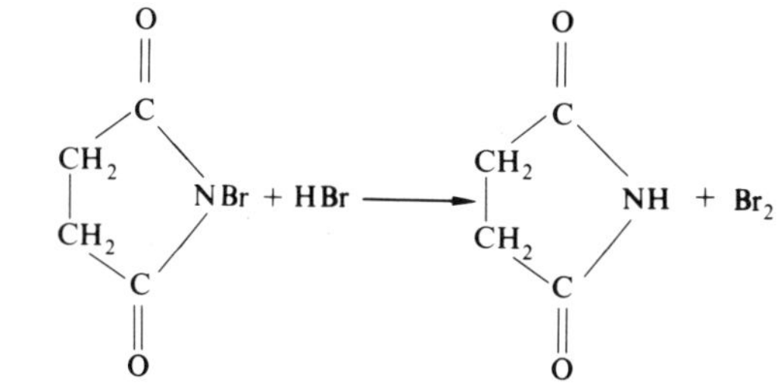

N – Bromosuccinimide   Succinimide

$$Br_2 + RCH=CH\dot{C}R_2 \longrightarrow RCH=CHCR_2Br + Br\cdot \text{ etc.}$$

6. *Addition of Acids.* Other acids add in a way similar to the hydrogen halides, and these additions also follow Markovnikov's Rule, i.e. the negative part of the molecule adds to the carbon atom with least hydrogen. In the case of sulphuric acid, an ester of sulphuric acid can be formed at low temperature, which can be hydrolysed to an alcohol; ethylene gives a primary alcohol; all other alkenes give secondary or tertiary alcohols. The mechanism is as in the preceding example:

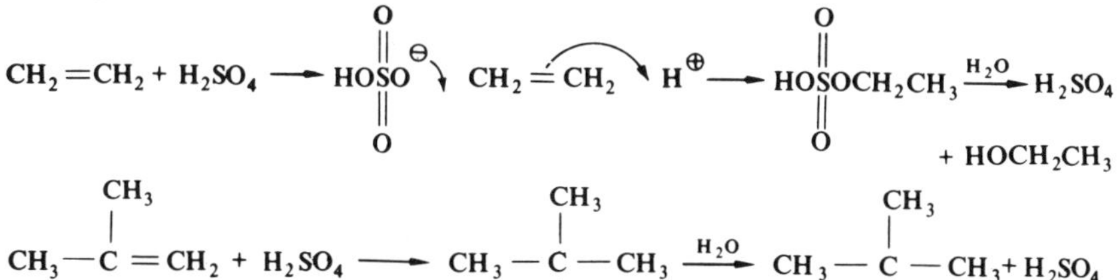

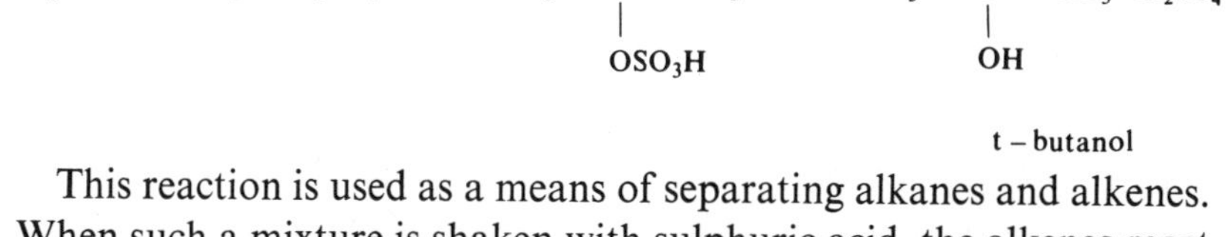

t – butanol

This reaction is used as a means of separating alkanes and alkenes. When such a mixture is shaken with sulphuric acid, the alkenes react and dissolve in the acid, but the alkanes are unchanged.

7. *Addition of Borane.* A reaction of alkenes that has proved very useful in recent years is the addition of borane (BH_3) to double

bonds. Borane does not exist freely as BH_3, but its dimer diborane (B_2H_6) prepared in solution from sodium borohydride and boron trifluoride is used as its source:

$$3NaBH_4 + 4BF_3 \longrightarrow 3NaBF_4 + 2 B_2H_6$$

Diborane reacts with an alkene to give first a monoalkylborane and then by further reaction, a trialkylborane. The trialkylborane can be hydrolysed to an alkane and boric acid or oxidized with hydrogen peroxide to an alcohol. The advantage is that H and OH are added by this reagent in the opposite sense to the normal hydration of a double bond (i.e. addition is apparently anti-Markovnikov) and provides a source of primary alcohols from olefins. This apparent exception to the rule is because the borane is the electron deficient species, being attacked by the nucleophilic double bond. As the C—B bond is formed, so a B—H bond must be broken and $H^\ominus$ is in this case transferred to carbon:

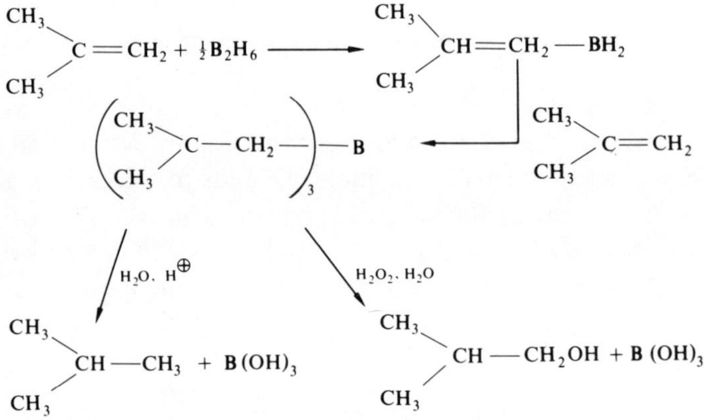

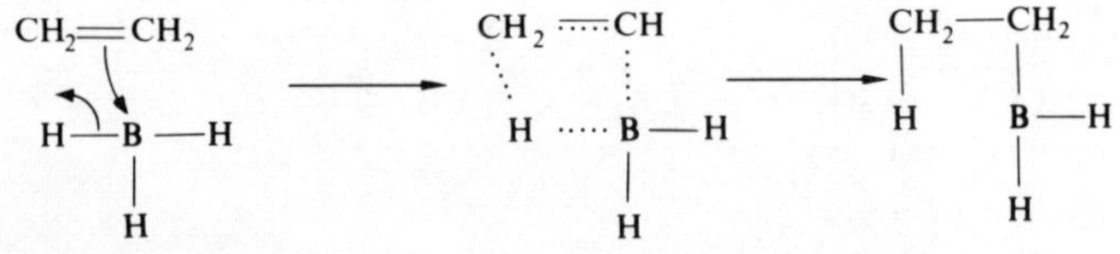

8. *Addition of Alkanes.* A reaction not often used in the laboratory, but of considerable importance in the petroleum industry, is the addition of alkanes to alkenes. Acidic catalysts such as sulphuric acid or aluminium chloride are used, and a complex mixture of products is formed. For example, if a mixture of butanes and butenes is

treated with sulphuric acid, a mixture of hydrocarbons valuable as a motor fuel is formed.

The reaction proceeds through the formation of carbonium ions, which can themselves add to alkenes. Carbonium ions can rearrange by the migration of a hydrogen atom or a methyl group from one carbon atom to another, and in this way more stable ions are formed:

$$CH_3CH{=}CHCH_3 \xrightarrow{H^\oplus} CH_3\overset{\oplus}{\underset{|}{C}}H{-}CH_2 \rightleftharpoons CH_3\overset{CH_3}{\underset{|}{C}}{-}\overset{\oplus}{C}H_2 \rightleftharpoons {=}CH_3\overset{CH_3}{\underset{\oplus}{C}}CH_3$$

The hydrogen atoms or methyl groups migrate with their electron pairs. Notice that the more highly branched tertiary carbonium ions tend to be formed through these migrations. The mixture of carbonium ions thus formed, adds to the mixture of butenes, for example:

$$CH_3\overset{CH_3}{\underset{CH_3}{C^\oplus}} \quad CH{=}CHCH_3 \longrightarrow CH_3\overset{CH_3}{\underset{CH_3}{C}}{-}\overset{\oplus}{C}H{-}CHCH_3 \xrightarrow{-H^\oplus} CH_3\overset{CH_3}{\underset{CH_3}{C}}{-}C{=}CHCH_3$$

The products formed tend to be highly branched alkenes, and these after hydrogenation to alkanes are valuable for addition to motor fuel to raise its octane rating.

Isobutene is self-condensed by sulphuric acid to di-isobutene, which was formerly important in the preparation of synthetic detergents. The reaction entails the formation of a t-butyl carbonium ion followed by attack of an isobutene double bond on the ion to form another carbonium ion which is stabilized by loss of $H^\oplus$ to give a new alkene. Some trimer is formed by attack on the dimeric ion by another molecule of isobutene:

$$CH_3\overset{CH_3}{\underset{CH_3}{C}}{=}CH_2 \xrightarrow{H^\oplus} CH_3\overset{CH_3}{\underset{CH_3}{C}}{\oplus} \quad CH_2{=}C\underset{CH_3}{\overset{CH_3}{<}} \longrightarrow CH_3\overset{CH_3}{\underset{CH_3}{C}}{-}CH_2{-}\overset{\oplus}{C}\underset{CH_3}{\overset{CH_3}{<}}$$

$$CH_3\overset{CH_3}{\underset{CH_3}{C}}{-}CH{=}C\underset{CH_3}{\overset{CH_3}{<}} \xleftarrow{-H^\oplus}$$

di–isobutene or
2, 4, 4–trimethylpent –2–ene

9. *Isomerization*. At high temperatures (500–700°) thermal iso-merization of alkene takes place with a tendency to form more highly branched chains and the double bond moves towards the middle of the chain:

$$CH_3CH_2CH_2CH=CH_2 \longrightarrow CH_3CH_2CH=CHCH_3 + CH_3CH_2\underset{\underset{CH_3}{|}}{C}=CH_2 + CH_3CH=\underset{\underset{CH_3}{|}}{C}-CH_3$$

Isomerization can take place at 200–300° on various solid catalysts.

10. Cis-trans *Isomerization*. The methods of preparing alkenes discussed so far give essentially *trans*-alkenes. How then can *cis*-alkenes be made and are the isomers interconvertible? *Trans*-alkenes are generally more stable isomers, with the larger groups on opposite sides of the molecule. *Cis*-alkenes are less stable because the larger groups are close together and may even overlap, preventing the molecule from completely attaining the planar arrangement that the bonds require (Fig. 11).

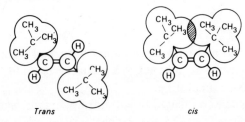

Trans   *cis*

Fig. 11   Crowding of alkyl groups in a cis-*alkene. The extreme case of di-*t-*butylethylene is shown*

Interconversion of isomers to an equilibrium mixture occurs on heating with iodine. A reversible addition of iodine takes place with rotation about the single C—C bond in the intermediate:

$$I\cdot + \underset{H}{\overset{R}{>}}C=C\underset{R}{\overset{H}{<}} \rightleftharpoons I-\overset{R}{\underset{H}{C}}\cdot\big)\overset{H}{\underset{R}{C}\cdot} \rightleftharpoons I-\overset{R}{\underset{H}{C}}-\overset{R}{\underset{H}{C}}\cdot \rightleftharpoons \underset{H}{\overset{R}{>}}C=C\underset{H}{\overset{R}{<}} + I\cdot$$

trans   *cis*

Irradiation with visible or ultraviolet light is often a more efficient means of isomerization, and good yields of the less stable *cis*-isomer are obtained in this way:

$$\underset{H}{\overset{R}{>}}C=C\underset{R}{\overset{H}{<}} \xrightarrow{\;h\nu\;} \underset{H}{\overset{R}{>}}C=C\underset{H}{\overset{R}{<}}$$

In the next chapter a method is described for the preparation of essentially pure *cis*-alkenes from acetylenes by hydrogenation.

11. Many metal ions form complexes with alkenes, for example, ethylene is absorbed by a solution of silver nitrate, from which it can be recovered by heating. Transition metals form a wide range of alkene complexes, e.g. $PdCl_2(C_2H_4)_2$. We are only beginning to understand the reactions of these complexes and they probably offer some exciting discoveries for the future.

12. *Aluminium Alkyls*. Finely divided aluminium reacts with hydrogen and alkenes under pressure to give aluminium alkyls. These compounds form part of the Ziegler catalyst system for polymerizing ethylene and propylene (see next section). A typical Ziegler catalyst is a mixture of aluminium triethyl and titanium tetrachloride.

The aluminium alkyls are very sensitive to oxygen; they ignite in air, giving alumina: they react with water (violently) or alcohols to give respectively aluminium hydroxide or alkoxides, and alkanes.

$$Al + 3CH_2 = CH_2 + 1\tfrac{1}{2}H_2 \longrightarrow Al(CH_2CH_3)_3$$

$$Al(CH_2CH_3)_3 + 3ROH \longrightarrow Al(OR)_3 + CH_3CH_3$$

Polymerization

We have seen (in paragraph 8) that a molecule of di-isobutene (a *dimer*) is produced by the joining of two molecules of isobutene (a *monomer*). Given the right conditions, the dimer can react with another molecule of isobutene to form a *trimer*, and so on, until a large number of monomer molecules have been joined together to give a *polymer*. The linking together of small molecules to form a very large molecule is called *polymerization*. Polymerization of olefins has been known for a long time, but the early polymers were not very useful because they were high-boiling oils or sticky semi-solids. In 1934 a method was discovered by Imperial Chemical Industries for polymerizing ethylene at 100–200° and pressures of 3000 atmospheres. This method gave a solid polymer called polyethylene of m.p. 110° to 120°. This polyethylene is a white wax-like solid, tough and flexible, insoluble in all common solvents, and with excellent electrical insulating properties. With these properties, added to its low cost, it is natural that today we see polyethylene in thousands of articles in daily use.

We can represent this polymerization formally as follows, where the reaction is a radical reaction, initiated by some radical R. The polymer produced contains an average 100 to 1000 ethylene units per molecule:

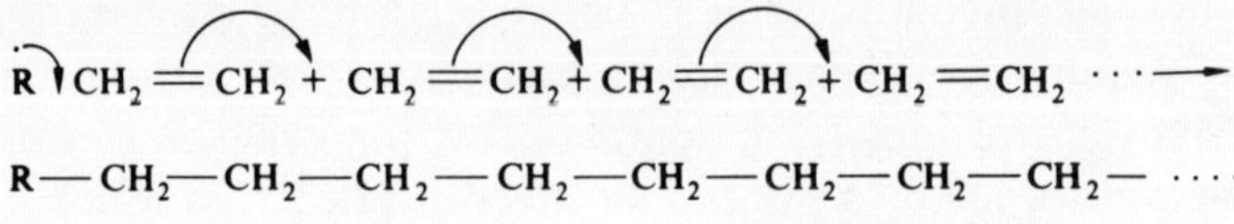

$$R\text{)}CH_2=CH_2 + CH_2=CH_2 + CH_2=CH_2 + CH_2=CH_2 \cdots \longrightarrow$$

$$R—CH_2—CH_2—CH_2—CH_2—CH_2—CH_2—CH_2—CH_2— \cdots$$

polyethylene

The polymerization of propylene and higher alkenes to solid polymers was not possible by the high pressure method. This was first accomplished with the catalysts discovered by Ziegler. These consist of mixture of aluminium alkyls and titanium tetrachloride. When these are mixed, the $TiCl_4$ is first reduced to $TiCl_3$, and this remarkable catalyst system can then polymerize ethylene, propylene and higher 1-alkenes to useful solid polymers at temperatures and pressures only a little above normal. Just how this occurs is not yet clear, but it seems the growing alkyl chain is attached to the surface of titanium trichloride crystals and associated with aluminium in some way. The process is illustrated below, for propylene, assuming aluminium as the active metal atom.

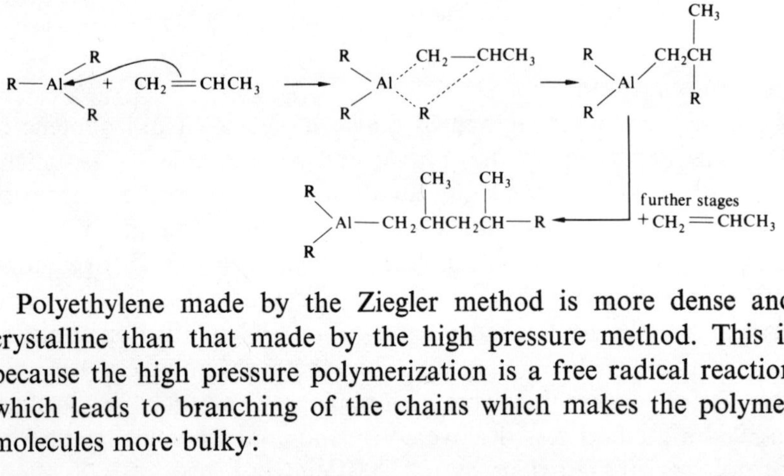

Polyethylene made by the Ziegler method is more dense and crystalline than that made by the high pressure method. This is because the high pressure polymerization is a free radical reaction which leads to branching of the chains which makes the polymer molecules more bulky:

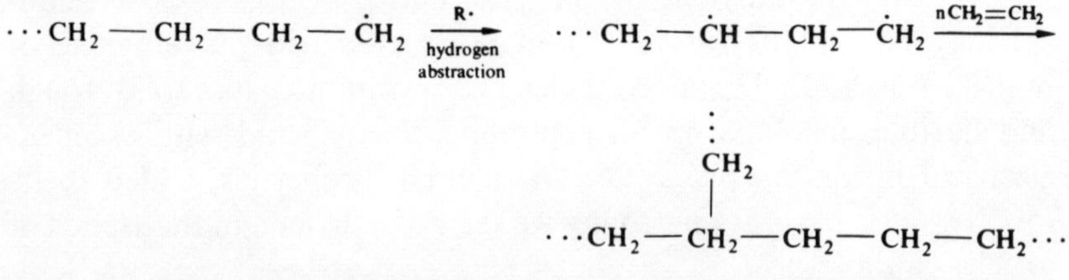

The mild conditions of the Ziegler reaction lead to very uniform linear chains which pack together into a uniform dense solid.

Polypropylene is a very tough and higher melting material. Both types of polyethylene and polypropylene are prepared commercially in thousands of tons per year.

Isobutylene has been polymerized commercially to a rubber-like material called butyl rubber since the beginning of World War II. This is done with a catalyst such as aluminium chloride, zinc chloride or boron trifluoride at $-100°C$. This polymerization is by still another mechanism, ionic polymerization.

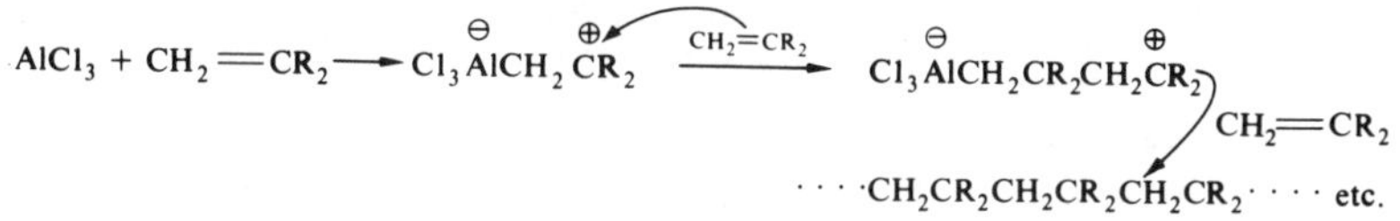

$$AlCl_3 + CH_2{=}CR_2 \longrightarrow Cl_3\overset{\ominus}{Al}CH_2\overset{\oplus}{C}R_2 \xrightarrow{CH_2=CR_2} Cl_3\overset{\ominus}{Al}CH_2CR_2CH_2\overset{\oplus}{C}R_2$$

$$\overset{CH_2=CR_2}{\diagdown}$$

$$\cdots CH_2CR_2CH_2CR_2\overset{\cdot}{C}H_2CR_2\cdots \text{ etc.}$$

here $R{=}CH_3$

Other commercially important polymers are based on alkene derivatives. Polyvinyl chloride made by radical polymerization is a brittle, glassy substance. It can be softened (plasticized) by adding viscous high molecular weight liquids (e.g. dibutyl phthalate) to break down its crystalline structure.

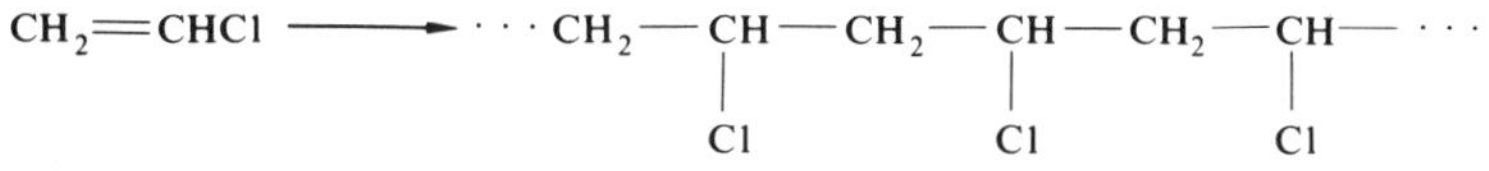

$$CH_2{=}CHCl \longrightarrow \cdots CH_2{-}\underset{\underset{Cl}{|}}{CH}{-}CH_2{-}\underset{\underset{Cl}{|}}{CH}{-}CH_2{-}\underset{\underset{Cl}{|}}{CH}{-}\cdots$$

vinyl chloride   polyvinyl chloride (PVC)

Styrene, a derivative of benzene, is easily polymerized by ionic or radical catalysts to a clear glassy solid:

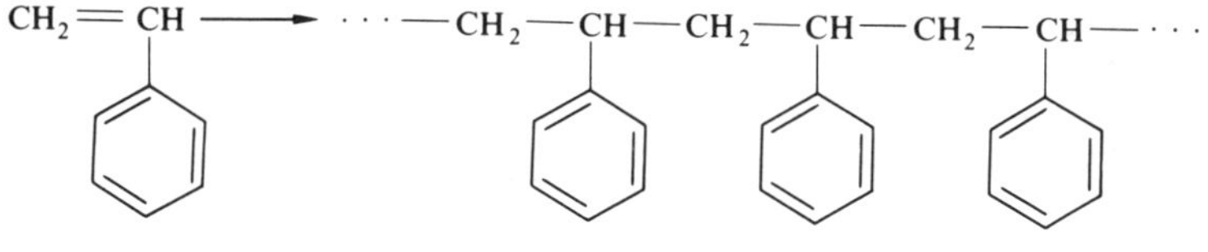

styrene   polystyrene

Methyl methacrylate can be polymerized by radical or ionic initiators to polymethyl methacrylate (known commonly as Perspex or Plexiglas). The polymer can be depolymerized by heating above its melting point.

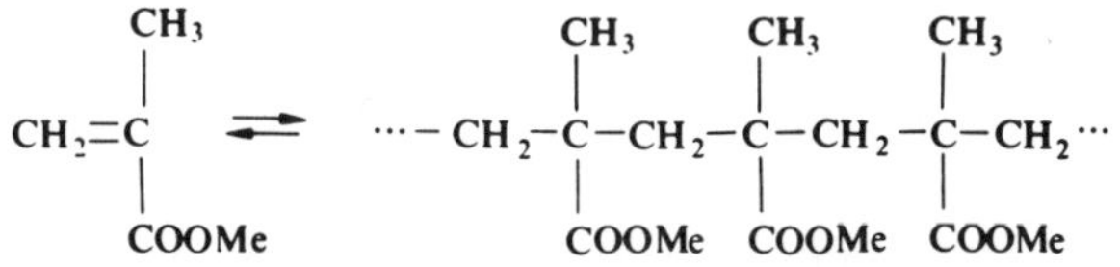

methyl methacrylate   polymethyl methacrylate (perspex)

This by no means exhausts the list of polymers or methods by which they are made. The above are all synthetic polymers, but protein (wool, silk, horn), cellulose (wood, cotton) and rubber are all natural polymers; and others, made by modifying natural polymers, e.g. cellophane and rayon from cellulose, are semi-synthetic.

Polymers can be divided into two groups, *thermoplastic*—those that melt on heating and can be cast into moulds or drawn into fibres (polyethylene, nylon), and *thermosetting*—those where the monomers react on heating, usually, to form a solid which does not melt, but decomposes if heated high enough (e.g. bakelite). Thermoplastic polymers usually consist of linear chains and usually the polymer is soluble in one or more solvents. At lower temperatures the thermoplastic polymer molecules will arrange themselves in an orderly array resembling the ordered arrangement of small molecules in a crystal. At higher temperatures, this ordered arrangement is broken down. The crystalline polymer may have a definite melting point or may slowly alter to a plastic form. Thermosetting polymers are made by condensing dissimilar molecules, containing two or more function groups, into a rigid, three-dimensional structure:

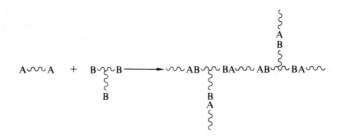

This rigid structure cannot be loosened by solvent molecules or by heating, so the polymer is insoluble and has no melting or softening point. If heated strongly enough the bonds begin to rupture and the polymer is thermally decomposed.

A schematic two-dimensional representation of thermoplastic and thermosetting polymers is given in Fig. 12.

Alkadienes

Alkenes containing two double bonds are called *dienes*, or *alkadienes*, three double bonds, *trienes*, and so on. The double bonds can be arranged in three ways, called *cumulative*, *conjugated* or *isolated*. Cumulative double bonds are immediately adjacent to each other, conjugated double bonds are separated by one single bond, and

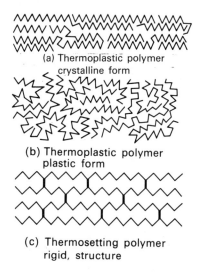

(a) Thermoplastic polymer
crystalline form

(b) Thermoplastic polymer
plastic form

(c) Thermosetting polymer
rigid, structure

Fig. 12   Schematic representation of thermoplastic and thermosetting polymers (a) Thermoplastic polymer, crystalline form (b) Thermoplastic polymer plastic form (c) Thermosetting polymer, rigid structure

isolated double bonds are separated by more than one single bond.

$$\overset{\diagdown}{\diagup}C\!=\!C\!=\!C\overset{\diagup}{\diagdown} \qquad \overset{\diagdown}{\diagup}C\!=\!\overset{|}{C}\!-\!\overset{|}{C}\!=\!C\overset{\diagup}{\diagdown} \qquad \overset{\diagdown}{\diagup}C\!=\!\overset{|}{C}\!-\!(CH_2)_n\!-\!\overset{|}{C}\!=\!C\overset{\diagup}{\diagdown}$$

cumulative   conjugated   isolated

If the double bonds are isolated, their properties are simply those of any alkene. If they are conjugated, they have enhanced reactivity and in addition reactions the two double bonds usually, but not invariably, react as a single unit:

$$CH_2 = CH - CH = CH_2 + HBr \longrightarrow BrCH_2 - CH = CH - CH_3$$

buta-1,3-diene   1-bromobut-2-ene

We have learned that electrons in a double bond are more mobile and more easily attacked than in a single bond. When two double bonds are separated by one single bond, the opportunities for mobility are much greater and the two double bonds are near enough for a reagent attacking one bond to disturb the electrons of the second. Indeed, measurement of the length of the single bond between two

71

double bonds shows that it is intermediate in length between that of a single and double bond.

Table of bond lengths

Bond	Length (Å)
C—C	1·54
C=C	1·34
C≡C	1·20
C—C in butadiene	1·46

In the example above, the addition of hydrogen bromide, the approach of a hydrogen ion with its positive charge causes the electrons of the conjugated double bond to move towards the charge:

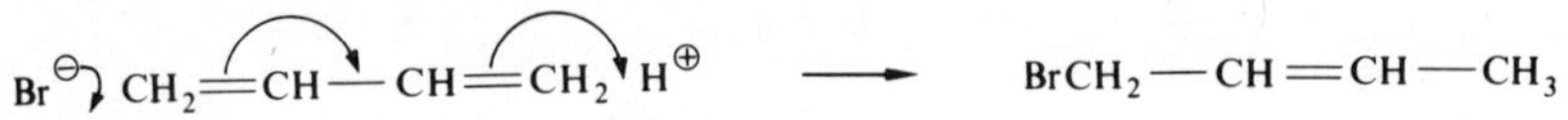

$$Br^{\ominus}\ CH_2\!\!=\!\!CH\!-\!CH\!=\!CH_2\ H^{\oplus} \longrightarrow BrCH_2\!-\!CH\!=\!CH\!-\!CH_3$$

Butadiene, the simplest conjugated alkadiene, is made industrially on a large scale for the manufacture of synthetic rubber. There are a number of technical processes, all using high temperature and catalysts. The starting material depends upon what is cheaply available in a particular area.

1. By dehydrating butan-1,4-diol (which in turn is made in several steps from acetylene) at 270° over a phosphoric acid catalyst.

$$\underset{\underset{OH}{|}}{CH_2}-CH_2-CH_2-\underset{\underset{OH}{|}}{CH_2} \longrightarrow CH_2\!=\!CH\!-\!CH\!=\!CH_2\ +\ 2H_2O$$

2. By catalytic dehydrogenation of butane and a mixture of butenes from petroleum or natural gas at 600–650° with a chromium-aluminium oxide catalyst.

$$CH_3CH_2CH_2CH_3 \xrightarrow{-H_2} \begin{array}{c} CH_3CH\!=\!CHCH_3 \\ + \\ CH_3CH_2CH\!=\!CH_2 \end{array} \xrightarrow{-H_2} CH_2\!=\!CH\ CH\!=\!CH_2$$

3. Vapour phase dimerization of ethanol, or better, the reaction of ethanol and acetaldehyde over a dehydrating catalyst at 325–350° gives butadiene. This method has been used where alcohol is very cheap.

$$CH_3CH_2OH\ +\ CH_3\overset{\overset{\displaystyle H}{|}}{C}\!=\!O \longrightarrow CH_2\!=\!CH\ CH\!=\!CH_2\ +\ 2H_2O$$

Butadiene is a **gas** (b.p. $-4.5°$). It is more reactive than simple alkenes, but undergoes the same type of addition reactions. It is easily polymerized to a variety of products depending upon the catalyst used; with sodium, a synthetic rubber is produced (page 77) and with nickel and cobalt complexes, cyclic dimers and trimers are formed (Chapter 10).

The addition of halogens and hydrogen halides to butadiene mentioned above, gives a mixture of 1,2 and 1,4-addition products. These may be represented thus:

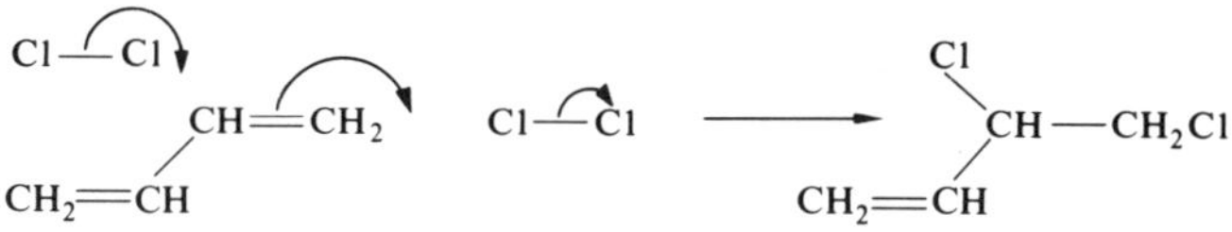

$$CH_2{=}CH{-}CH{=}CH_2 \qquad Cl{-}Cl \longrightarrow$$

and:

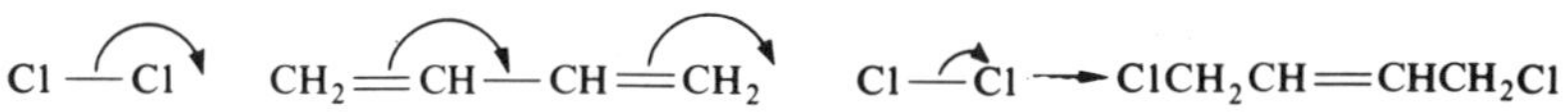

The formation of 1,4-dichlorobut-2-ene is a good example of the mobility of double bonds and the way they can behave as a single unit in conjugated dienes. Which product predominates depends upon the reaction conditions and the solvent. Low temperature favours 1,2 and high temperature 1,4-addition. Either product will react with excess reagent to give a fully saturated product.

$$\left. \begin{array}{l} CH_2{=}CHCHClCH_2Cl \\[1em] ClCH_2CH{=}CHCH_2Cl \end{array} \right\} \xrightarrow{Cl_2} ClCH_2CHClCHClCH_2Cl$$

1,2,3,4 –tetrachlorobutane

Two reactions characteristic of conjugated dienes are the Diels–Alder reaction (see Chapter 11) and the addition of sulphur dioxide to give sulpholenes. Butadiene is commercially converted to sulpholene and then hydrogenated to sulpholane. This is used as a solvent because it has the useful property of dissolving aromatic hydrocarbons but not aliphatic hydrocarbons. It is used to achieve this separation in petroleum refineries.

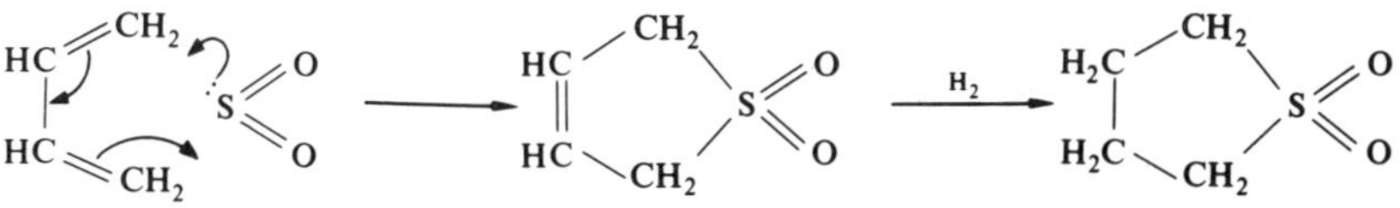

sulpholene   sulpholane

73

Isoprene (2-methylbutadiene) is an important diene. It is produced in small quantities in cracking processes in petroleum refining, but it is not easy to isolate the mixture of pentanes and pentenes, which by dehydrogenation will give isoprene, by analogy with the preparation of butadiene.

One method used has been to condense acetone and acetylene, using sodamide as catalyst, and reducing and dehydrating the resulting product:

$$CH_3-\underset{\underset{CH_3}{|}}{C}=O \ + \ HC\equiv CH \xrightarrow{NaNH_2} CH_3-\underset{\underset{OH}{|}}{\overset{\overset{CH_3}{|}}{C}}-C\equiv CH \xrightarrow{H_2} CH_3-\underset{\underset{OH}{|}}{\overset{\overset{CH_3}{|}}{C}}-CH=CH_2$$

$$CH_2=\underset{\underset{CH_3}{|}}{C}-CH=CH_2 \ \longleftarrow \ \Big\uparrow Al_2O_3$$

isoprene

A more recent method is to dimerize propylene (from 'cracked' petroleum) to 2-methylhexene. This is then pyrolysed and isomerized to isoprene and methane:

$$2CH_3CH=CH_2 \longrightarrow CH_3CH_2CH_2\underset{\underset{CH_3}{|}}{C}=CH_2 \longrightarrow CH_4 \ + \ CH_2=CH-\underset{\underset{CH_3}{|}}{C}=CH_2$$

Isoprene is a liquid (b.p. 42°). It is the basic unit of natural rubber, from which it was first obtained by destructive distillation. Isoprene can be regarded as the basic unit of the *terpenes*, a large and diverse group of naturally occurring substances. The isoprene 'units' are joined head to tail in these natural products, but they are not formed by polymerizing isoprene, rather they come from acetate groups through an intermediate C_6 compound, mevalonic acid.

Allene

The simplest example of a *cumulative* diene is *allene*, a reactive gas (b.p. −34°) which can be made by the action of zinc dust on dibromopropene, itself available from glycerol.

$$\underset{\underset{Br}{|}}{CH_2}-\underset{\underset{Br}{|}}{CH}-\underset{\underset{Br}{|}}{CH_2} \xrightarrow{KOH} CH_2=\underset{\underset{Br}{|}}{C}-\underset{\underset{Br}{|}}{CH_2} \xrightarrow{Zn} CH_2=C=CH_2$$

allene

74

It adds bromine readily to give tetrabromopropane and is hydro-lysed (with sulphuric acid as catalyst) to acetone. With sodium in ether it rearranges to methylacetylene. Though cumulative dienes are very reactive and are infrequently encountered in laboratory work, they do occur in a few products made by nature, and par-ticularly in certain fungi.

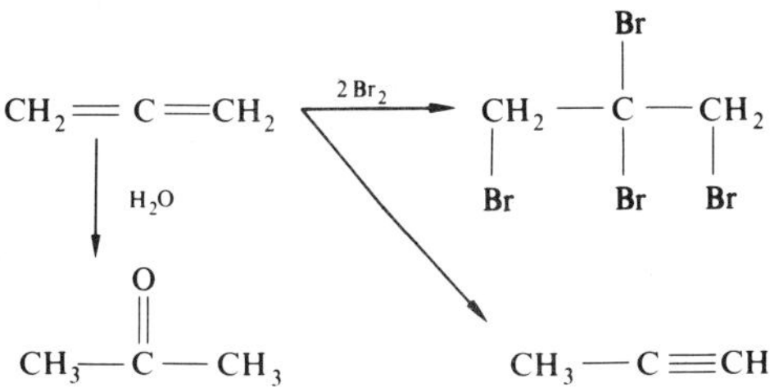

The next member of the series is butatriene, $CH_2=C=C=CH_2$, which together with higher homologues has been prepared in recent years.

Alkenes with many double bonds are known; if sufficient double bonds are in a conjugated system the products are coloured yellow, orange or red. Lycopene ($C_{40}H_{56}$) the red colouring matter of to-matoes, has thirteen double bonds, eleven of them conjugated and the other two isolated.

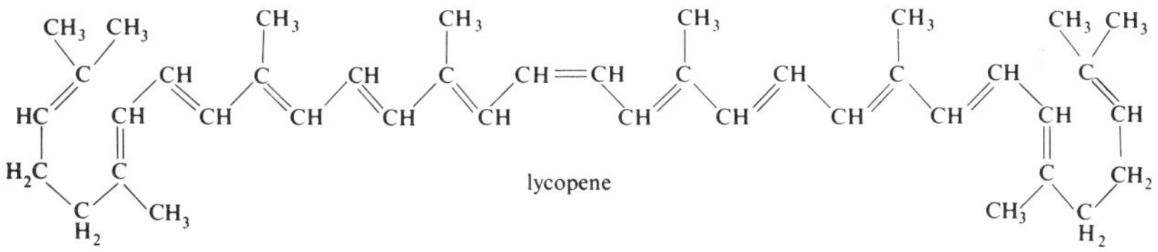

lycopene

Rubber

Natural rubber can be regarded as a head-to-tail polymer of isoprene units. This is known from the ozonolysis of a solution of rubber in chloroform or ethyl acetate. Decomposition of the ozonide gives almost pure laevulinic aldehyde ($CH_3COCH_2CH_2CHO$). X-Ray analysis of stretched rubber shows that the double bonds are all *cis*-substituted.

$$n \quad \underset{\substack{CH_2 \quad CH_2 \\ \text{isoprene}}}{\overset{CH_3}{\underset{\parallel}{C}-CH}} \xrightarrow[\text{catalysts}]{\text{various}} \quad \underset{\substack{\cdots CH_2 \\ }}{\overset{CH_3}{C=CH}} \quad \underset{\substack{CH_2-CH_2}}{\overset{CH_3}{C=CH}} \quad \underset{\substack{CH_2-CH_2}}{\overset{CH_3}{C=CH}} \quad \underset{\substack{CH_2-CH_2 \\ \text{rubber}}}{\overset{CH_3}{C=\cdots}}$$

$$\big\downarrow O_3$$

laevulinic aldehyde

The other form of polyisoprene, with all *trans* double bonds is also known and occurs in nature as gutta-percha. It does not have the elastic properties of rubber at normal temperature, but above 70° it becomes elastic.

gutta-percha

Just why rubber should have the remarkable properties of stretching and snapping back is not clear, but probably in unstretched rubber the long polyisoprene chains are folded and coiled and in stretching, these straighten out.

Natural rubber is soft, soluble in several organic solvents (e.g. chloroform, ether, benzene) and very susceptible to atmospheric oxidation. *Vulcanized* rubber is tougher, insoluble in these solvents and not as liable to oxidation in the air. The discovery of this process changed rubber from a curiosity into a useful industrial product. In vulcanization the rubber is treated with sulphur (usually 5–8% of the weight of rubber) or a variety of sulphur-containing chemicals; the result is the formation of C—S—C bonds between the linear chains. These are called *cross-links*. They convert rubber into a three-dimensional polymer, like thermosetting polymers. The more sulphur that is used, the more cross-linking is produced and the harder the product. In vulcanizing, some of the double bonds react with sulphur, but even in very hard rubber up to 15% of the original double bonds may still be present. Thus rubber can 'perish' (oxidize) in air.

Synthetic Rubbers

Neoprene (*neo*=new+*prene*, from isoprene) was the first synthetic rubber produced on a large scale. The addition of hydrogen chloride to vinyl acetylene gives chloroprene (2-chlorobutadiene) which polymerizes spontaneously to neoprene.

$$CH_2=CH-C\equiv CH \xrightarrow{HCl} CH_2=CH-\underset{\underset{Cl}{|}}{C}=CH_2$$

$$\cdots-CH_2-CH=\underset{\underset{Cl}{|}}{C}-CH_2-CH_2-CH=\underset{\underset{Cl}{|}}{C}-CH_2-\cdots$$

Neoprene is vulcanized best with sulphur or zinc oxide. It compares well with natural rubber and is superior in its resistance to oxidation and many chemicals, but is damaged by organic solvents.

Buna rubber was developed in Germany and used in place of natural rubber during World War II. It is produced by the polymerization of butadiene with sodium (*Buna*=from *bu*tadiene+*Na*). Also during World War II, the U.S.A. developed a butadiene rubber GR-S (Government Rubber, Styrene) which contained a little styrene incorporated into the butadiene polymer. This material is still of great industrial importance.

The presence of double bonds is not necessary for elastic properties, since Butyl rubber (polyisobutylene), mentioned earlier has no double bonds. Butyl rubber is therefore chemically more inert than other rubbers, but it cannot be vulcanized. Therefore, it is mixed with natural rubber or a small amount of butadiene is added to the isobutene before polymerization which produces some double bonds in the final product, which can then be vulcanized in the normal way.

The Ziegler catalysts have given a new impetus to rubber chemistry, and as a result isoprene can now be polymerized to all *cis*-polyisoprene that is practically identical with natural rubber. A co-polymer (i.e. a polymer made by mixing two different monomers and polymerizing them together) of ethylene and propylene made with Ziegler catalysts is also an excellent rubber. This too contains no double bonds. Rubber-like properties seem to be associated with long chains which do not lie in parallel lines but are twisted and convoluted. Stretching straightens them out partially. Relaxation allows the chains to regain a random, twisted state (Fig. 13).

Molecular Orbitals

The simple picture of chemical bonding as electron pair sharing between atoms, though adequate for many of the needs of the organic chemist, is surpassed in usefulness and elegance by the more detailed picture provided by molecular orbital theory which though theoret-

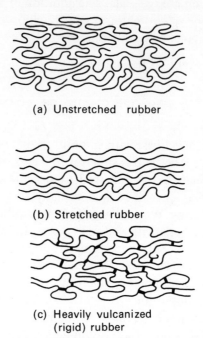

(a) Unstretched rubber

(b) Stretched rubber

(c) Heavily vulcanized
(rigid) rubber

*Fig. 13   Schematic representation of relaxed and stretched rubber, and hard
rubber (a) unstretched rubber (b) stretched rubber (c) heavily
vulcanized (rigid) rubber*

ically difficult to derive, in application is extremely useful and provides clear explanations of many phenomena that would otherwise
baffle us.

Schrödinger in 1928 showed that mathematically, electrons in
atoms can be represented as three-dimensional standing waves. The
calculations required to obtain these wave shapes from Schrödinger's
wave equation are beyond the scope of this book and the mathematical ability of many chemists. For a non-mathematical account
of the theory, the reader is referred to a simple text, such as that of
Audrey L. Companion's *Chemical Bonding*, published by McGraw-
Hill or P. L. Lynch's *Orbitals and Chemical Bonding* published by
Longmans. The following paragraphs are only a summary of the
points we must understand to make molecular orbitals useful in
organic chemistry.

From Schrödinger's wave equation we can plot the amplitude of
the wave (represented by the Greek letter psi, ψ) at various distances
from the nucleus. Just what this represents is difficult to say. We can

understand one-dimensional waves in a violin string or a two-dimensional wave in a drum head. Just as these waves have amplitude and nodes—points or lines of no vibration—and as intensity of sound is proportional to the square of the amplitude of vibration, so with the electron waves there are nodal planes where the value of ψ is zero, and ψ^2 is found to represent a probability of the electron being at any given point, if we regard it as a particle. Diagrams are commonly drawn to show the electron density, as represented by ψ^2 in the various sorts of standing waves (Fig. 14a and b).

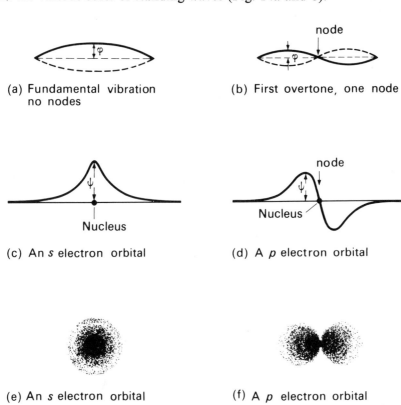

(a) Fundamental vibration
no nodes

(b) First overtone, one node

(c) An *s* electron orbital

(d) A *p* electron orbital

(e) An *s* electron orbital

(f) A *p* electron orbital

Fig. 14   Pictorial representation of electron density in atomic orbitals. A vibrating string showing the amplitude Φ (a) Fundamental vibration, no nodes (b) First overtone, one node. One-dimensional representation of electron orbitals, showing how the amplitude varies with distance from the atomic nucleus (c) An s electron orbital (d) A p electron orbital. The charge cloud of (e) an s orbital (f) a p orbital

1. *The shapes of atomic orbitals.* For a given quantum level the regions of electron density or electron *orbitals* have either no nodes, one, two or three nodes, and so on in order of increasing energy and

are given the symbols *s, p, d, f, g*, respectively. Here we need consider only the simplest kinds of *s* and *p*. The *s* orbital is a spherically symmetrical cloud of electron density, with the nucleus of the atom at its centre. Because the value of ψ only falls to zero at infinite distance from the nucleus (Fig. 14c), we cannot draw a neat figure of the electron density; it can be represented either as a diffuse cloud (Fig. 14e) or else by drawing, for simplicity, a sphere that encloses say 95% of the electron density, as in Fig. 15. Figure 15 represents the hydrogen atom with its one electron in an *s* orbital.

The next higher group of orbitals in terms of potential energy are the *p* orbitals. There are three such orbitals, each directed along an axis in space (Fig. 16).

Fig. 15   *The spherically symmetrical* s *orbital*

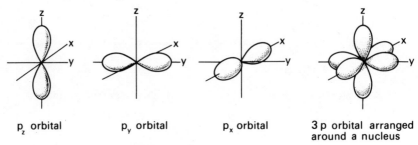

| p_z orbital | p_y orbital | p_x orbital | 3 p orbital arranged around a nucleus |

Fig. 16·   *The three* p *orbitals arranged in space*

2. *Hybrid orbitals.* The carbon atom in its simplest form has four valency electrons, one in an *s* orbital, and one in each of the three *p* orbitals. These orbitals can combine with each other to give new hybrid orbitals, having some of the characteristics of their parents. For example, one *s* orbital and a *p* orbital can give two *sp* hybrid orbitals (Fig. 17). Note the hybrids retain some of the spherical character of the *s* and the two-lobed nature of the *p* orbitals.

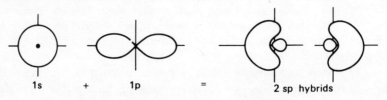

1s   +   1p   =   2 sp hybrids

Fig. 17   *Formation of* sp *hybrid orbitals*

Putting these hybrid orbitals into a single atom, their lobes overlap and it would be very difficult to draw, so they are represented in a simplified way, showing only the large lobes. Calculations from the wave equation show that carbon can exist in three hybridized forms, as follows. One *s* and three *p* orbitals can give:

1. Two *sp* hybrid orbitals and two *p* orbitals (Fig. 18a)
2. Three *sp*² hybrid orbitals and one *p* orbital (Fig. 18b)
3. Four *sp*³ hybrid orbitals (Fig. 18c)

The three forms differ slightly in energy and the last is the most stable. Note that four *sp*³ hybrid orbitals have a tetrahedral arrangement in space.

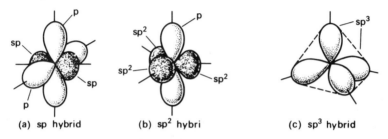

Fig. 18 *Hybridized orbitals in the carbon atom*

3. *Orbitals in molecules.* Molecular orbital theory says that when two atoms approach each other so that their atomic orbitals overlap, they can combine to give a new molecular orbital in which electron density is concentrated between the two nuclei and a covalent bond is formed. We can consider this concentration of electron density between two nuclei as the real force that cements nuclei together into molecules. Since any orbital (atomic or molecular) can accommodate only two electrons, we are chiefly considering those atomic orbitals, such as in carbon that contain only one and which by overlap can lead to a new molecular orbital containing two electrons. If two atoms combine so that the orbitals overlap and combine *along* their axes (the orbitals can be *s*, *p* or hybrid), we form a sigma (σ) molecular orbital. If they overlap and combine with their axes *parallel* (this can only occur for *p* orbitals in our simplified treatment) a pi (π) molecular orbital (Fig. 19) is formed.

By a combination of Figs 18c and 19c we can picture the methane molecule in molecular orbital dress. It is not significantly different from that we have been using.

Combination of two *sp*² hybrid carbon atoms (Fig. 18b) along the axes of any pair of hybrid orbitals will cause the *p* orbitals to overlap

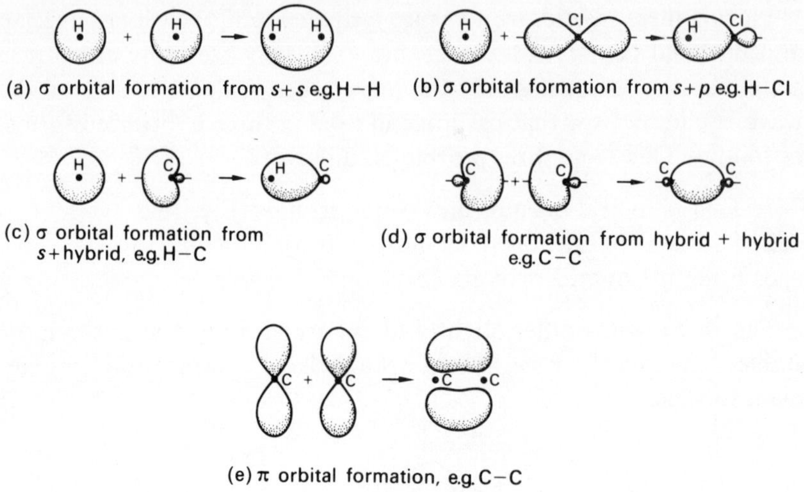

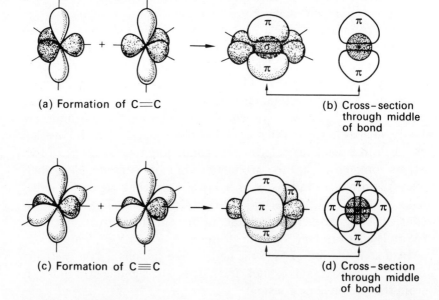

(a) σ orbital formation from s + s e.g. H−H

(b) σ orbital formation from s + p e.g. H−Cl

(c) σ orbital formation from s + hybrid, e.g. H−C

(d) σ orbital formation from hybrid + hybrid e.g. C−C

(e) π orbital formation, e.g. C−C

Fig. 19   Molecular orbital formation. For simplicity, only the relevant orbitals are shown in each atom

so we have σ and π bonds formed together (Fig. 20a). This is the molecular orbital picture of the carbon-carbon double bond. The C=C bond then is not two bonds of equal kind, one is an 'ordinary' σ bond and the other is a π bond. The π bond electron density cloud

(a) Formation of C=C

(b) Cross−section through middle of bond

(c) Formation of C≡C

(d) Cross−section through middle of bond

Fig. 20   Formation of multiple carbon-carbon bonds in sp² and sp hybrid carbon atoms

stretches further out from the nuclei than the σ bond does, it is less concentrated between the nuclei, and this can help us to understand why a C=C bond is more easily distorted or polarized than a single C—C bond and explains why the double bond is a nucleophile and more easily attacked by many reagents. Note that the remaining sp^2 hybrid orbitals are all in the same plane and this conforms to the picture we have of C=C compounds such as ethylene.

Combination of two sp hybrid carbon atoms along the axis of a pair of hybrid orbitals allows both p orbital pairs to overlap and we have a carbon-to-carbon triple bond formed (Fig. 20c), consisting of one σ bond and two π bonds. If the two remaining hybrid orbitals are combined with hydrogen, we have the linear molecule acetylene (Chapter 4).

A σ bond is symmetrical about the axis joining the two atoms, so atoms can rotate about a σ bond, but a π bond is not symmetrical about the axis joining the two atoms (Fig. 20b). If we try to rotate one carbon atom against the other, it would involve twisting and breaking the π bond, but not the σ bond. This is in accordance with our experience of rotation about C—C single bonds and lack of rotation around C=C double bonds.

Figure 21 shows why conjugated double bonds form a special class. From the representation of butadiene in Fig. 21a we can see that all the p orbitals are lined up with axes parallel, so that in addition to π bond formation between carbon atoms 1 and 2, and 3 and 4, we can have overlap between carbons atoms 2 and 3 as well, in fact the whole system becomes one π-bonded unit. But rotation about the central carbon-carbon bond (Fig. 21b) would destroy this

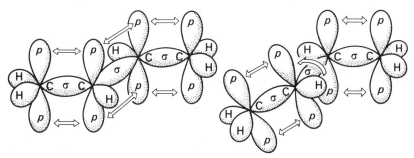

(a) All orbitals overlapping   (b) Overlap prevented by central bond rotation

Fig. 21   Formation of π orbitals in butadiene

simple system by preventing the overlap between carbon atoms 2 and 3, and then the system behaves as two isolated double bonds. Butadiene is slightly more stable as shown in Fig. 21a and normally exists in this form. The implication of this explanation on the length of the central C—C bond and the fate of the π bond system when C_1 is converted to a tetrahedral sp^3 carbon atom by reaction with HBr should be apparent.

Identification

Alkenes can be distinguished from alkanes and halo-alkanes by the fact that they dissolve in concentrated sulphuric acid, and they react rapidly with dilute solutions of bromine in water or carbon tetrachloride to decolorize them and add bromine to the double bond. They react with dilute neutral permanganate solution giving a precipitate of brown (hydrated) manganese dioxide.

The C—H bond stretching vibration of hydrogen attached to C=C occurs at only a slightly different value from alkane C—H, but these values are sufficiently constant and characteristic to be very useful in identifying alkenes by their infrared spectra. The =C—H group absorbs at 3100 to 3010 cm^{-1}, i.e. alkene C—H just above 3000 cm^{-1} and alkane C—H just below 3000 cm^{-1}. The alkene absorption is rather weaker than that of alkanes in this region. The C=C stretching occurs at 1680–1620 cm^{-1} and is often not a strong band but can vary with the groups attached to it. A completely symmetrically substituted double bond (as in ethylene or tetra-methylethylene) will not have a C=C stretching band visible. The C—H bending vibrations of alkene occur below 1000 cm^{-1} and vary with the number and arrangement of alkyl groups attached to the double bond. For example *cis*-alkenes absorb around 700 cm^{-1} and *trans*-alkenes near 970 cm^{-1}. Double bonds at the end of a chain absorb at 990 and 910 cm^{-1}. With a little experience, bands in this region can be useful to get information about the arrangement of groups around the double bond.

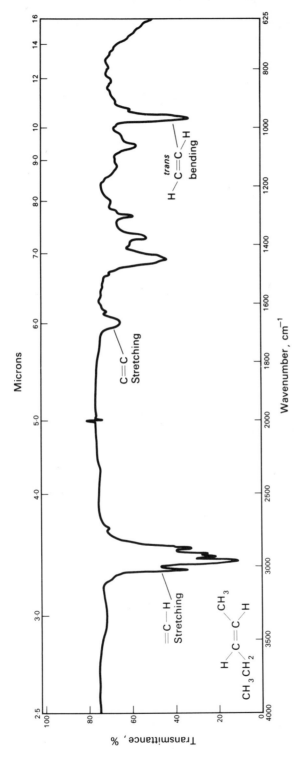

Fig. 22 Infrared spectrum of trans-pent-2-ene

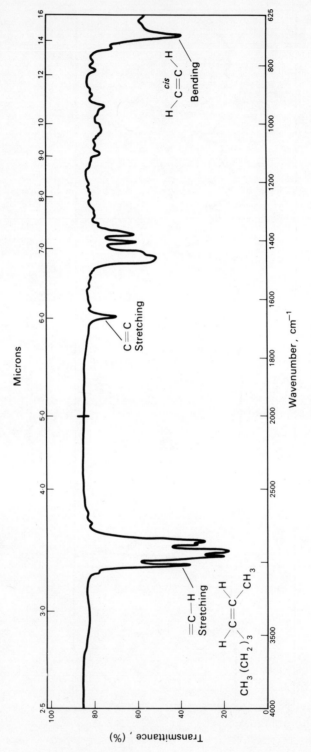

Fig. 23   Infrared spectrum of cis-hept-2-ene

4
Acetylenes

Among the groups of hydrocarbons conforming to the general formula C_nH_{2n-2}, we find the alkadienes mentioned in the previous chapter. Another group with this general formula is also highly unsaturated, exhibits many of the reactions of alkenes, but also possess highly characteristic properties. The first member of the series is acetylene, C_2H_2, the methods of preparation and the reactions of which have led to the conclusion that the hydrocarbon contains a carbon-carbon *triple* bond, that is the two carbon atoms share *six* valence electrons. Using the tetrahedral carbon picture, two faces of the tetrahedra are joined (Fig. 24), or in the language of molecular orbitals, the carbon atoms are in an *sp* hybrid state, with one σ bond and two π bonds joining the carbon atoms (Fig. 20). Either picture gives us a linear molecule, with one hydrogen atom at each end; the molecular orbital picture shows that there is a large, roughly cylindrical region of π electron density in the molecule, which we might predict will react with electron-deficient reagents.

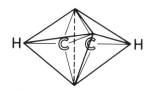

Fig. 24   The triple bond of acetylene shown as two tetrahedral carbon atoms face to face

The high electron density between the carbon atoms holds them closely together and the carbon-carbon bond length in acetylene (1·20 Å) is shorter than that of either ethylene (1·34 Å) or ethane (1·54 Å).

Nomenclature
The higher homologues can be regarded as substituted acetylenes, in which case the names of the substituting alkyl groups are given, e.g

methylethylacetylene $CH_3C \equiv CCH_2CH_3$. The IUPAC names for acetylenes indicate the chain length with the suffix *-yne* to denote a triple bond. Thus, methylacetylene is propyne. Some simple members of the series are listed in Table 7.

Table 7.   Alkynes

Systematic name	Trivial name	Formula	m.p. °C	b.p. °C
Ethyne	Acetylene	$HC \equiv CH$	−82	−75
Propyne	Methylacetylene	$CH_3C \equiv CH$	−101	−23
But–1–yne	Ethylacetylene	$CH_3CH_2C \equiv CH$	−122	9
But–2–yne	Dimethylacetylene	$CH_3C \equiv CCH_3$	−24	27

For complex molecules containing both double and triple bonds, the suffix indicating a double bond comes first:

$$CH_2 = CH-C \equiv CH \qquad CH_3-CH = CH-CH_2C \equiv CH$$

butenyne
vinylacetylene   hex-2-en-5-yne

$$CH_2 = CH-C \equiv C-C \equiv CH$$

hex-1-en-3,5-diyne

Preparation of Acetylenes

1. The addition of water to metal carbides gives acetylene. Calcium carbide, made commercially by heating lime and coke in an electric furnace, is commonly used. This constitutes a synthesis from inorganic materials:

$$CaO + 3C \xrightarrow{2000-3000°} CaC_2 + CO$$

$$CaC_2 + 2H_2O \longrightarrow Ca(OH)_2 + HC \equiv CH$$

2. Dehydrohalogenation, the removal of the elements of a hydrogen halide. This method is similar to that used for preparing alkenes from alkyl halides. An alcoholic solution of potassium hydroxide eliminates two molecules of hydrogen halide from a dihalide, to give an acetylene. The halogen atoms are usually on adjoining carbon atoms. The intermediate halo-alkane formed by loss of one HX can usually be obtained as an intermediate product:

$$CH_3CHBrCH_2Br \xrightarrow[\text{EtOH}]{\text{KOH}} CH_3CHBr = CHBr$$

Halogen on doubly-bonded carbon is less reactive than alkyl halogen so the second stage does not go readily with alcoholic KOH. Sodamide, a very strong base, can be used instead:

$$CH_3CBr=CH_2 \quad + \quad CH_3CH=CHBr \xrightarrow{\text{NaNH}_2} \quad CH_3 \, C\equiv CH$$

An alkene can thus be converted to an acetylene:

$$RCH=CHR' \xrightarrow{X_2} RCHX-CHXR' \xrightarrow[\text{EtOH}]{\text{KOH}} RC\equiv CR'$$

3. A reactive metal, such as zinc, or zinc–copper couple, removes halogen from a vicinal tetrahalide to give an acetylene:

$$CHBr_2CHBr_2 \xrightarrow{Zn} 2ZnBr_2 + HC\equiv CH$$

4. Cracking of methane from natural gas or petroleum at a temperature of at least 1200° gives acetylene, plus carbon, hydrogen, ethylene and diacetylene ($HC\equiv C-C\equiv CH$) as by-products. The high temperature is produced by an electric arc and the products are cooled rapidly to prevent decomposition of the acetylene.

$$2CH_4 \longrightarrow HC\equiv CH + 3H_2$$

5. Combustion of methane or lower hydrocarbons with a controlled amount of oxygen gives acetylene. The heat of combustion of part of the hydrocarbons provides the energy for the pyrolysis of another part to acetylene, and carbon monoxide and hydrogen are produced in the proportions necessary for the synthesis of methanol. Both methods, 4 and 5, are used in the petroleum chemicals industry,

$$6CH_4 + O_2 \xrightarrow{1500°} 2HC\equiv CH + 2CO + 2H_2$$

Physical Properties

Acetylene itself is a colourless gas with a characteristic ether-like odour when pure, but as produced from carbide it has an unpleasant smell due to impurities. Acetylene is a highly endothermic compound, i.e. decomposition to its elements liberates energy, and this fact relates its preparation at high temperature, and its instability under some conditions. It is sparingly soluble in water but very soluble in acetone. When compressed it can decompose explosively; cylinders of compressed acetylene therefore consist of a solution of acetylene in acetone (at 10 atmospheres pressure), absorbed on a porous solid. Methylacetylene is also a gas, but higher homologues are colourless liquids. Acetylene is of great importance as an industrial chemical; the higher homologues, however, are not so readily available and are not commercial products.

Reactions of Acetylenes

Acetylene burns with a smoky, very brilliant flame. The type of flame can be used as a rough guide to distinguish groups of hydrocarbons. The higher the ratio of H to C, the bluer the flame, the lower the ratio of H to C, the more luminous and smoky the flame. Methane (ratio $H:C=4:1$) burns with an almost invisible blue flame. Acetylene (ratio $1:1$) or benzene (ratio $1:1$) give very sooty flames. Alkanes and alkenes fall between these extremes. Acetylene forms a highly explosive mixture with air. Mixed with oxygen it burns with a very hot flame ($3000°C$); the oxyacetylene flame is used for welding and cutting metals (and safe-breaking!).

1. *Oxidation.* Mild oxidation of acetylene with nitric acid and mercuric nitrate gives oxalic acid:

$$HC\equiv CH \quad \xrightarrow[Hg(NO_3)_2]{HNO_3} \quad HO-\overset{O}{\overset{\|}{C}}-\overset{O}{\overset{\|}{C}}-OH$$

2. *Reduction.* Acetylenes react with hydrogen in the presence of catalysts, usually platinum, to give alkanes. Alkenes are more readily reduced than acetylenes, so usually the reduction cannot be stopped at the alkene stage. With a specially prepared palladium catalyst (Lindlar's catalyst) hydrogenation of acetylenes can give an olefin. However, the hydrogenation of vinylacetylene to butadiene is not possible by this method.

$$HC\equiv CCH_3 \xrightarrow[Pt]{2H_2} CH_3CH_2CH_3$$

$$HC\equiv CCH_3 \quad \xrightarrow[\text{Lindlar's Pd}]{H_2} \quad CH_2=CHCH_3$$

As would be expected from the mechanism of hydrogenation given in Chapter 3, the addition of hydrogen is *cis*, so a *cis*-olefin is always obtained by this reduction.

Lindlar's catalyst is made by precipitating palladium onto calcium carbonate which acts as an inert support and then 'poisoning' the catalyst with a little lead acetate. Why it is a selective reducing agent is not understood.

3. *Addition of Halogens.* In addition reactions, acetylenes react similarly to alkenes. For alkylacetylenes, addition of hydrogen halide follows Markovnikov's rule. Impure acetylene and chlorine react explosively, but with pure acetylene the reaction is very slow. Under

suitable conditions, reaction can be stopped after one molecule of halogen or hydrogen halide has been added:

$$CH \equiv CH + Cl_2 \longrightarrow CHCl = CHCl \xrightarrow{Cl_2} CHCl_2CHCl_2$$

dichloroethylene   tetrachloroethane

and:

$$H \overset{\oplus}{\underset{}{}} HC \equiv CH \overset{\ominus}{Br} \longrightarrow H_2C = CHBr \xrightarrow{HBr} CH_3CHBr_2$$

vinyl bromide

Though acetylene might be expected to react more rapidly than ethylene in addition reactions because it is more unsaturated, in fact alkynes undergo addition more slowly than corresponding alkenes. For example one equivalent of bromine reacts with vinylacetylene (page 96) to saturate the double, not the triple bond:

$$HC \equiv CCH = CH_2 \xrightarrow{Br_2} HC \equiv CCHBrCH_2Br$$

Just why this is so is not clear.

Vinyl chloride, prepared by passing acetylene into concentrated hydrochloric acid, with cuprous ammonium chloride or mercuric chloride as catalyst, is produced in large quantities for polymerizing to polyvinyl chloride (page 69) which is an extremely valuable industrial material used for example for insulating electric cables, for mouldings, flooring and fabrics.

$$HC \equiv CH + HCl \xrightarrow{HgCl_2} H_2C = CHCl$$

4. *Hydration.* Addition of water to the triple bond occurs in acid solution. As with halogenation, this reaction goes less readily for alkynes than for alkenes, and the hydration of alkynes must be catalysed by mercuric ions. Many reactions of alkynes are catalysed by cuprous, mercuric or nickel ions, further examples are given in this chapter. Again just why these ions have this effect is not clear. In the case of hydration, at least part of the effect must be by increasing the solubility of the acetylene in aqueous solvent by complex formation. Acetylene with sulphuric acid and mercuric sulphate given initially vinyl alcohol, but this immediately rearranges to a more stable isomer, acetaldehyde:

$$H^{\oplus} \quad CH \equiv CH \quad :O—H \xrightarrow[H_2SO_4]{HgSO_4} \left[HC = C—OH \right] \longrightarrow CH_3—C=O$$

91

This represents the largest use of industrial acetylene; the acetaldehyde is produced in good yield and is easily oxidized to acetic acid. Substituted acetylenes give ketones; note that the addition again follows Markovnikov's rule.

$$CH_3 \rightarrow C \equiv CH \quad H^{\oplus} \quad OH^{\ominus} \xrightarrow[H_2SO_4]{HgSO_4} \left[CH_3C = CH_2 \atop OH \right] \longrightarrow CH_3 - \overset{O}{\overset{\|}{C}} - CH_3$$

Methanol adds to acetylene in a similar way, to give methyl vinyl ether, which by treatment with dilute acid is hydrolysed to give acetaldehyde:

$$CH_3OH + HC \equiv CH \longrightarrow \underset{\text{methyl vinyl ether}}{CH_3O - CH = CH_2} \xrightarrow[H_2O]{H^{\oplus}} CH_3OH + CH_3CHO$$

5. *Addition of Acetic Acid.* The addition of acetic acid is also catalysed by mercuric salts. The product, vinyl acetate, is important in the manufacture of synthetic resins, rubbers and emulsion paints.

$$CH_3 - \overset{O}{\overset{\|}{C}} - OH \quad + \quad HC \equiv CH \xrightarrow{Hg^{\oplus\oplus}} CH_3\overset{O}{\overset{\|}{C}} - O - CH = CH_2$$

6. *Formation of Metal Acetylides.* The most characteristic reaction of acetylene and alkynes containing the $\equiv CH$ groups is the formation of metal derivatives. These can be subdivided into three groups, alkali metal derivatives, heavy metal derivatives and magnesium halide derivatives (Grignard reagents, page 36).

Sodium dissolves in liquid ammonia (at about $-30°$) to give a solution of sodamide, a very strong base:

$$Na + NH_3 \longrightarrow Na^{\oplus} NH_2^{\ominus} + \tfrac{1}{2}H_2$$

Adding acetylene or a mono-alkylacetylene to this gives a sodium acetylide:

$$HC \equiv CH + Na^{\oplus}NH_2^{\ominus} \rightleftharpoons HC \equiv C^{\ominus}Na^{\oplus} + NH_3$$

If sodium acetylide is dissolved in water, acetylene is regenerated:

$$HC \equiv C^{\ominus}Na^{\oplus} + H_2O \rightleftharpoons HC \equiv CH + Na^{\oplus} OH^{\ominus}$$

Acetylene is evidently behaving as a very weak acid. On the principle that a weak acid is regenerated from its salts by the action of a stronger acid, then acetylene is intermediate in acidity between NH_3 and H_2O.

The acid strength of water is given from the equilibrium of the reaction:

$$H_2O \rightleftharpoons H^{\oplus} + OH^{\ominus}$$

from which we obtain the acid dissociation constant K_a.

$$K_a = \frac{[H^{\oplus}][OH^{\ominus}]}{[H_2O]} = 10^{-16}$$

The more familiar value of 10^{-14} neglects the $[H_2O]$ factor, but for comparison with organic compounds measured in other solvents, it is more correct to include it here. To measure the acid dissociation of acetylene is very difficult, it is too weakly acidic to ionize in water, but indirectly we can calculate it:

$$HC\equiv CH \rightleftharpoons HC\equiv C^{\ominus} + H^{\oplus} \qquad K_a = 10^{-22}$$

And for ammonia:

$$NH_3 \rightleftharpoons NH_2^{\ominus} + H^{\oplus} \qquad K_a = 10^{-35}$$

If we compare these with acetic acid $K_a = 10^{-5}$ and hydrogen cyanide $K_a = 10^{-10}$, we see that acetylene is indeed a very weak acid.

The action of water on sodium or potassium acetylide, or calcium carbide, is simply the hydrolysis of the salt of a very weak acid:

$$HC\equiv C^{\ominus} Na^{\oplus} + 2H_2O \longrightarrow NaOH + HC\equiv CH$$

A possible explanation why the hydrogen of **acetylene** is acidic, can be seen if we consider that two electron pairs joining the carbon atoms are in π bonds which are located further from the carbon nuclei than they would be in σ bonds, so the carbon nuclei are left somewhat electropositive and attract the electron pair in the σ bond holding hydrogen rather more strongly than in, say ethane. This in turn makes the hydrogen more easily removed by a strong base.

$$\equiv C \longleftarrow H \overset{\ominus}{\frown} :NH_2 \longrightarrow \equiv C:^{\ominus} + NH_3$$

It is interesting to compare acetylene with the stronger acidity of $H-C\equiv N$; the electronegativity of the nitrogen atom draws electron density away from carbon still more strongly in the latter.

The salts of weak acids are themselves strong bases; for example in NaOH, which is the sodium salt of the weak acid H_2O, $OH^{\ominus}$ is a strong base. The acetylide ion $HC\equiv C^{\ominus}$ is also a very strong base, and undergoes a number of reactions. For example, with alkyl halides it gives substitution (by the S_N2 mechanism) or elimination (by E2) according to the structure of the alkyl group, as described in

Chapter 3. Primary halides give substitution, forming alkyl substituted acetylenes:

$$RCH_2X \quad + \quad HC\equiv C^{\ominus} Na^{\oplus} \longrightarrow RCH_2 C\equiv CH \quad + \quad Na^{\oplus}X^{\ominus}$$

This provides a way of making higher alkynes. Secondary and tertiary halides undergo elimination more readily so the method is not useful for them.

Passing acetylene into an ammoniacal solution of cuprous or silver salts gives a precipitate of cuprous or silver acetylide:

$$HC\equiv CH + CuCl_2 + H_2O \longrightarrow CuC\equiv CCu \cdot H_2O$$

These heavy metal acetylides (chiefly confined to copper, mercury and silver) are covalent compounds, not hydrolysed by water, unlike the alkali metal and calcium acetylides. Cuprous acetylide is a red-brown powder, very explosive when dry, and easily detonated by shock or friction. It is usually decomposed while still moist with dilute acid. Cuprous acetylide is a useful synthetic intermediate. For example, it is oxidized by ferricyanide ion in alkaline solution to diacetylene:

$$2HC\equiv CCu \longrightarrow HC\equiv C\!-\!C\equiv CH$$

The formation of insoluble silver or copper compounds can also be used as a test for the $-C\equiv CH$ group.

7. *Grignard Reagents.* Acetylenes containing at least one hydrogen attached to the triply bonded carbon react with Grignard reagents to give acetylenic Grignard reagents. These are extremely useful tools for synthesis, as illustrated by a few examples below:

$$HC\equiv CH \quad + \quad RMgBr \longrightarrow HC\equiv C\!-\!MgBr \quad + \quad RH$$

Excess Grignard reagent reacts further:

$$HC\equiv CMgBr + RMgBr \longrightarrow BrMgC\equiv CMgBr \quad + \quad RH$$

$$R\!-\!C\equiv C\!-\!MgBr \xrightarrow{R'X} R\!-\!C\equiv C\!-\!R' \quad + \quad MgBrX$$

$$\xrightarrow[H_2O]{CO_2} R\!-\!C\equiv C\!-\!COOH \quad + \quad MgBrOH$$

8. *With carbonyl compounds.* Acetylene reacts with formaldehyde, using copper acetylide as catalyst to give propargyl alcohol and

butynediol. Both of these are of industrial interest, propargyl alcohol for making allyl alcohol and glycerol, butynediol for making butadiene.

$$HC\equiv CH + \overset{H}{HC}=O \xrightarrow{Cu_2C_2} HC\equiv CCH_2OH \xrightarrow{H_2} H_2C=CHCH_2OH$$

propargyl alcohol   allyl alcohol

$$HC\equiv CH + 2\overset{H}{HC}=O \xrightarrow{Cu_2C_2} HOCH_2C\equiv CCH_2OH \xrightarrow{H_2} HOCH_2CH_2CH_2CH_2OH$$

butynediol   butanediol

The addition of acetone to acetylene is catalysed by potassium hydroxide or sodamide. From the resulting acetylenic alcohol, isoprene (page 74) can be produced by hydrogenation and dehydration:

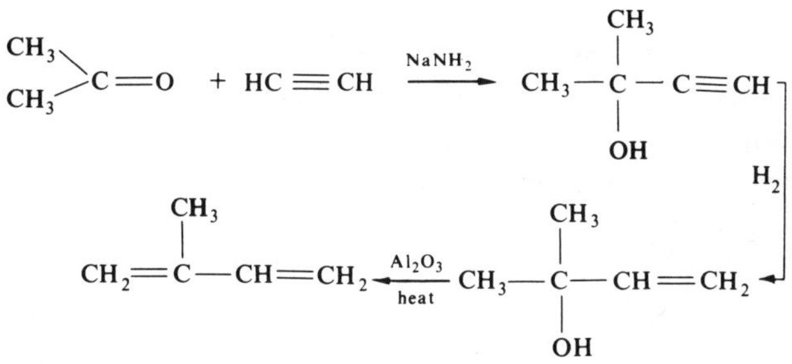

9. *Polymerization.* Acetylene polymerizes in a variety of ways, depending upon conditions and catalysts. Acetylene passed through a heated tube forms some benzene, and methylacetylene forms mesitylene:

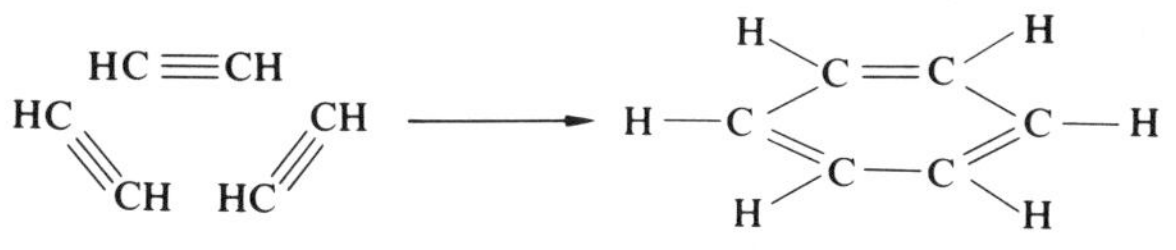

benzene

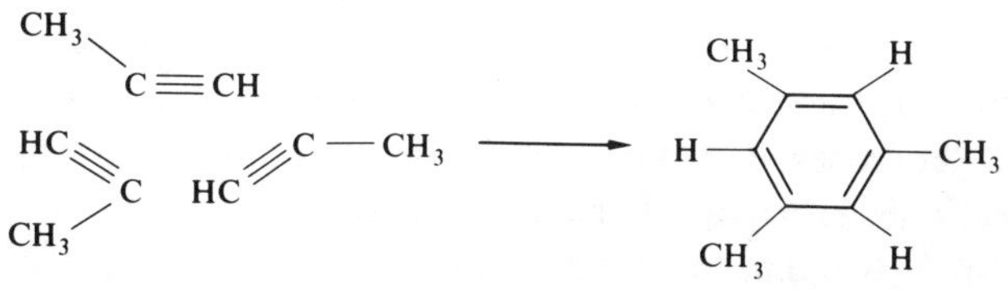

mesitylene

Using nickel cyanide as catalyst, cyclo-octatetraene is formed (see Chapter 11):

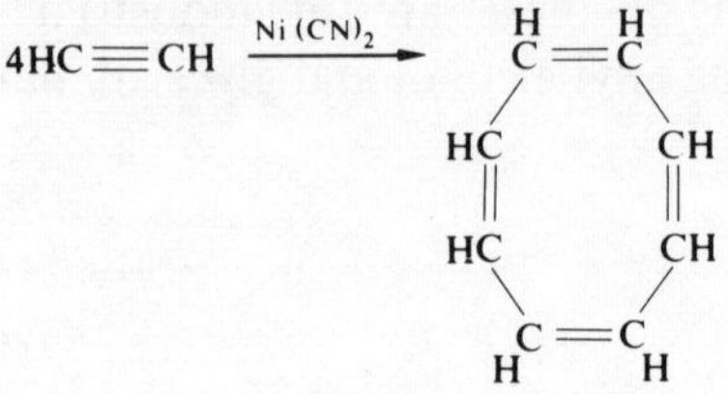

$$4HC \equiv CH \xrightarrow{Ni(CN)_2}$$

cyclo- octatetraene

In a solution of cuprous chloride and ammonium chloride, acetylene gives a mixture of vinylacetylene and divinylacetylene:

$$HC \equiv CH \xrightarrow[NH_4Cl]{Cu_2Cl_2} H_2C = CH - C \equiv CH + H_2C = CH - C \equiv C - CH = CH_2$$

vinylacetylene   divinylacetylene

Addition of hydrogen chloride, catalysed by cuprous chloride, to vinylacetylene gives a 1,4-addition product which rearranges to 2-chlorobutadiene, known commercially as chloroprene, which is polymerized to the synthetic rubber Neoprene:

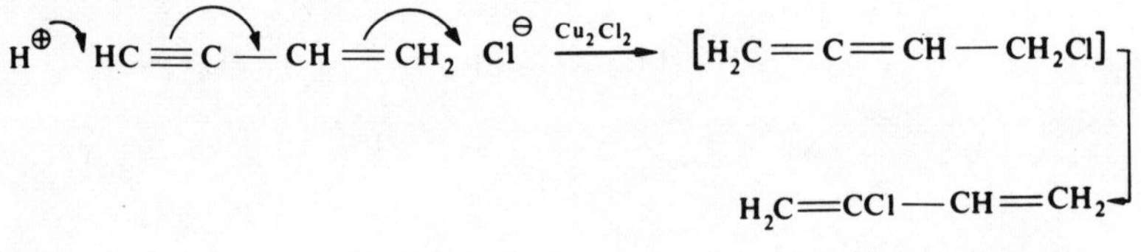

$$H^{\oplus} \; HC \equiv C - CH = CH_2 \; Cl^{\ominus} \xrightarrow{Cu_2Cl_2} [H_2C = C = CH - CH_2Cl]$$

$$H_2C = CCl - CH = CH_2$$

chloroprene

10. *Addition of Hydrogen Cyanide.* The addition of hydrogen cyanide to acetylene, catalysed by barium cyanide is similar to the formation of vinylacetylene. The product, acrylonitrile is used for the manufacture of Buna N synthetic rubber:

$$HC \equiv CH + HC \equiv N \xrightarrow{Ba(CN)_2} H_2C = CH - C \equiv N$$

acrylonitrile

Though the importance of acetylene as the starting point for industrial chemicals is declining, it is interesting in this case to notice how many of the reactions described above are actually used in industrial processes. Table 8 gives the most important products made from acetylene. Horizontal arrows lead to intermediates and vertical arrows, final use. The numbers over arrows refer to the numbered paragraphs on the reactions of acetylenes above. World production of acetylene (excluding Communist countries) is over 850 000 tons per year.

96

Table 8. Uses of acetylene

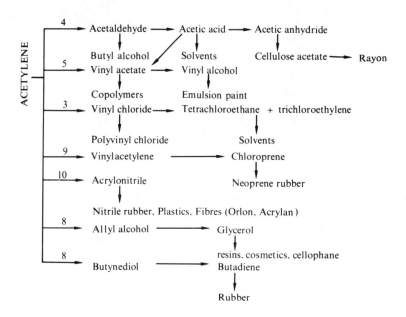

Identification

Terminal acetylenes, i.e. those containing the $\equiv$CH group, are easily identified by their insoluble copper or silver compounds, mentioned earlier. When the C$\equiv$C is in the middle of an alkyl chain, an alkyne might be confused with an alkene, since it would decolorize bromine solution and reduce permanganate. However, treatment with dilute sulphuric acid and mercuric sulphate (reaction 4) would give a ketone or mixture of ketones which could easily be identified.

The infrared absorption due to acetylenic C—H stretching is a sharp band at approximately 3300 cm^{-1}, even higher than that of alkenes; and the C$\equiv$C stretching absorption is at 2200–2100 cm^{-1}. It can be a medium or weak absorption depending upon other attached groups, but occurs in a region of the spectrum where little else, except C$\equiv$N and C$\equiv$O absorb so it is not easily missed. As in the case of the C=C stretching absorption, this will only be seen if the molecule is asymmetric, i.e. acetylene or dimethylacetylene will not give a C$\equiv$C absorption.

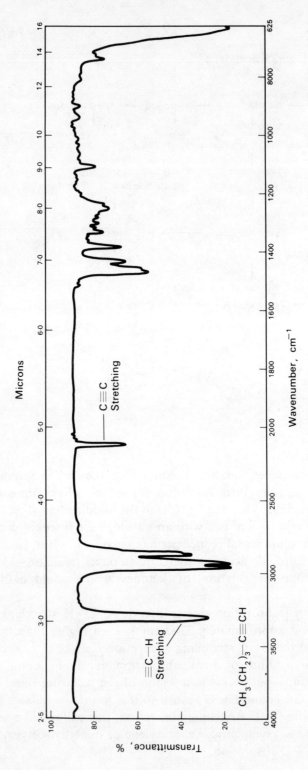

Fig. 25 *Infrared spectrum of hex-1-yne*

5

Alcohols

Alcohols are compounds represented by the general formula ROH where R can be an alkyl or substituted alkyl group; OH is termed the *hydroxyl* functional group. This chapter deals with alcohols, ROH, where R is a simple alkyl group. These alcohols have considerable importance as industrial chemicals and in the biochemistry of living organisms. Apart from hydrocarbon fuels, ethyl alcohol has the distinction of being the organic chemical produced in the largest volume annually.

Ethanol, or ethyl alcohol, long known as a product of the fermentation of sugar has the molecular formula C_2H_6O, and giving to carbon, hydrogen and oxygen their usual valencies of four, one, and two, respectively, there are only two possible structures that can be assigned to the molecule, i.e.:

$$H-\underset{\underset{H}{|}}{\overset{\overset{H}{|}}{C}}-O-\underset{\underset{H}{|}}{\overset{\overset{H}{|}}{C}}-H \quad \text{and} \quad H-\underset{\underset{H}{|}}{\overset{\overset{H}{|}}{C}}-\underset{\underset{H}{|}}{\overset{\overset{H}{|}}{C}}-OH$$

It was an important task for early chemists to decide which of these was the correct structure, and this they did by examining the reactions of the compound and its relation to other substances. For example, sodium will dissolve in ethanol, releasing hydrogen and forming C_2H_5ONa. Only one hydrogen atom per molecule can be replaced, which at once suggests that one hydrogen atom is bound in a different way from the other five. Phosphorus pentachloride acts on ethanol, giving the compound C_2H_5Cl, thus one hydrogen and one oxygen atom are together replaced by a chlorine atom, and the compound C_2H_5Cl can be reconverted to ethanol by treatment with an aqueous solution of sodium hydroxide. These, and many other reactions are consistent only with the second of the above formulae. The behaviour of methyl alcohol or methanol (CH_4O) fits in very well with the explanation. On valency arguments, CH_4O can only be H_3COH and here again it is found that one H is replaceable by Na and one OH by

Cl. To clinch the matter, another compound is known which has the formula C_2H_6O, with different physical and chemical properties. This is dimethyl ether, and this has all the properties consistent with the formula $CH_3.O.CH_3$.

The same reasoning can be applied to other alcohols and we can write the expression $C_nH_{2n+1}OH$ for the whole homologous series of the saturated aliphatic alcohols. Alcohols, then, are hydroxy-hydrocarbons, or looked at in another way, they are monoalkyl derivatives of water. The dialkyl derivatives of water, the ethers, will be considered in the next chapter.

$$H-O-H \qquad R-O-H \qquad R-O-R$$

water   alcohols   ethers

Recognition of this relation in the early nineteenth century led to the theory of types, in this case the *water* type, which was then written:

$$\left.\begin{array}{c}H\\H\end{array}\right\}O \qquad \left.\begin{array}{c}R\\H\end{array}\right\}O \qquad \left.\begin{array}{c}R\\R\end{array}\right\}O$$

and this was one of the foundations of modern structural theory.

The above explanation follows the classical approach to structure, but today, with physical methods, we can immediately recognize and distinguish hydroxyl and ether groups, for example by their infrared spectra, as discussed at the end of the chapter.

Nomenclature

The trivial names of the alcohols follow directly from the trivial names of the halides (Chapter 2), e.g. ethyl bromide becomes ethyl alcohol; isobutyl bromide, isobutyl alcohol. The systematic names include the radical names given in Table 3 (Chapter 1) and used for halides, alkenes and acetylenes. The characteristic suffix for saturated alcohols is *-anol*, *-an-* to show it is saturated, and *-ol* for an alcohol, e.g. the systematic name for ethyl alcohol is ethanol; an alcohol also containing a double bond would have the suffix *-enol*, etc. As in the case of halides, there are three types, namely primary, secondary (s-) and tertiary (t-) alcohols, of somewhat different properties, depending upon the number of alkyl groups attached to C—OH. This can be illustrated by the four isomeric butyl alcohols or butanols:

100

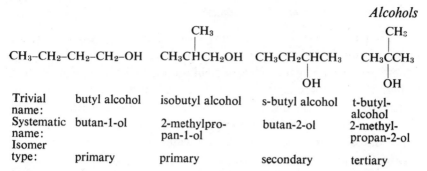

Trivial name:	butyl alcohol	isobutyl alcohol	s-butyl alcohol	t-butyl-alcohol
Systematic name:	butan-1-ol	2-methylpro-pan-1-ol	butan-2-ol	2-methyl-propan-2-ol
Isomer type:	primary	primary	secondary	tertiary

Iso- refers to an alkyl group with a methyl branch furthest from the functional group, as explained earlier and amyl is a trivial name used frequently for *pentyl*, e.g.:

$$CH_3CH_2CH_2CH_2CH_2OH$$
amyl alcohol

$$\underset{\text{isoamyl alcohol}}{CH_3\overset{\overset{\textstyle CH_3}{|}}{C}HCH_2CH_2OH}$$

Hydrogen Bonds

The alkanes, haloalkanes, alkenes, etc., form series with boiling points ascending smoothly with molecular weight, and all are only very slightly soluble in water. It may seem surprising at first that the lower alcohols boil about 100° higher than the corresponding alkanes, and are readily soluble in water, though they differ from alkanes by only one oxygen atom per molecule.

If we look at some inorganic hydrides we find that water (b.p. 100°) and ammonia (b.p. −33°) have much higher boiling points than their near relatives hydrogen sulphide (b.p. −61°) and phosphine (b.p. −88°) respectively, in spite of the higher molecular weights of the latter. This phenomenon is confined only to the strongly electronegative elements (F, O, N and sometimes Cl) and hydrogen.

The molecules of water, in addition to being held together by van der Waals forces, (the weak forces that hold all molecules together, and enable us to liquefy even helium), are also held together by a stronger force, called *hydrogen bonding*. This results in the liquid having a reduced vapour pressure and consequently raises its boiling point.

Because oxygen is a strongly electronegative element, the O—H bonds of water are polarized, that is, the valence electrons are held more firmly by oxygen, the net result is:

$$\delta\ominus \qquad \delta\oplus$$

$$O \leftarrow H$$

Hydrogen has no other electrons with which to shield its nuclear charge. But divalent oxygen has two unshared pairs of electrons, and these produce a negative electric field on one side of the oxygen atom (Fig. 26), and there is an attraction between this negative field of the unshared pairs and the $\delta^{\oplus}$ hydrogen atom. By associating with the unshared pairs of another oxygen atom, hydrogen can 'clothe its nakedness'. Thus the molecules of water are held together by an attraction greater than the van der Waals forces, but weaker than a normal covalent bond (Fig. 27). The same effect operates in HF, giving rise to the stable $(F—H \cdots F)^{\ominus}$ ion, and to a lesser extent in NH_3. Hydrogen bonds are usually represented by a dotted line.

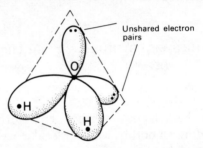

Unshared electron pairs

Fig. 26   The water molecule showing oxygen as an sp³ *hybrid, with two of the orbitals filled with unshared electron pairs and the other two orbitals used in bonding to hydrogen*

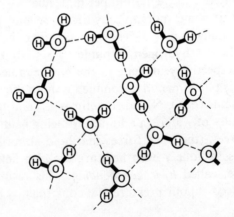

Fig. 27   Random hydrogen bonding in liquid water

Some idea of the strength of hydrogen bonding forces can be found by comparing some *bond energies*. It is evident that the formation and breaking of chemical bonds involves uptake or release of energy. By indirect methods the amount of energy released in forming one mole of water from *atoms* of hydrogen and oxygen can be calculated (one must know, for example, the energy absorbed in the reactions $H_2 \rightarrow 2H$ and $O_2 \rightarrow 2O$) and by dividing the result by 2, the energy released in forming the O—H bond is obtained, which can be expressed in kilojoules per mole (kJ/mole). This and some other covalent bond energies are given in Table 9. together with some hydrogen bond energies. Covalent bond energies vary from 150 kJ/mole for I—I upwards; the hydrogen bond energies are all much less than this.

Table 9. Bond energies

Bond	Energy (kJ/mole)
F · · · H hydrogen bond	42
O · · · H hydrogen bond	29
N · · · H hydrogen bond	8
C—H covalent bond	410
N—H covalent bond	385
O—H covalent bond	460
F—H covalent bond	565
C—C covalent bond	348
C—O covalent bond	360

Hydrogen bonding can and does occur in alcohols too, reducing their vapour pressures and raising their boiling points. In addition, alcohols can hydrogen-bond with water molecules and so the alcohols dissolve readily in water (Fig. 28). As we ascend the homologous series of alcohols the size of the alkyl group becomes dominant, and the effect of the hydroxyl group less important, and so their solubility in water decreases rapidly (see Table 10) and their boiling points approach more closely to those of the corresponding alkanes.

Many further examples of hydrogen bonding will arise in organic chemistry; above all, it plays a vital part in the complex molecules of biological substances.

Preparation of Alcohols

1. *From an Alkyl Halide.* Treating a primary alkyl halide, preferably a chloride, with aqueous alkali gives an alcohol by an S_N2 mechanism. Secondary alkyl halides react less well, and tertiary alcohols cannot

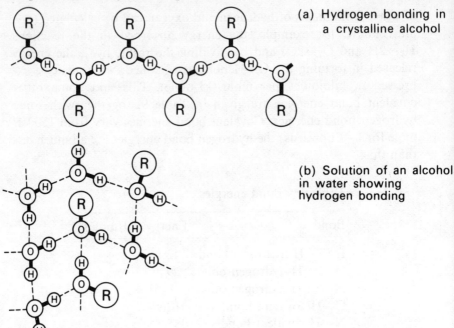

(a) Hydrogen bonding in a crystalline alcohol

(b) Solution of an alcohol in water showing hydrogen bonding

Fig. 28   Hydrogen bonding in alcohols (a) Hydrogen bonding in a crystalline alcohol (b) Solution of an alcohol in water showing hydrogen bonding

be obtained by this method. The competing effects of substitution and elimination have already been discussed in Chapter 2.

$$RCH_2Br + KOH \xrightarrow{H_2O} RCH_2OH + KBr$$

A very mild reagent for converting alkyl halides to alcohols is a suspension of moist silver oxide in ether, which is effective in the cold or on slight warming, and elimination to an olefin is avoided. This seems not to be a simple S_N1 or S_N2 reaction between RBr and OH⁻. The heavy silver ion may take part by helping to pull off the bromine atom:

$$OH^{\ominus} + RBr + Ag^{\oplus} \longrightarrow OH \cdots R \cdots Br \cdots Ag \longrightarrow ROH + AgBr$$

Other mild reagents are hot aqueous sodium carbonate, or suspensions of magnesium oxide or lead oxide (PbO) in water, or even water alone at 100° or under pressure at higher temperatures industrially. Because conversion of a halide to an alcohol is, in essence, a reaction with water it is referred to as *hydrolysis* or a hydrolytic reaction:

104

$$RX + H_2O \longrightarrow ROH + HX$$

2. *Hydrolysis of Esters.* An ester (see Chapter 9) is a compound formed from an acid and an alcohol by elimination of water. For example, acetic acid and ethanol when condensed together with loss of H_2O form ethyl acetate, an ester:

$$\underset{\displaystyle \|}{\overset{O}{CH_3COH}} + HOC_2H_5 \rightleftharpoons \underset{\displaystyle \|}{\overset{O}{CH_3COC_2H_5}} + H_2O$$

This is an equilibrium reaction, so that esters can be decomposed by water; a catalyst, either an acid or an alkali is usually added. The acid and alcohol are regenerated. It should be noted that the alkyl halides are esters of the inorganic acids HX. Many alcohols occur in nature as esters, and can be obtained from them in this way.

$$RCOOR' + H_2O \xrightarrow[orOH^\ominus]{H^\oplus} RCOOH + R'OH$$

Esters from tertiary alcohols and organic acids can usually be hydrolysed smoothly to give the alcohol. Alkyl halides react with silver acetate to give acetic esters, so tertiary alcohols can be made from the halides by first converting the halides to the acetates.

$$\underset{\underset{\displaystyle CH_3}{\displaystyle |}}{\overset{\overset{\displaystyle CH_3}{\displaystyle |}}{CH_3Cl}} + CH_3COOAg \longrightarrow \underset{\underset{\displaystyle CH_3}{\displaystyle |}}{\overset{\overset{\displaystyle CH_3}{\displaystyle |}}{CH_3COOCCH_3}} \xrightarrow{H_2O} \underset{\underset{\displaystyle CH_3}{\displaystyle |}}{\overset{\overset{\displaystyle CH_3}{\displaystyle |}}{CH_3COH}} + CH_3COOH$$

3. *Hydration of Alkenes.* As already stated in Chapter 3, alkenes add sulphuric acid to give alkyl sulphates which are esters of sulphuric acid. These are easily hydrolysed by boiling water to give alcohols. Ethylene affords a primary alcohol; higher alkenes give the secondary or tertiary alcohols that are predicted by Markovnikov's Rule:

$$CH_2{=}CH_2 + H_2SO_4 \longrightarrow \underset{\underset{\displaystyle O}{\displaystyle \|}}{\overset{\overset{\displaystyle O}{\displaystyle \|}}{CH_3CH_2-OSOH}} \xrightarrow{H_2O} CH_3CH_2OH + H_2SO_4$$

$$CH_3CH{=}CH_2 + H_2SO_4 \longrightarrow \underset{\underset{\displaystyle OSO_3H}{\displaystyle |}}{CH_3CHCH_3} \xrightarrow{H_2O} \underset{\underset{\displaystyle OH}{\displaystyle |}}{CH_3CHCH_3} + H_2SO_4$$

To obtain primary alcohols from higher alkenes, the alkene is treated with aluminium and hydrogen under pressure to give aluminium alkyls (Chapter 3). These react vigorously with oxygen in the air to give aluminium alkoxides, the oxygen inserting itself between aluminium and carbon. Treatment of the alkoxides with water and dilute acid yields the alcohols.

High molecular weight (C_{12}—C_{18}) alcohols made in this way are important intermediates for synthetic detergent manufacture (page 199).

$$RCH=CH_2 \ + \ Al \xrightarrow{H_2} Al(CH_2CH_2R)_3 \xrightarrow{O_2} Al(OCH_2CH_2R)_3$$

$$\downarrow H_2O . H_2SO_4$$

$$RCH_2CH_2OH \longleftarrow$$

Another recent industrial method is to pass the alkene with steam over phosphoric acid supported on silica, at moderately high temperature and pressure:

$$CH_2=CH_2 \ + \ H_2O \xrightarrow{cat.} CH_3CH_2OH$$

Anti-Markovnikov hydration of alkenes is best achieved in the laboratory by hydroboration followed by oxidation, as described in Chapter 3. The yields in this reaction are high, and because carbonium ions are not involved, no rearrangement occurs:

$$\underset{\substack{| \\ CH_3}}{\overset{\substack{CH_3 \\ |}}{CH_3-C}}=CH_2 \xrightarrow{BH_3} (CH_3CHCH_2-)_3B \xrightarrow[OH^{\ominus}]{H_2O_2} \underset{\substack{| \\ CH_3}}{CH_3CHCH_2OH} \ + \ B(OH)_3$$

4. *Reduction.* Several groups of compounds can be reduced to alcohols and the following are among the general methods for this purpose. The groups of compounds are aldehydes, ketones and esters. Aldehydes and esters give primary alcohols; ketones give secondary alcohols.

$$\underset{\text{acetaldehyde}}{\overset{\substack{H \\ |}}{CH_3-C}}=O \ + \ 2[H] \longrightarrow CH_3CH_2OH$$

$$\underset{\text{acetone}}{\overset{\substack{O \\ ||}}{CH_3-C}-CH_3} \ + \ 2[H] \longrightarrow \underset{\substack{| \\ OH}}{CH_3CHCH_3}$$

$$\underset{\text{ethyl pentanoate}}{\overset{\substack{O \\ ||}}{CH_3CH_2CH_2CH_2COC_2H_5}} \ + 4[H] \longrightarrow \underset{\text{pentanol b.p. } 138°}{CH_3(CH_2)_3CH_2OH} + \underset{\text{ethanol b.p. } 78°}{CH_3CH_2OH}$$

In the case of the ester, the products can easily be separated by fractional distillation.

The more important general methods of reduction are (a) catalytic, (b) with dissolving metals and (c) with complex hydrides.

(a) Aldehydes, ketones and esters are reduced to alcohols under various conditions of temperature and pressure by hydrogen in the presence of a metal catalyst, such as copper, nickel or platinum. The yields can be good, but more vigorous reduction gives hydrocarbons. For example, nickel catalysts above 250° give hydrocarbons. Note that double bonds are catalytically hydrogenated under much milder conditions (at room temperature and pressure) so unsaturated groups would also be reduced by this method.

(b) By treating esters with sodium dissolved in alcohol (the Bouveault method), or zinc in acetic or hydrochloric acid, good yields of alcohol are obtained. The Bouveault method has been widely used for the reduction of esters.

The reduction comes not from the hydrogen produced in the reaction but from donation of electrons from the metal to the organic compound, followed by donation of protons from the solvent to neutralize the charge created:

$$Zn \longrightarrow Zn^{\oplus\oplus} + 2e^{\ominus}$$

$$\begin{array}{c} R \\ \diagdown \\ R \diagup \end{array} C{=}O + 2e^{\ominus} + 2H_2O \longrightarrow \overset{\displaystyle R}{\underset{\displaystyle R}{\overset{\displaystyle |}{\underset{\displaystyle |}{HC}}}}{-}OH + 2OH^{\ominus}$$

(c) Complex metal hydrides used for selective reduction are exemplified by lithium aluminium hydride ($LiAlH_4$) and sodium borohydride ($NaBH_4$). They are made commercially by the following reactions:

$$4LiH + AlCl_3 \longrightarrow LiAlH_4 + 3LiCl$$

$$\underset{\substack{\text{sodium} \\ \text{hydride}}}{4NaH} + \underset{\substack{\text{trimethyl} \\ \text{borate}}}{B(OCH_3)_3} \longrightarrow NaBH_4 + 3NaOCH_3$$

Lithium aluminium hydride and sodium borohydride are powerful yet selective reducing agents. They both reduce aldehydes and ketones to alcohols rapidly at room temperature and in excellent yield, but do

107

not affect double bonds. Lithium aluminium hydride will reduce free acids, which are not affected by methods (a) and (b) above, and esters. Sodium borohydride does not reduce acid or ester groups.

$$CH_3(CH_2)_7 CH{=}CH(CH_2)_7 COOH \xrightarrow{LiAlH_4} CH_3(CH_2)_7 CH{=}CH(CH_2)_7 CH_2OH$$

oleic acid   oleyl alcohol

$$\overset{\displaystyle O}{\overset{\|}{CH_3 CCH_2 COOC_2 H_5}} \xrightarrow{NaBH_4} \underset{\underset{HO}{|}}{CH_3 CHCH_2 COOC_2 H_5}$$

ethyl acetoacetate   ethyl 3-hydroxybutyrate

Although lithium aluminium hydride reacts violently with water and must be used in hydroxyl-free solvents such as dry ether, sodium borohydride which is stable to water can be employed in aqueous or alcoholic solution.

The mechanism of the reaction involves the intermediate formation of complex aluminium compounds, which can then be decomposed by dilute acid.

$$4\,R{-}\overset{\overset{\displaystyle H}{|}}{C}{=}O + LiAlH_4 \longrightarrow (R{-}\underset{\underset{\displaystyle H}{|}}{\overset{\overset{\displaystyle H}{|}}{C}}{-}O)_4\, Al^{\ominus}\,Li^{\oplus} \xrightarrow{4HCl} 4RCH_2 OH + LiCl + AlCl_3$$

Diborane (page 63) also reduces aldehydes, ketones, acids and esters.

5. *By the Grignard Reaction.* Grignard reagents (page 36) can be used to form all three types of alcohol by reaction with suitable reagents.

Addition of formaldehyde to a Grignard reagent gives a primary alcohol.

$$RMgX + HC{\overset{\overset{\displaystyle H}{|}}{=}}O \longrightarrow RCH_2 OMgX \xrightarrow{H_2O} RCH_2OH + MgXOH$$

The mechanism of this reaction will be discussed in Chapter 8.

Higher aldehydes with one equivalent of a Grignard reagent (or formic esters with two equivalents) give secondary alcohols:

$$\underset{H}{\overset{H}{\underset{|}{CH_3C}}}=O + RMgX \longrightarrow CH_3\underset{|}{\overset{R}{C}}HOMgX \longrightarrow R-\underset{|}{\overset{H}{\underset{OH}{C}}}-CH_3$$

$$\overset{O}{\overset{||}{HCOEt}} + RMgX \longrightarrow R-\underset{|}{\overset{H}{\underset{OMgX}{C}}}-OEt \longrightarrow R\underset{\downarrow RMgX}{\overset{H}{C}}=O + MgXOEt$$

ethyl formate

$$\underset{OMgX}{\overset{H}{\underset{|}{RCR}}} \overset{H_2O}{\longrightarrow} \underset{OH}{\overset{H}{\underset{|}{RCR}}} + MgXOH$$

Ketones and higher esters give tertiary alcohols.

$$\overset{O}{\overset{||}{CH_3CCH_3}} + RMgX \longrightarrow \underset{OMgX}{\overset{R}{\underset{|}{CH_3CCH_3}}} \overset{H_2O}{\longrightarrow} \underset{OH}{\overset{R}{\underset{|}{CH_3CCH_3}}} + MgXOH$$

If dry air is blown through a solution of a Grignard reagent, reaction takes place, similar to that between aluminium alkyls and oxygen. Subsequent hydrolysis gives an alcohol of the same number of carbon atoms as the starting halide.

$$RBr \overset{Mg}{\longrightarrow} RMgBr \overset{O_2}{\longrightarrow} ROMgBr \overset{H_2O}{\longrightarrow} ROH + MgBrOH$$

6. *The 'Oxo' Process.* The Oxo or hydroformylation process is used on a large scale for making alcohols from alkenes. The alkenes are mixed with carbon monoxide and hydrogen and passed through a reactor filled with a cobalt catalyst at about 150° and 200 atmospheres. An aldehyde is produced, which can then be treated with more hydrogen to produce alcohols. The method is used for making butanol from propylene and for making higher alcohols (C_{10}—C_{19}) from alkenes produced by cracking petroleum fractions.

Under the conditions of the reaction, the cobalt is converted to $HCo(CO)_3$ which adds to the double bond. This product (a) takes up more carbon monoxide and rearranges to insert one CO between carbon and cobalt (b). Reduction by hydrogen gives an aldehyde and regenerates $HCo(CO)_3$ for further catalytic activity.

$$RCH=CH_2 + HCo(CO)_3 \longrightarrow \underset{(a)}{RCH_2CH_2Co(CO)_3} \overset{CO}{\searrow}$$

$$RCH_2CH_2Co(CO)_4$$

$$RCH_2CH_2\overset{H}{\underset{|}{C}}=O + HCo(CO)_3 \overset{H_2}{\longleftarrow} \underset{(b)}{RCH_2CH_2\overset{O}{\overset{||}{C}}Co(CO)_3}$$

By modifying the reaction and adding more hydrogen, the aldehyde is reduced to an alcohol. The overall reaction then is:

$$RCH = CH_2 + CO + H_2 \xrightarrow{\text{Co catalyst}} RCH_2CH_2CHO \xrightarrow{H_2} RCH_2CH_2CH_2OH$$

Physical Properties of Alcohols

Aliphatic alcohols are colourless liquids or white, waxy solids. The lower alcohols resemble water in their physical properties, and are completely miscible with water, but with increasing molecular weight the resemblance to water quickly diminishes, their solubility in water falls, and they become oily liquids soluble only in organic solvents. The lower alcohols have characteristic odours and tastes, the higher ones are tasteless and odourless.

The alcohols have higher melting and boiling points than the corresponding alkanes, due to hydrogen bonding. As in other homologous series, branched isomers have lower boiling points than the straight chain isomers.

Table 10. Alcohols

Alcohol	m.p. °C	b.p. °C	Solubility in H_2O (g/100 g at 20°)
Methyl	— 97	64·5	miscible
Ethyl	—115	78·3	miscible
Propyl	—126	97	miscible
Isopropyl	— 86	82·5	miscible
Butyl	— 90	118	7·9
Isobutyl	—108	107	9·6
s-Butyl	—114	99·5	12·5
t-Butyl	26	83	miscible
Amyl (pentyl)	— 78	138	2·6
Hexyl	— 52	157	0·6
Hexadecyl	49	180*	insoluble

Reactions of Alcohols

1. *Oxidation.* Alcohols may be regarded as the first oxidation products of hydrocarbons. Primary alcohols are easily oxidized to aldehydes which are themselves very susceptible to oxidation, so they are usually then oxidized further to carboxylic acids with excess

* At 10 mm pressure.

of the reagent. For example, potassium dichromate, chromic acid (H_2CrO_4), chromic oxide (CrO_3) or alkaline permanganate usually give the corresponding acid from a primary alcohol.

$$RCH_2OH \xrightarrow{KMnO_4} [R\overset{H}{\underset{}{C}}=O] \longrightarrow RCOOH$$

Secondary alcohols with these reagents give ketones, which in contrast to aldehydes are resistant to further oxidation. Tertiary alcohols are resistant to mild oxidants, but more powerful agents, for example chromic acid break down tertiary alcohols into mixtures of ketones and acids.

$$R-\overset{\overset{\displaystyle R'}{|}}{\underset{\underset{\displaystyle OH}{|}}{C}}-CH_2R' \xrightarrow{[O]} R-\overset{\overset{\displaystyle O}{||}}{C}-R' + R''COOH$$

The C—H bonds of alkanes are comparatively resistant to oxidation (see Chapter 1), but once an oxygen atom has been introduced into the molecule, the carbon atom to which the oxygen is attached becomes susceptible to further oxidation, and the hydrogen atoms attached to that carbon atom are easily replaced by OH or OR. Once these hydrogen atoms have all been removed, the molecule again becomes resistant to oxidation, and a strong oxidizing agent is required to break a C—C bond. That is why primary and secondary alcohols and aldehydes are oxidizable, whereas acids, ketones and tertiary alcohols, containing no O—C—H, are resistant. Thus acetic acid and acetone are good solvents for carrying out dichromate and permanganate oxidations of organic compounds.

The first stage in the chromic acid oxidation of an alcohol is the formation of an unstable ester between the chromic acid and alcohol, this is followed by a slow step in which water attacks the α-hydrogen atom:

$$\underset{R}{\overset{R}{\diagdown}}\!\!\!\underset{OH}{\overset{H}{C}}\!\!\!\diagup \quad + \quad H_2CrO_4 \rightleftharpoons \underset{R}{\overset{R}{\diagdown}}\!\!\!\underset{O-CrO_3H}{\overset{H}{C}}\!\!\!\diagup \quad + \quad H_2O$$

$$\underset{R}{\overset{R}{\diagdown}}\!\!\!\underset{O-CrO_3H}{\overset{H\; :O<^H_H}{C}}\!\!\!\diagup \quad \longrightarrow \quad \underset{R}{\overset{R}{\diagdown}}C=O \quad + \quad H_3O^{\oplus} \quad + \quad HCrO_3^{\ominus}$$

By this electron-pair shift, chromium has gone from an oxidation state of $+6$ to $+4$, but is is not stable in the latter state and then undergoes further reactions to give Cr^{+3} and Cr^{+6}.

111

Alcohols may be converted to aldehydes or ketones in a vapour-phase reaction over a copper catalyst. The alcohol vapour is passed through a tube filled with copper turnings at 200–300°. The reaction is a *dehydrogenation*, removal of two hydrogen atoms, rather than an oxidation, but some air can be admitted to oxidize the hydrogen formed:

$$RCH_2OH \xrightarrow{Cu} RCHO + H_2$$

$$R_2CHOH \xrightarrow{Cu} R_2C=O + H_2$$

$$RCH_2OH + O_2 \xrightarrow{Cu} RCHO + H_2O$$

2. *Dehydration.* Alcohols lose the elements of water, either by heating alone to a very high temperature (400–800°) or at lower temperatures with catalysts such as alumina or silica. Sulphuric and phosphoric acids also dehydrate alcohols, tertiary alcohols most easily and primary least readily. Thus heating with dilute sulphuric acid is sufficient to convert a tertiary alcohol to an alkene. These reactions of E2 and E1 type have already been discussed in Chapter 3.

When dehydration can take place in two or more ways to give different alkenes, hydrogen is eliminated from the adjacent carbon atom having least hydrogen (Saytzev's rule) as illustrated below. The compound underlined is the major or only product.

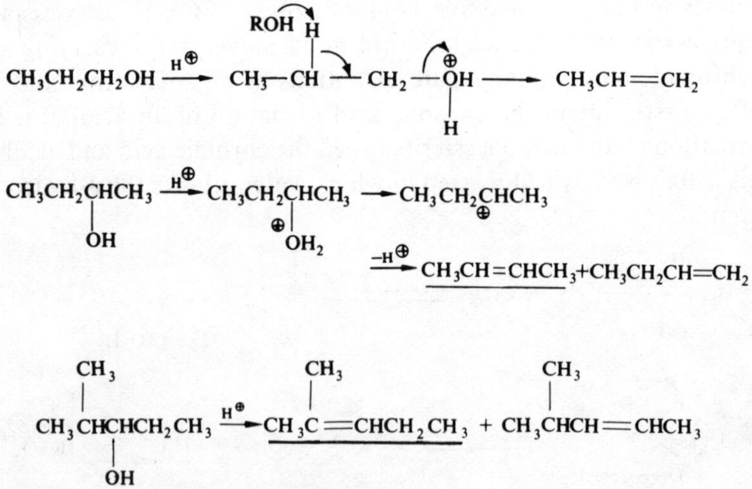

$$CH_3CH_2CH_2OH \xrightarrow{H^\oplus} CH_3-CH \overset{ROH\ H}{\curvearrowright} CH_2-\overset{\oplus}{O}H \longrightarrow CH_3CH=CH_2$$

$$CH_3CH_2CHCH_3 \xrightarrow{H^\oplus} CH_3CH_2\overset{\oplus}{C}HCH_3 \longrightarrow CH_3CH_2\overset{\oplus}{C}HCH_3$$
$$\underset{|}{OH} \qquad \overset{\oplus}{\underset{OH_2}{|}}$$
$$\xrightarrow{-H^\oplus} \underline{CH_3CH=CHCH_3} + CH_3CH_2CH=CH_2$$

$$CH_3\underset{\underset{OH}{|}}{\overset{\overset{CH_3}{|}}{CHCHCH_2CH_3}} \xrightarrow{H^\oplus} \underline{CH_3\overset{\overset{CH_3}{|}}{C}=CHCH_2CH_3} + CH_3\overset{\overset{CH_3}{|}}{C}HCH=CHCH_3$$

With some alcohols, the high temperature required for dehydration causes isomerization of the alkene formed, so the double bond does not appear in the predicted place.

3. *With Metals.* Reactive metals (e.g. Li, Na, K, Mg, Al) dissolve in alcohols to give metal alkoxides: e.g.

$$Na + CH_3CH_2OH \longrightarrow CH_3CH_2ONa + \tfrac{1}{2}H_2$$

sodium ethoxide

The order of reactivity of the alkali metals with alcohols is the same as with water, but the reaction is less vigorous. The reactivity of the alcohols decreases in the order primary > secondary > tertiary, and also decreases with rise in molecular weight. Sodium dissolves vigorously in methanol or ethanol. Potassium ignites in these alcohols but dissolves quietly in t-butanol. The metal alkoxides can be obtained as white hygroscopic solids, but they retain traces of alcohol very tenaciously.

The alkali metal compounds are essentially ionic, e.g. $RO^\ominus Na^\oplus$, the ionic character diminishes for lithium and the alkaline earth metals, and the aluminium compounds are essentially covalent.

The alcohols are amphoteric compounds like water, in the elimination reaction (paragraph 2) above, they are behaving as bases (HB represents any acid):

$$ROH + H^\oplus B^\ominus \rightleftharpoons ROH_2^\oplus + B^\ominus$$

Here they are behaving as weak acids, weaker than water (K_a for water is 10^{-16} (page 93), for ethanol K_a is 10^{-18}):

$$ROH + M \longrightarrow RO^\ominus M^\ominus + \tfrac{1}{2}H_2$$

Since the salts of weak acids are themselves strong bases, we find as we would expect, that the alkoxide ions $RO^\ominus$ are stronger bases than $OH^\ominus$ and are frequently used in organic reactions where a strong base is required.

A list of weak acids and strong bases that have been discussed so far can be drawn up thus:

Strong acid	$\longrightarrow$			Weak acid
HX	> H_2O	> ROH	>	$RC{\equiv}CH$ > NH_3
$X^\ominus$	< $OH^\ominus$	< $RO^\ominus$	<	$RC{\equiv}C^\ominus$ < $NH_2^\ominus$
Weak base	$\longleftarrow$			Strong base

113

4. *Esters*. Alcohols form esters with both organic and inorganic acids, e.g.

$$C_2H_5OH \ + \ HO\overset{\overset{\textstyle O}{\|}}{C}CH_3 \ \rightleftharpoons \ C_2H_5O\overset{\overset{\textstyle O}{\|}}{C}CH_3 \ + \ H_2O$$

The reaction is an equilibrium reaction, the mechanism is discussed in Chapter 9: in order to obtain a good yield of the ester, some means of displacing the equilibrium is usually necessary, such as using a large excess of one component, or of removing water as it is formed, and to speed up the rate of reaction, a catalyst is needed, usually a strong acid. The common method of preparing esters is to heat under reflux a mixture of the acid (which is usually the more valuable component) with about five equivalents of the alcohol in the presence of a little strong acid as catalyst.

Esters of sulphuric acid can be prepared from the acid and alcohol. Fuming sulphuric acid and methanol heated together give dimethyl sulphate:

$$2C_2H_3OH \ + \ SO_3 \longrightarrow CH_3\,O\overset{\overset{\textstyle O}{\|}}{\underset{\underset{\textstyle O}{\|}}{S}}OCH_3$$

dimethyl sulphate

Fuming acid in the cold, or warm concentrated acid and higher alcohols give alkyl hydrogen sulphates. The latter reaction is of S_N1 or S_N2 type, depending on the nature of the alcohol, similar to the formation of alkyl halides:

$$CH_3CH_2OH + H^{\oplus} \longrightarrow CH_3CH_2\overset{\oplus}{O}H_2 \xrightarrow[S_{N2}]{HSO_4^{\ominus}} CH_3CH_2O\overset{\overset{\textstyle O}{\|}}{\underset{\underset{\textstyle O}{\|}}{S}}OH \ + \ H_2O$$

Stronger heating of ethanol and higher alcohols with sulphuric acid gives mixtures of esters, ethers and olefins.

Alkyl phosphates are made from alcohol and phosphoryl chloride in the presence of pyridine.

$$3ROH \ + \ POCl_3 \xrightarrow{C_5H_5N} RO\overset{\overset{\textstyle OR}{|}}{\underset{\underset{\textstyle OR}{|}}{P}}{=}O \ + \ 3HCl$$

114

Other esters of inorganic acids are usually prepared by indirect methods, e.g. alkyl nitrates are prepared from alkyl halides and silver nitrate. They are dangerously unstable to heat or shock.

$$RX + AgNO_3 \longrightarrow RONO_2 + AgX$$

5. *Formation of halides.* The action of phosphorus halides on alcohols to give the corresponding alkyl halides has already been discussed in Chapter 2. Hydrogen halides also convert alcohols to halides but at widely different rates, tertiary alcohols most readily. Thus t-butyl alcohol is converted to the chloride on shaking with concentrated hydrochloric acid *via* a S_N1 reaction. A secondary alcohol is converted slowly to the chloride on heating with zinc chloride and concentrated hydrochloric acid: a primary alcohol is hardly affected. This reaction is used to distinguish the three types of alcohols (Lucas test). Hydrogen bromide in concentrated aqueous or acetic acid solution is much more reactive and converts secondary and primary alcohols to alkyl bromides, at 50–90°.

$$RCH_2OH \xrightarrow{H^{\oplus}} RCH_2\overset{\oplus}{O}H_2 \xrightarrow[S_N2]{Br^{\ominus}} RCH_2Br + H_2O$$

Free halogens do not replace the hydroxyl, but oxidize alcohols. From ethanol and chlorine, acetaldehyde is formed which is then substituted with three atoms of chlorine to give trichloroacetaldehyde or *chloral*, used medically as a sedative and illicitly in 'knockout drops':

$$CH_3CH_2OH + 4Cl_2 \longrightarrow CCl_3\overset{\overset{\displaystyle H}{|}}{C}=O + 5HCl$$

chloral

Several alcohols are of such industrial importance that their technical preparations and uses deserve consideration.

Methanol

Methanol was for a long time prepared by the 'destructive distillation' of wood, that is, heating wood with the exclusion of air. This provides a distillate called pyroligneous acid which is a complex mixture containing acetic acid, acetone, ethanol, allyl alcohol (CH_2=$CHCH_2OH$), acetaldehyde and other minor constituents. Wood consists chiefly of cellulose and lignin, a substance of high molecular weight containing methoxyl (—OCH_3) groups. The methanol is produced from the lignin methoxyls during pyrolysis. This method gave methanol the name 'wood alcohol'. Impurities in the alcohol caused it to have a very unpleasant odour. Today wood is much too valuable for this use and coke or hydrocarbons are the raw material for methanol production.

The starting point for methanol is usually synthesis gas ($CO+H_2$) made by passing steam over heated coke:

$$C + H_2O \longrightarrow CO + H_2$$

This gas is not rich enough in hydrogen for methanol synthesis so more hydrogen is produced by a 'shift' reaction by adding more steam:

$$CO + H_2O \xrightarrow{cat.} CO_2 + H_2$$

The carbon dioxide can easily be removed from the gas mixture by dissolving it in water under pressure. The carbon monoxide-hydrogen mixture is then compressed and passed over a zinc oxide-chromic oxide catalyst at 300°. The yield is very high and the product pure.

$$CO + 2H_2 \underset{200\,atm.}{\overset{ZnO\cdot Cr_2O_3}{\longrightarrow}} CH_3OH$$

The synthesis gas can alternatively be obtained from natural gas:

$$CH_4 + H_2O \longrightarrow CO + 3H_2$$

This mixture is too rich in hydrogen and more CO can be made and added in a variety of ways to give the $CO:H_2$ ratio of $1:2$ necessary for methanol. Direct catalytic oxidation of methane to methanol with air at 200° is also used.

$$2CH_4 + O_2 \xrightarrow{cat.} 2CH_3OH$$

The U.K. production of methanol was 350 000 tons a year in 1973. The chief uses are for the preparation of formaldehyde $H_2C{=}O$ (Chapter 8), as a solvent particularly in the manufacture of vitamins and pharmaceuticals, in methyl methacrylate for polymers, and as an intermediate in a wide range of other processes.

Methanol burns quietly with a pale blue flame. Over alumina at 350° it is dehydrated to dimethyl ether; under the same conditions ethanol and higher alcohols give olefins. Physiologically, it is a very dangerous compound: drinking it causes permanent injury to the brain and death, and as little as 10 ml is sufficient to cause blindness.

116

Since it is also a low boiling solvent, one should take care not to breathe the vapours of boiling methanol in the laboratory. It is a good solvent for many organic compounds and is miscible with most other organic solvents, and dissolves many inorganic salts.

Ethanol

Ethanol must share with acetic acid the distinction of being the earliest industrial organic chemical. Today it is the most important organic chemical judged by volume of production. It is found free in plants in traces, but commonly as a metabolic product of micro-organisms.

Ethanol is produced industrially from four different kinds of material.

1. *Sugars, such as cane sugar (molasses) or sugar beet.* Yeast contains enzymes which convert the sugars (maltose, sucrose and others) to glucose and an enzyme zymase, which converts glucose ($C_6H_{12}O_6$) to ethanol with considerable efficiency; abut 94–95% of the sugar is converted to ethanol.

$$C_6H_{12}O_6 \xrightarrow{zymase} 2C_2H_5OH + 2CO_2$$

Fermentation takes place in a dilute aqueous solution at 20–30°; if the alcohol content rises above 12% or the temperature to 60°, the fermentation stops. When fermentation is complete, the water and alcohol are distilled and a residue is left called fusel oil, derived chiefly from the protein of the plant debris by removal of nitrogen. Fusel oil consists chiefly of higher-boiling alcohols.

2. *Starch as contained in grain or potatoes.* The starch has first to be digested with acid or malted, i.e. the natural enzymes of the grain are used to break down the starch into sugars, which are then fermented with yeast. After distillation of the alcohol, a mash is left behind, which is pressed free of water and provides a highly nutritious cattle feed.

3. *Cellulose, from wood or agricultural waste, such as straw.* This must first be digested with acid to break down the cellulose to sugar, which is then fermented.

4. *Ethylene from petroleum refining.* The ethylene is hydrated either with sulphuric acid at 85° or as a vapour at 300° over a catalyst of phosphoric acid on silica.

The method which is used is largely governed by the raw materials available at the place of production. In Great Britain about 75% of ethanol production is by synthesis from ethylene. It is interesting that in some countries it is economical to dehydrate fermentation alcohol to make ethylene industrially. The total world production of industrial ethanol was about 2 million tons in 1973, worth $240 million of which the U.S.A. produced 900 000 and the U.K. about 150 000 to 200 000 tons.

Ethanol is used as a solvent in the preparation of esters, ether, and acetic acid and as an intermediate in a range of other chemicals. It can be added to petrol for motor fuel. Larger quantities of alcohol are consumed in the form of beer, wines and spirits. Such potable forms of alcohol usually carry a heavy tax. Denatured alcohol is untaxed alcohol for industrial use to which about 5% of a substance (methanol, benzene, petrol, etc.) has been added to make it unfit for drinking. Industrial methylated spirit consists of 90% ethanol, 5% methanol and 5% water (by volume). For domestic use, pyridine (to given an offensive odour) and a violet dye are added as well.

Aqueous alcohol mixtures are measured in degrees proof. In Great Britain 100° proof is a solution that is 50% by weight of ethanol, and in the U.S.A., 100° proof is 50% by volume or 42% by weight of ethanol. In this, as in certain other measures, the British unit is more generous than the American. A discerning chemist, offered U.K. or U.S. gin of 98° proof at the same price would take the U.K. bottle. Pure alcohol is 76° over proof (U.K. standard) or 200° U.S. proof. When ethanol and water are mixed, there is a contraction in volume, so that more than 50 ml of ethanol can be added to 50 ml of water to make 100 ml solution.

Fractional distillation of an alcohol–water mixture cannot give alcohol purer than 95% because ethanol and water form a constant boiling mixture or *azeotrope* of this composition, with a boiling point (78·15°) lower than that of either pure component. To remove the last 5% of water, a third component (benzene, ether or trichloroethylene) is added. In the case of benzene, at first a ternary azeotropic mixture of benzene, ethanol and water distils (b.p. 64·8°); when all the water has been removed a binary mixture of ethanol and benzene distils (b.p. 68·2°) until all the benzene is removed, and then pure ethanol distils (b.p. 78·3°). Ethanol cannot be dried with the common laboratory drying agent, calcium chloride, because the two substances form a complex ($CaCl_2 . 3C_2H_5OH$). Potassium carbonate can be used as a drying agent, but better results are obtained by distilling from calcium, barium oxide, or magnesium turnings (the magnesium is converted to magnesium ethoxide which in turn reacts with any water present to give magnesium hydroxide and ethanol). Dry ethanol absorbs water rapidly from the air. Ethanol is a good solvent for many inorganic and organic compounds and dissolves gases to a greater extent that does water. It is an excellent antiseptic. It has a pleasant odour and a sharp burning taste. The physiological effect of ethanol is to depress the control of the central nervous system over the body, causing a feeling of stimulation or euphoria. In the body, ethanol is rapidly oxidized to carbon dioxide and water.

n-Propanol

Propyl alcohol or propan-1-ol is obtained from fusel oil, or by oxidation of propane-butane mixture with air; propanol is one product in the mixture produced. It is formed in the catalytic reduction of propylene oxide.

$$CH_3-CH-CH_2 \quad \xrightarrow{[H]} \quad CH_3CH_2CH_2OH$$
$$\diagdown_O\diagup$$

It is only slightly soluble in water saturated with calcium chloride, whereas ethanol is miscible with this solution. It is used as a solvent for resins and cellulose derivatives.

Isopropyl Alcohol (Propan–2–ol)

Isopropyl alcohol is made industrially by the reaction of propylene with sulphuric acid followed by hydrolysis of the sulphuric ester produced. Propylene can be directly hydrated over a tungsten-zinc oxide catalyst. It is used widely as a solvent particularly in quick-drying oils and inks, and cosmetics, in anti-freeze compositions and by oxidation or catalytic dehydrogenation over copper provides the chief source of acetone. Like ethanol, it forms an azeotrope with water, the mixture containing 87.7% of isopropyl alcohol. The pure alcohol is obtained from the azeotrope by addition of toluene or xylene and distilling. Both the propanols are more intoxicating and more toxic than ethanol. The toxicity of alcohols increases from ethanol up to C_6 or C_8 and then falls; the n-C_{16} alcohol is essentially physiologically inert.

Butanols

n-Butanol, butan-1-ol, is prepared by fermentation of starch or sugar material with special bacteria, such as *Clostridium acetobutylicum* which degrades the carbohydrate to a mixture of butanol, acetone and a little ethanol (the Fernbach–Weizmann method). Butanol is also made by a series of reactions from acetylene.

$$HC{\equiv}CH \quad \xrightarrow{H_2O} \quad CH_3CHO \quad \xrightarrow{NaOH} \quad CH_3CH{=}CHCHO$$

119

$$CH_3CH=CHCHO \xrightarrow{[H]} CH_3CH_2CH_2CH_2OH$$

The Hydroformylation or Oxo process (page 109) can also be used to produce butanol from propylene, carbon monoxide and hydrogen and is rapidly becoming the most important source of this alcohol.

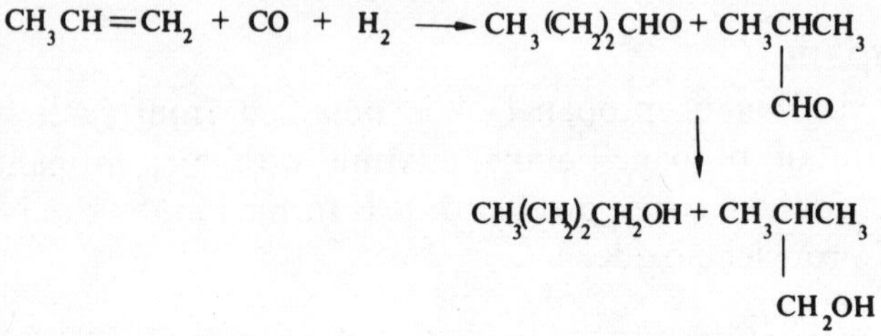

$$CH_3CH=CH_2 + CO + H_2 \longrightarrow CH_3(CH_2)_2CHO + CH_3CHCH_3$$
$$\underset{\displaystyle CHO}{|}$$

$$\downarrow$$

$$CH_3(CH_2)_2CH_2OH + CH_3CHCH_3$$
$$\underset{\displaystyle CH_2OH}{|}$$

Butanol is used as a solvent, especially for lacquers, in artificial flavours as butyl acetate, and in perfumery.

Secondary butyl alcohol or butan-2-ol is prepared by the hydration of the mixture of butenes obtained as a by-product in petroleum refineries. It is used to make methyl ethyl ketone, an important solvent.

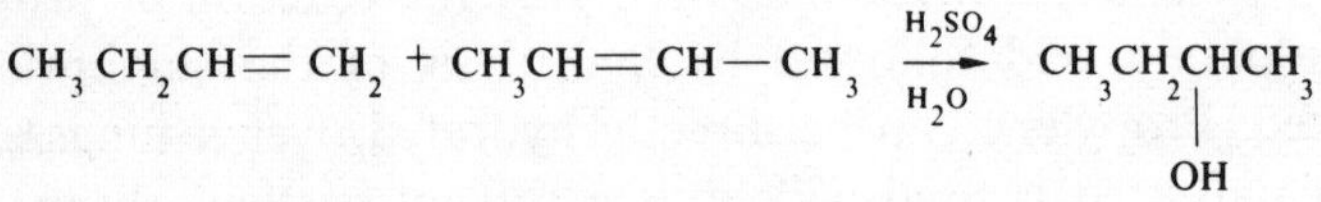

$$CH_3CH_2CH=CH_2 + CH_3CH=CH-CH_3 \xrightarrow[H_2O]{H_2SO_4} CH_3CH_2CHCH_3$$
$$\underset{\displaystyle OH}{|}$$

Isobutyl alcohol or 2-methylpropan-1-ol is obtained from fusel oil, or as a by-product in the production of methanol from carbon monoxide and hydrogen. In the Oxo reaction on propylene, it is obtained, together with n-butanol. Though it can be produced cheaply, it has not found many industrial applications.

Tertiary butyl alcohol or t-butanol is the first alcohol of the homologous series that is solid at room temperature. It is prepared by the hydration of isobutylene with 50% sulphuric acid at 10–30°.

Pentanols or Amyl Alcohols

There are eight isomeric amyl alcohols. A mixture of three isomers, including n-pentanol is obtained from fusel oil. A mixture of six primary and secondary isomers is obtained by chlorinating a mixture of pentane and isopentane and treating the chloropentanes with sodium hydroxide solution containing soap, to emulsify the insoluble chloropentane mixture. These mixtures of amyl alcohols are used as paint solvents.

Higher molecular weight alcohols are found in certain natural waxes, for example, carnaubyl alcohol or n-tetracosanol ($C_{24}H_{49}OH$) occurs in carnauba wax, and wool wax, and melissyl alcohol (n-triacontanol $C_{30}H_{61}OH$) as esters in beeswax.

The preparation of alcohols in the range of C_{12} to C_{18} has become important because the sodium sulphate esters of these alcohols, $ROSO_3^{\ominus}Na^{\oplus}$, are very useful synthetic detergents. Because the calcium salts of these sulphates are soluble in water, they do not form hard water scum, and because they are biologically 'soft', i.e. they are decomposed by sewage bacteria, the problem of contamination of rivers and water supplies that was becoming serious with the earlier detergents is avoided. These 'detergent range' alcohols are prepared from acids of natural oils or fats by reduction, or from alkenes from cracked petroleum.

Polyhydroxy alcohols

Compounds containing more than one hydroxyl group are very common. The simplest of these is ethylene glycol ($HOCH_2CH_2OH$). Many of them are important natural products, such as glycerol, sugars and polysaccharides (derivatives of sugars, such as starch and cellulose), reserved for detailed study at a more advanced stage.

Identification

No single chemical test readily distinguishes alcohols as a class. The molecular weight and the nature of the OH group both affect their behaviour in tests. As seen from Table 10, the lowest members of the series are all soluble in water giving solutions that are neutral to litmus. Alcohols lower than hexyl are all soluble in concentrated hydrochloric acid. For these the Lucas test (page 115) is a convenient way of distinguishing between primary, secondary and tertiary types. With zinc chloride as catalyst, tertiary alcohols react at once with the hydrochloric acid to give an alkyl chloride which appears as oily drops. Secondary alcohols react similarly but slowly, usually within five minutes, and primary alcohols do not react at room temperature.

Another useful test is to add a small lump of sodium metal provided the substance being tested is thoroughly dry. Sodium dissolves rapidly in lower alcohols, evolving bubbles of hydrogen. Higher alcohols and particularly tertiary alcohols react very slowly. Alcohols together with many other classes of compounds dissolve in concentrated sulphuric acid, but they do not react with bromine in carbon tetrachloride as alkenes do.

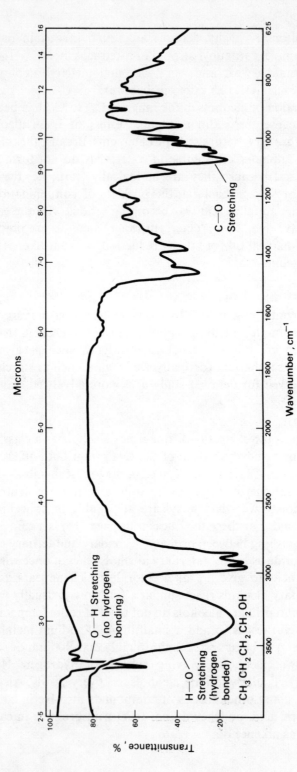

Fig. 29 Infrared spectrum of n-butanol. Inset shows O—H stretching absorption measured on a 0·4% solution of the alcohol in carbon tetrachloride

Identification of alcohols from their infrared spectra is simpler and more certain. All alcohols show a strong band between 3650 cm^{-1} and 3200 cm^{-1} due to the O—H stretching vibration (Fig. 29). The alcohols are recognized by the broad shape of this band which is caused by hydrogen bonding. In the absence of other functional groups the band is between 3400 and 3200 cm^{-1}. If the spectrum is measured in a very dilute solution in a neutral solvent (such as carbon tetrachloride) the alcohol molecules are too far apart for hydrogen bonding and then the O—H vibration produces a sharp band at 3650 to 3590 cm^{-1} (see inset in Fig. 29).

Strong absorption due to C—O stretching can also be seen between 1150 and 1000 cm^{-1} and in favourable cases the position of this band can be used to distinguish between primary, secondary and tertiary alcohols, but so many other groups have absorption in this region, it cannot be relied upon for positive identification.

6
Ethers

Just as alcohols can be considered the monoalkyl derivatives of water, so ethers are the dialkyl derivatives of water.

$$\underset{\text{water}}{\overset{\displaystyle O}{\underset{\displaystyle H \qquad H}{\diagdown}}} \qquad \underset{\text{ethyl alcohol}}{\overset{\displaystyle O}{\underset{\displaystyle C_2H_5 \qquad H}{\diagdown}}} \qquad \underset{\text{diethyl ether}}{\overset{\displaystyle O}{\underset{\displaystyle C_2H_5 \qquad C_2H_5}{\diagdown}}}$$

But the student handling diethyl ether (or simply 'ether') in the laboratory is aware that ether bears little resemblance to either water or ethanol in physical and chemical properties. Diethyl ether, in spite of a higher molecular weight, has a much lower boiling point than either water or ethanol. It is not completely miscible with water but dissolves many non-polar organic compounds that do not dissolve in water or ethanol.

The essential point is that ethers do not have hydroxyl groups, which as we have seen, confer characteristic properties on alcohols and water because of hydrogen bonding. The electrons of the C—H bond are displaced to a small extent towards the C atom, C←H, whereas in the hydroxyl group they are displaced to a larger extent towards the O atom, O←H. Although oxygen in ethers still possesses two lone pairs of electrons, no hydrogen bond is formed, because there are no electron-deficient hydrogen atoms. The molecules of the liquid are not associated and so the compounds are more volatile. In fact, the boiling points of ethers are in line with the hydrocarbons. Diethyl ether, molecular weight 74, b.p. 35° compares well with n-pentane, molecular weight 72, b.p. 36°.

Nomenclature

Ethers can be of two types, *simple* or symmetrical ethers, where the two alkyl groups are the same, and *mixed* or unsymmetrical ethers, where the alkyl groups are different. Unsymmetrical ethers can be named in two ways. The IUPAC has adopted the system whereby the ether is considered an alkoxy derivative of an alkane; the larger

124

of the two groups is considered to be the alkane. In the examples below the IUPAC names are given in brackets.

Symmetrical Ethers	Unsymmetrical Ethers
CH_3OCH_3	dimethyl ether
$CH_3CH_2OCH_2CH_3$	diethyl ether
$CH_3CH_2CH_2OCH_2CH_2CH_3$	dipropyl ether

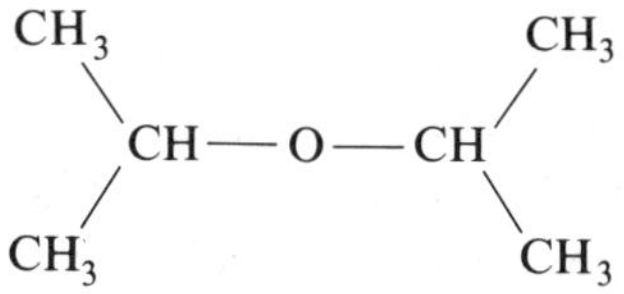

di-isopropyl ether

$CH_3OCH_2CH_3$	methyl ethyl ether (methoxyethane)
$CH_3CH_2OCH_2CH_2CH_3$	ethyl propyl ether (1-ethoxypropane)

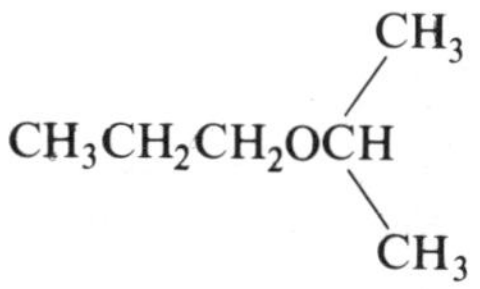

propyl isopropyl ether (2-propoxypropane)

Preparation of Ethers

1. *From alcohols by removal of the elements of water.* This is probably the first synthetic organic method. Basil Valentine in the early fifteenth century described how he prepared a wonderful medicine with an agreeable and stupefying odour by the reaction of spirits of wine (ethanol) and oil of vitriol. The method is used today for the commercial production of diethyl ether from ethanol. The alcohol is heated with sulphuric or phosphoric acid. The conditions must be controlled to favour formation of ether and not ethylene. Lower temperature favours ether formation and higher temperature gives mainly olefins. In the sulphuric acid method, a mixture of ethanol and concentrated sulphuric acid is first heated to 130–140° and then more ethanol is added at a rate equal to the rate at which ether distils. Ethyl hydrogen sulphate is first formed which then reacts with more ethanol:

125

$$C_2H_5OH \; + \; H_2SO_4 \; \rightleftharpoons \; C_2H_5O\overset{\overset{\textstyle O}{\|}}{\underset{\underset{\textstyle O}{\|}}{S}}OH \; + \; H_2O$$

$$C_2H_5O\overset{\overset{\textstyle O}{\|}}{\underset{\underset{\textstyle O}{\|}}{S}}OH \; + \; C_2H_5OH \; \rightleftharpoons \; C_2H_5\overset{\overset{\textstyle H}{|}}{\underset{}{O}}{}^{\oplus}\!\!-C_2H_5 \; + \; HSO_4^{\ominus}$$

$$\longrightarrow C_2H_5OC_2H_5 \; + \; H^{\oplus}$$

The reaction is driven forward by distilling out the ether, which is the most volatile component. Some ethylene is also unavoidably formed. Primary alcohols can be used in this method, secondary alcohols react but in lower yield, tertiary alcohols give alkenes too readily. Di-isopropyl ether is obtained as a by-product in the hydration of propylene to isopropanol with sulphuric acid.

2. *Williamson Synthesis.* The reaction of a sodium or potassium alkoxide and an alkyl halide can be used to form simple or mixed ethers by an S_N2 or S_N1 reaction:

$$RONa \; + \; R'X \longrightarrow ROR' \; + \; NaX$$

This reaction is historically important, since it led to the establishment of the relation between alcohols and ethers in 1852, and became one of the foundations of the theory of organic chemistry. Williamson made mixed ethers, and these were only credible on the 'water-type' theory. When either R or R' is to be a secondary or tertiary alkyl group, it is better to prepare that portion of the molecule from the alcohol, because the alkoxides of secondary or tertiary alcohols are less readily dehydrated to olefins than the corresponding halides.

3. *By heating an alcohol with silver oxide and a simple halide*, e.g. methyl or ethyl iodide, the corresponding ethers can be formed. This is a special case of the Williamson synthesis. The method is applicable to a wide range of alcohols but the yields are not always good; using dimethylformamide as solvent for the reaction, the yield can be greatly improved.

$$2\,ROH \; + \; Ag_2O \; + \; 2CH_3I \longrightarrow 2\,ROCH_3 \; + 2AgI \; + \; H_2O$$

Methyl iodide and silver oxide are together often referred to as Purdie's reagent, because Purdie developed this method for the preparation of methyl derivatives of the hydroxyl groups of carbohydrates (e.g. sugar ethers), and it is an important method in carbohydrate chemistry. If an alkyl iodide is heated with silver oxide but without an alcohol the symmetrical ether is formed:

$$2RI \ + \ Ag_2O \ \longrightarrow \ ROR \ + \ 2AgI$$

4. *A method used on a large scale industrially is the addition of an alcohol to an alkene with sulphuric acid as catalyst* This is really a variation of (1) above, but in this case unsymmetrical ethers can be produced.

$$CH_3CH = CH \ + H_2SO_4 \longrightarrow \underset{\underset{SO_4H}{|}}{CH_3CHCH_3} + \ CH_3CH_2OH \longrightarrow \underset{\overset{|}{CH_3}}{CH_3CHOCH_2CH_3} + \ H_2SO_4$$

5. *A newer method for making methyl ethers from alcohols is to add diazomethane with fluoroboric acid as catalyst.*

$$ROH \ + \ CH_2N_2 \ \xrightarrow{\ HBF_4\ } \ ROCH_3 \ + \ N_2$$

Properties of Ethers

The lower members of the series, except dimethyl ether, which is a gas, are volatile, pleasant-smelling, inflammable liquids. Their vapours form explosive mixtures with air over a wide concentration range. Many laboratory accidents have been caused by diethyl ether because of its high volatility and inflammability. The ethers are less dense than water and only slightly soluble in it (100 g water dissolves 7·5 g diethyl ether at 16°). As noted earlier, the boiling points of

Table 11.   Alkyl ethers

Compound	Formula	m.p. °C	b.p. °C
Dimethyl ether	CH_3OCH_3	-140	-25
Methyl ethyl ether	$CH_3OC_2H_5$	–	8
Diethyl ether	$C_2H_5OC_2H_5$	-116	34·5
Methyl propyl ether	$CH_3OC_3H_7$	–	39
Methyl butyl ether	$CH_3OC_4H_9$	-116	70
Dipropyl ether	$C_3H_7OC_3H_7$	-122	90
Di-isopropyl ether	$(CH_3)_2CHOCH(CH_3)_2$	-60	68
Dibutyl ether	$C_4H_9OC_4H_9$	–	141

ethers closely resemble the corresponding hydrocarbons with an oxygen atom in place of $-CH_2-$, e.g. $CH_3CH_2CH_2OCH_2CH_2CH_3$ b.p. 90°, $CH_3(CH_2)_5CH_3$ b.p. 98°.

Ethers are chemically rather inert. They are not attacked by sodium (which is used to dry ethers), mild oxidizing agents or alkalis, but they are decomposed slowly by strong acids.

1. Heating an ether with dilute sulphuric acid and water under pressure hydrolyses it to the corresponding alcohol:

$$ROR + H^\oplus \longrightarrow \overset{\oplus}{R}\underset{H}{O}R \xrightarrow{H_2O} ROH + \overset{\oplus}{R}\underset{H}{O}H \longrightarrow 2ROH + H^\oplus$$

With concentrated sulphuric acid, alkyl hydrogen sulphates are formed:

$$ROR + H^\oplus \longrightarrow \overset{\oplus}{R}\underset{H}{O}R \xrightarrow{HSO_4^\ominus} ROH + ROSO_3H$$

Tertiary ethers are decomposed more readily than secondary or primary ethers by acid:

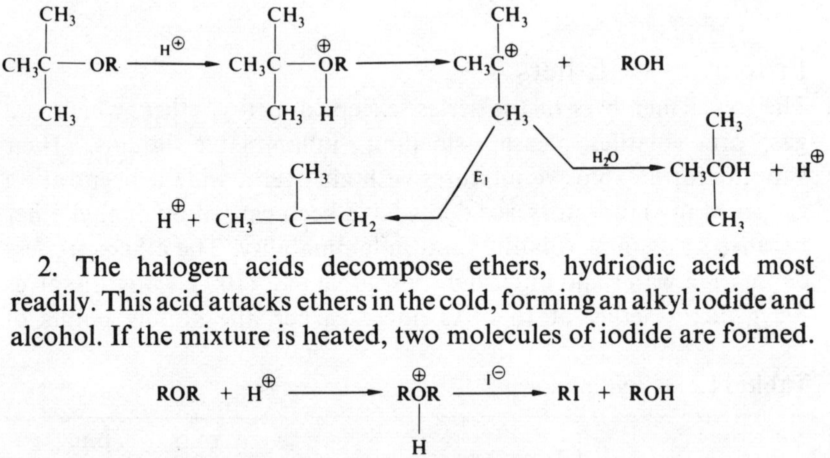

2. The halogen acids decompose ethers, hydriodic acid most readily. This acid attacks ethers in the cold, forming an alkyl iodide and alcohol. If the mixture is heated, two molecules of iodide are formed.

$$ROR + H^\oplus \longrightarrow \overset{\oplus}{R}\underset{H}{O}R \xrightarrow{I^\ominus} RI + ROH$$

In the case of mixed ethers, the iodide is formed from the simpler alkyl group:

$$CH_3OCH_2CH_2CH_3 + HI \longrightarrow CH_3I + CH_3CH_2CH_2OH$$

At moderate temperatures hydrobromic acid reacts with ethers only partially and hydrochloric acid hardly at all. The reaction with hydrogen iodide, however, is so complete that use is made of it in the *Zeisel* alkoxy determination. By this method methoxy ($-OCH_3$) and ethoxy ($-OC_2H_5$) groups can be quantitatively determined in an unknown compound. The volatile methyl or ethyl iodide is distilled out of a mixture of hydriodic acid and the unknown compound, and

128

the iodide passed into an alcoholic solution of silver nitrate when a $CH_3I.AgNO_3$ complex is precipitated.

3. Ethers can be chlorinated in the dark to give mono- or di-chloroethers, and in light to give fully chlorinated ethers.

$$CH_3CH_2OCH_2CH_3 + Cl_2 \longrightarrow CH_3\underset{\underset{Cl}{|}}{C}HOCH_2CH_3 \xrightarrow{Cl_2} CH_3\underset{\underset{Cl}{|}}{C}HOCH\underset{\underset{Cl}{|}}{C}HCH_3$$

$$CH_3CH_2OCH_2CH_3 + Cl_2 \xrightarrow{h\nu} CCl_3CCl_2OCCl_2CCl_3$$

4. Although ethers cannot form hydrogen bonds, as discussed at the beginning of this chapter, the oxygen atom still possesses two pairs of unshared electrons, one of which enables ethers to form complexes with strong acids, called *oxonium* salts, by analogy with ammonium salts; they are also comparable with the hydrated proton $H_3O^{\oplus}$:

$$\underset{H}{\overset{H}{\diagdown}}O + HCl \longrightarrow \underset{H}{\overset{H}{\diagdown}}O^{\oplus} - H\ Cl^{\ominus}$$

$$\underset{R}{\overset{R}{\diagdown}}O + HCl \longrightarrow \underset{R}{\overset{R}{\diagdown}}O^{\oplus} - H\ Cl^{\ominus}$$

compare :

$$\underset{\underset{R}{|}}{\overset{\overset{R}{|}}{R}N} + HCl \longrightarrow \underset{\underset{R}{|}}{\overset{\overset{R}{|}}{R}NH}^{\oplus}\ Cl^{\ominus}$$

This property is used as a qualitative test for ethers and also to separate ethers from alkanes and alkyl halides. Ethers dissolve in concentrated sulphuric acid, and so by passing a mixture of an ether and alkane, or alkyl halide, through concentrated sulphuric acid, the ether dissolves and the other components do not.

The formation of ether complexes is not restricted to acids containing a proton; Lewis acids also form complexes, e.g. boron trifluoride forms a complex with diethyl ether which is a liquid and can be distilled unchanged. It is easier to handle and has many of the catalytic uses of the more corrosive gas BF_3.

$$\underset{CH_3}{\overset{CH_3}{\diagdown}}\overset{..}{O} \diagdown BF_3$$

boron trifluoride etherate

Ethers are excellent solvents for preparing Grignard reagents because they do not react together, but the ethers *solvate* the Grignard reagent by forming complexes by electron donation to the positive magnesium. Crystalline complexes of the form $RMgBr$. $(C_2H_5OC_2H_5)_2$ have been isolated from diethyl ether solutions, and the structure examined by X-ray diffraction.

$$\overset{\delta\ominus}{R}\!-\!\overset{\delta\oplus\oplus}{Mg}\!-\!\overset{\delta\ominus}{Br} \quad\xrightarrow{\ 2R'OR'\ }\quad \begin{array}{c} R' \quad R' \\ \diagdown\,\diagup \\ \ddot{O}\!: \\ \mid \\ R\!-\!Mg\!-\!Br \\ \mid \\ \ddot{O}\!: \\ \diagup\,\diagdown \\ R' \quad R' \end{array}$$

5. Ethers form unstable peroxides on long contact with air. If a sample of ether has been standing for some time, and is then distilled to dryness, the residue of peroxide left at the end of the distillation can explode violently. The presence of peroxides can be detected by adding a little of the ether to a solution of potassium iodide in glacial acetic acid, and shaking. A brown iodine colour indicates peroxides present. The peroxides can be removed by shaking the ether with a saturated aqueous solution of ferrous sulphate, or by passing the ether through a column of activated alumina, when the peroxides become adsorbed on to the alumina.

Dimethyl ether is a gas at normal temperatures, easily liquefied under pressure, so it can be used commercially as a refrigerant. It is made by dehydrating methanol over an aluminium phosphate catalyst.

Diethyl ether, usually known simply as 'ether', is of considerable importance as an extracting solvent because it is a good solvent for fats, oils, some resins and many other types of organic compounds; it is low boiling and so can easily be distilled, with economy of fuel, and without damaging heat-sensitive compounds. Its chief disadvantages are its high vapour pressure at normal temperature and the danger of fires or explosions of the vapour in air; and its slight but significant solubility in water, by which on an industrial scale large quantities of the solvent can be lost, but this loss can be cut down by adding an inorganic salt such as NaCl to the water which depresses the solubility of ether in the water phase appreciably.

Ether has been much employed as a general anaesthetic since its first use in Boston, U.S.A. in 1840. Because of its explosiveness, it is no longer used. At the present time other general anaesthetics are preferred, notably fluothane ($CF_3CHBrCl$).

Ether is prepared industrially by dehydration of ethanol, either with sulphuric acid at 140° or with metal oxide catalysts. It is also made from ethylene and ethanol, using sulphuric acid as catalyst.

Di-isopropyl and dibutyl ethers are obtained as by-products in the formation of the alcohols from the corresponding alkenes, and both are used as industrial solvents. Di-isopropyl ether is also added to petrol, because it has a high octane rating, but it is dangerously liable to formation of explosive peroxides.

Cyclic Ethers

Some cyclic ethers are worthy of mention here. Their detailed description will come later. The simplest of these is ethylene oxide, which contains a three-membered ring, easily opened in a number of reactions not common to other ethers. Ethylene oxide is an important industrial intermediate for a number of products. Tetrahydrofuran is frequently used as a solvent for Grignard and LiAlH$_4$ reactions. The name is derived from the heterocyclic compound furan from which it is produced by catalytic hydrogenation. Dioxan is a diether, it is also used as a solvent where conditions demand a polar solvent which does not contain hydroxyl groups. Both tetrahydrofuran and dioxan are completely miscible with water.

$$CH_2 \!-\! CH_2 \diagdown_{O} \diagup$$

ethylene oxide
b.p. 13°

$$\begin{array}{c} CH_2 \!-\! CH_2 \\ | \qquad\quad | \\ CH_2 \quad CH_2 \\ \diagdown_{O}\diagup \end{array}$$

tetrahydrofuran
b.p. 66°

$$\begin{array}{c} \diagup^{O}\diagdown \\ CH_2 \qquad CH_2 \\ | \qquad\quad | \\ CH_2 \qquad CH_2 \\ \diagdown_{O}\diagup \end{array}$$

dioxan
b.p. 101°

Identification

Ethers are unreactive towards nearly all common reagents, and apart from their odours, which are usually stronger and sweeter than hydrocarbons, might easily be mistaken for hydrocarbons in qualitative tests. The one important test that distinguishes them is the solubility of ethers in cold, concentrated sulphuric acid, with formation of oxonium salts. Unsaturated hydrocarbons also dissolve in sulphuric acid, but they also react with bromine and reduce permanganate, whereas ethers are inert to both these reagents.

The C—O stretching vibration of ethers appears as a strong infrared band in the region 1060–1150 cm^{-1}, depending upon the kind of alkyl groups attached to oxygen. In simple compounds the appearance of such a band is helpful, but since many other functional groups also contain C—O bonds (e.g. alcohols, acids, esters) which absorb in this region, in complex compounds it must be interpreted cautiously (e.g. absence of O—H or C=O absorption). The unusual appearance of the C—H stretching absorption at 2860 cm^{-1} in Fig. 30 is due to the CH$_3$ group of ethyl ethers. Such unexpected effects occur at times in infrared spectra.

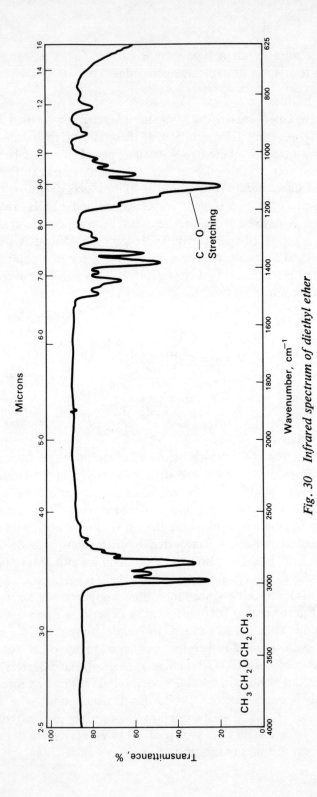

Fig. 30 Infrared spectrum of diethyl ether

7

Amines

In the last two chapters compounds formally derived from water by replacing one and then two hydrogen atoms by alkyl groups have been considered. A similar group of compounds are known which are formally derived from ammonia by replacing hydrogen atoms by alkyl groups. These compounds are collectively known as *amines*. They form the most important group of basic substances in organic chemistry.

Nomenclature

Amines are designated primary, secondary or tertiary depending upon the number of hydrogen atoms of ammonia that have been replaced by alkyl groups:

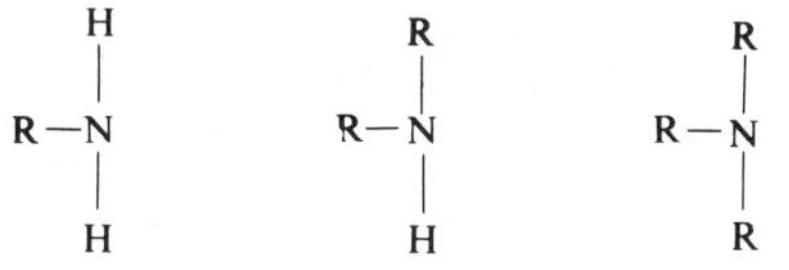

$$
\begin{array}{ccc}
\text{H} & \text{R} & \text{R} \\
| & | & | \\
\text{R}-\text{N} & \text{R}-\text{N} & \text{R}-\text{N} \\
| & | & | \\
\text{H} & \text{H} & \text{R} \\
\text{primary} & \text{secondary} & \text{tertiary}
\end{array}
$$

In naming individual amines the stem is prefixed by the names of the alkyl groups:

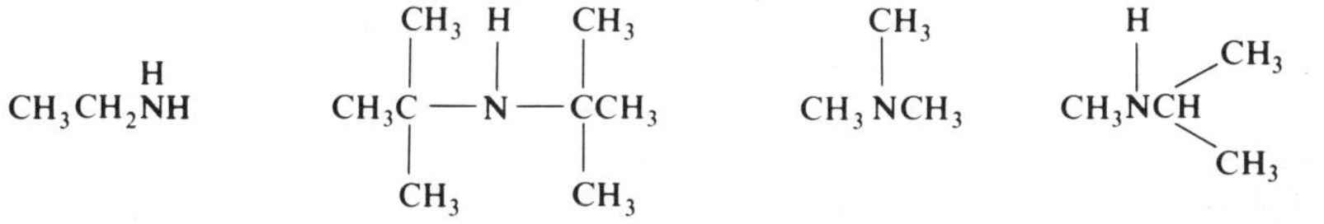

$$
\begin{array}{cccc}
\text{H} & \text{CH}_3\ \ \text{H}\ \ \ \text{CH}_3 & \text{CH}_3 & \text{H} \\
| & |\ \ \ \ |\ \ \ \ \ | & | & |\quad \diagup\text{CH}_3 \\
\text{CH}_3\text{CH}_2\text{NH} & \text{CH}_3\text{C}-\text{N}-\text{CCH}_3 & \text{CH}_3\text{NCH}_3 & \text{CH}_3\text{NCH} \\
& |\ \ \ \ \ \ \ \ \ | & & \diagdown\text{CH}_3 \\
& \text{CH}_3\ \ \ \ \text{CH}_3 & &
\end{array}
$$

ethylamine   di–t–butylamine   trimethylamine   methylisopropylamine

Amines form salts with acids as ammonia does, and such salts are correctly named as derivatives of the ammonium ion, but are commonly named by adding the acid name on to the amine:

$$
(\text{CH}_3)_3\text{NH}^{\oplus}\ \text{Cl}^{\ominus} \qquad \text{CH}_3\text{CH}_2\overset{\oplus}{\text{NH}}_3\ \ \text{HSO}_4^{\ominus}
$$

trimethylammonium chloride   ethylammonium hydrogen sulphate
or trimethylamine hydrochloride   or ethylamine hydrogen sulphate

In addition to the amines, compounds are known where the four hydrogen atoms of the ammonium ion are replaced by alkyl groups; these are known as quaternary ammonium salts and named as for the amine salts:

$$(CH_3)_4 N^{\oplus} \quad Cl^{\ominus}$$
tetramethylammonium chloride

Basicity

Ammonia has one unshared pair of electrons, which is available for bonding with a proton, i.e. ammonia is a proton acceptor (a base by the Brønsted–Lowry definition) or it is a donor of electrons (a base by the Lewis definition). This lone pair of electrons is available for salt formation in amines, and like ammonia, amines display basic properties. We have already met basic properties and oxonium salt formation in alcohols and ethers, but these are extremely weak bases, accepting protons only from very strong acids, such as concentrated sulphuric acid, or donating electrons to the very strong Lewis acid, boron trifluoride.

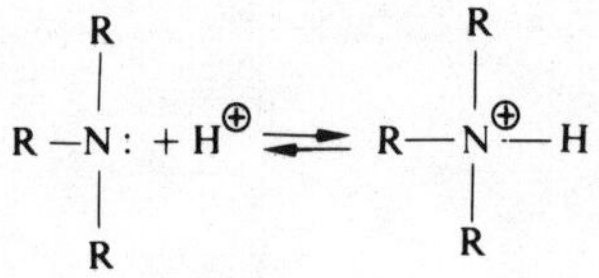

$$\begin{array}{ccc} & R & & R \\ & | & & | \\ R-N: & + H^{\oplus} \rightleftharpoons & R-N^{\oplus}-H \\ & | & & | \\ & R & & R \end{array}$$

The stability of the ammonium ion is a measure of its basic strength. Simple aliphatic amines are all stronger bases than ammonia itself because the electron-donating alkyl groups help to neutralize the positive charge on the nitrogen atom and stabilize the ion, just as alkyl groups help to stabilize carbonium ions; but there is not a simple relationship for amines between basicity and the number of substituting alkyl groups. The bulkiness of the alkyl groups also plays a part.

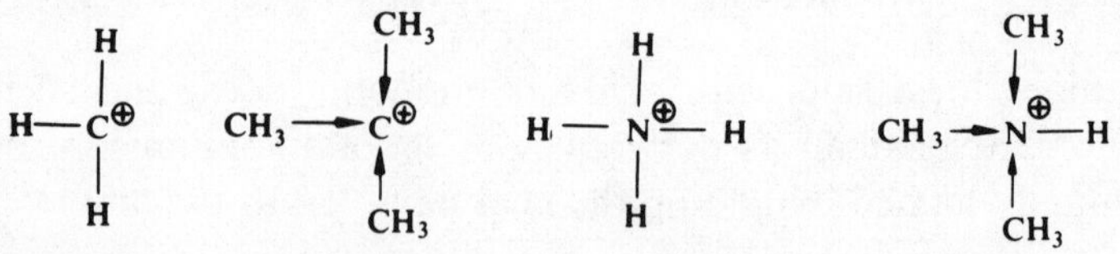

less stable   more stable   less stable   more stable

$K_b\, 1{\cdot}8 \times 10^{-5}$ $K_b\, 4{\cdot}4 \times 10^{-4}$

where K_b, the basicity constant is defined for the equation:

$$R_3N + H_2O \rightleftharpoons R_3NH^\oplus + OH^\ominus$$

$$\text{as } K_b = \frac{[R_3NH^\oplus][OH^\ominus]}{[R_3N]} \quad R=H \text{ or alkyl}$$

Preparation

1. *From Alkyl Halides and Ammonia.* This method has already been encountered (page 38) as one of the examples of nucleophilic substitution reactions of alkyl halides:

$$\overset{..}{N}H_3 \quad \overset{R}{\underset{|}{CH_2}}\!\!-\!\!X \xrightarrow{\ S_N2\ } RCH_2\overset{\oplus}{N}H_3 \quad X^\ominus$$

The reaction does not stop here because the amine salt formed can exchange with ammonia:

$$RCH_2\overset{\oplus}{N}H_3 \quad X^\ominus + NH_3 \rightleftharpoons RCH_2NH_2 + NH_4^\oplus \ X^\ominus$$

The primary amine can now compete with ammonia for attack on the alkyl halide:

$$RCH_2\overset{..}{N}H_2 \quad \overset{R}{\underset{|}{CH_2}}\!\!-\!\!X \longrightarrow (RCH_2)_2\overset{\oplus}{N}H_2 \quad X^\ominus$$

And after again exchanging with ammonia:

$$(RCH_2)_2\overset{..}{N}H \quad \overset{R}{\underset{|}{CH_2}}\!\!-\!\!X \longrightarrow (RCH_2)_3\overset{\oplus}{N}H \ X^\ominus$$

$$\text{and } (RCH_2)_3\overset{..}{N} + RCH_2X \longrightarrow (RCH_2)_4\,N^\oplus\,X^\ominus$$

In the presence of excess ammonia, the reaction gives a mixture of primary, secondary and tertiary amines and their salts, a quaternary ammonium halide and ammonium halide. By adding a strong alkali the free amines are liberated from their salts, and the primary, secondary and tertiary amines can be extracted into an organic solvent and separated by fractional distillation. The quaternary ammonium ion is unaffected by cold alkali and not extracted by an organic solvent.

The reaction may equally well be used for the alkylation of a primary, secondary or tertiary amine. When ammonia is used, to achieve a satisfactory rate of reaction, it is carried out under pressure at 100° or higher temperature.

2. *From Alcohols.* Industrially, simple alcohols and ammonia are

heated together under pressure with a catalyst (Al_2O_3, $SnCl_2$, or ThO_2) to 300–450°:

$$CH_3OH + NH_3 \longrightarrow CH_3NH_2 + H_2O$$
$$\longrightarrow (CH_3)_2NH, \text{ etc.}$$

The reaction is usually arranged for continuous operation by passing the reactants through a heated tube.

3. *Reduction of an alkyl cyanide can give a primary amine.* The reduction can be carried out by sodium in alcohol or catalytically with Ni or Pd and hydrogen. Indirectly this provides another route from alkyl halides, the amine having one more carbon atom than the starting halide:

$$RX \xrightarrow{CN^{\ominus}} RC\equiv N \xrightarrow{H_2/Ni} RCH_2NH_2$$

Many other methods of preparation are available and some will be encountered in later chapters.

Physical Properties

Nitrogen is less electronegative than oxygen but nevertheless is capable of hydrogen bonding. Thus the boiling point of ammonia ($-33°$) is lower than that of water, but a good deal higher than methane ($-161°$). Primary and secondary amines are also capable of hydrogen bonding, like alcohols. We would therefore expect primary and secondary amines to have higher boiling points than hydrocarbons of similar molecular weight but lower than similar alcohols; tertiary amines, which have no hydrogen bound directly to nitrogen, should have boiling points close to hydrocarbons of similar molecular weight. This is indeed found to be the case (Table 12).

Simple amines have pungent odours, methylamine is similar to ammonia, trimethylamine has a very unpleasant odour similar to rotting fish. The methylamines and ethylamine are gaseous at room temperature, the others are liquids. Again because of hydrogen bonding, the lower members of the series are freely soluble in water, but solubility falls off rapidly with increasing molecular weight. They are all soluble in common organic solvents, such as alcohols, ethers and chloroform.

The salts of amines and tetra-alkylammonium salts are water-soluble, crystalline solids, insoluble in non-polar organic solvents. This change of properties is very useful, for by converting an amine to its salt and extracting the salt with water, it can be separated from a mixture of organic compounds. Neutralizing the aqueous solution with alkali, liberates the amine which can then be extracted with a water-insoluble organic solvent, e.g. ether.

Table 12.   Amines

Name	b.p.	Solubility in H_2O	K_b
Ammonia	−33	very soluble	$1·8 \times 10^{-5}$
Methylamine	− 6	∞	$4·4 \times 10^{-4}$
Dimethylamine	7	∞	$5·4 \times 10^{-4}$
Trimethylamine	3	very soluble	$7·4 \times 10^{-5}$
Ethylamine	17	∞	$5·6 \times 10^{-4}$
Diethylamine	55	very soluble	$9·1 \times 10^{-4}$
Triethylamine	90	slightly soluble	$4·4 \times 10^{-4}$
n-Propylamine	49	∞	$5·1 \times 10^{-4}$
Isopropylamine	34	∞	—
n-Butylamine	78	very soluble	$5·9 \times 10^{-4}$

Reactions

1. *Salt Formation.* The basic nature of amines and their ability to form salts with acids have already been mentioned. The properties of the salts are similar to those of ammonium salts, i.e. ionic, crystalline and water-soluble. Many of them are deliquescent. The salts of weak acids (e.g. carbonates, acetates) are easily decomposed by heating. In aqueous solution, they are extensively hydrolysed, to the un-ionized acid and base:

$$R_3NH^{\oplus} CH_3COO^{\ominus} \rightleftharpoons R_3N + CH_3COOH$$

2. *Alkylation.* Primary, secondary and tertiary amines can all be alkylated with an alkyl halide. The ultimate product is a quaternary ammonium salt:

$$RNH_2 \xrightarrow{R'X} RNH \xrightarrow{R'X} RN \xrightarrow{R'X} RNR \; X^{\ominus}$$

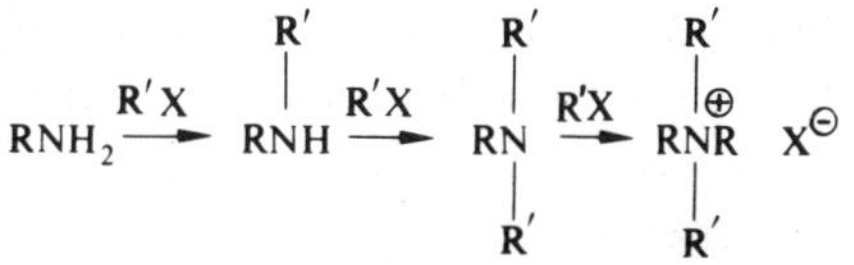

3. *Hofmann Elimination.* Quaternary ammonium salts are the salts of strong bases, the quaternary ammonium hydroxides, which can be

prepared from the halide salts by treatment with moist silver oxide, silver halide being precipitated:

$$R_4N^{\oplus} \ X^{\ominus} \ + \ AgOH \longrightarrow R_4N^{\oplus} OH^{\ominus} + AgX$$

Hofmann discovered in 1851 that heating a dry tetra-alkyl-ammonium hydroxide to 125° or higher caused it to decompose in a characteristic way. Tetramethylammonium hydroxide decomposes to trimethylamine and a mixture of methanol and dimethyl ether:

$$(CH_3)_4N^{\oplus} \ OH^{\ominus} \longrightarrow (CH_3)_3N \ + \ (CH_3)_2O + CH_3OH$$

Compounds with higher alkyl groups give a tertiary amine and an alkene, and if an ethyl group is present that group is preferentially eliminated as ethylene:

$$(CH_3)_3 \overset{\oplus}{N} CH_2CH_3 \quad \overset{\ominus}{OH} \longrightarrow (CH_3)_3N \ + \ CH_2{=}CH_2 + H_2O$$

$$(CH_3)_3 \overset{\oplus}{N} CH_2CH_2CH_3 \ \overset{\ominus}{OH} \longrightarrow (CH_3)_3N \ + \ CH_2{=}CHCH_3 + H_2O$$

The reaction is similar to the base-catalysed E2 dehydrohalogenation of an alkyl halide (page 51):

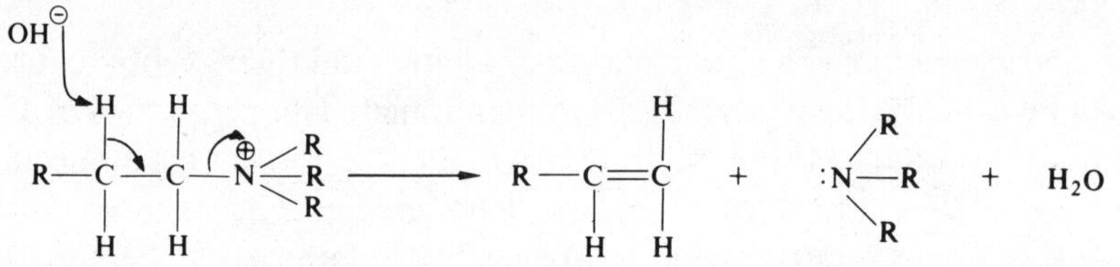

However, different controlling factors apply here; Hofmann elimination leads to the formation of the least alkylated olefin, whereas Saytzev's rule (page 50) predicts the most highly alkylated olefin.

Hofmann's method can be used as a synthesis of alkenes, but it has found much greater use in the study of the structure of complex amine bases found in plants, known as alkaloids (e.g. nicotine, strychnine, morphine, and quinine). By first treating the amine with methyl iodide until a quaternary ammonium iodide is formed ('exhaustive methylation') and heating the corresponding hydroxide, an olefin is produced with a double bond located (in general) where the nitrogen was formerly attached.

4. *Reaction with Nitrous Acid.* Primary, secondary and tertiary amines react differently with nitrous acid, and their different behaviour can be used as a means of telling them apart. Nitrous acid

is not very stable so it is made *in situ* by mixing hydrochloric acid and sodium nitrite. The resulting solution contains a number of species, including nitrite $NO_2^{\ominus}$ and nitrosonium $NO^{\oplus}$ ions.

The reaction with primary amines is complex. First a nitroso-amine is formed which is converted to an unstable diazonium salt, which loses nitrogen to give a carbonium ion. This in turn can undergo reaction in a number of ways, so that several products are formed.

$$CH_3CH_2CH_2NH_2 \xrightarrow{NO^{\oplus}} CH_3CH_2CH_2\overset{H}{\underset{H}{\overset{\oplus}{N}}}-N{=}O \longrightarrow CH_3CH_2CH_2\overset{\oplus}{N}{\equiv}N + H_2O$$

$$\longrightarrow CH_3CH_2\overset{\oplus}{CH_2} + N_2$$

The carbonium ion can react with the solvent to give an alcohol, or lose a proton to give an alkene, or first rearrange to a more stable secondary or tertiary carbonium ion and then in the same way react with the solvent or form an alkene.

$$CH_3CH_2\overset{\oplus}{CH_2} + H_2O \longrightarrow CH_3CH_2CH_2OH + H^{\oplus}$$

$$\longrightarrow CH_3CH{=}CH_2 + H^{\oplus}$$

$$CH_3\overset{\oplus}{CH}CH_3 + H_2O \longrightarrow CH_3\underset{OH}{CHCH_3} + H^{\oplus}$$

Small amounts of chloride are also formed by reactions of the carbonium ions with $Cl^{\ominus}$ in solution.

What is observed is the amine dissolves in the acidic solution, bubbles of nitrogen are formed and an oily layer consisting of alkenes, alkyl chlorides and possibly alcohols (depending upon chain length) separates.

With secondary amines, the nitroso-amine forms more slowly and the reaction stops at that stage. The nitroso-amines are not basic and are insoluble in the aqueous solution. A yellow oil is seen to separate and there is no evolution of hydrogen.

$$(CH_3CH_2)_2NH + \overset{\oplus}{N}{=}O \longrightarrow (CH_3CH_2)_2-\overset{H}{\overset{\oplus}{N}}-N{=}O \longrightarrow (CH_3CH_2)_2-N-N{=}O$$

$$\text{nitrosodiethylamine} \qquad + H^{\oplus}$$

Tertiary amines cannot form nitroso-amines, but dissolve in the acidic solution to give salts.

Identification

Amines are recognized chiefly by their basicity. The first test to perform with a substance suspected of being an amine is to test it with dilute hydrochloric acid. If it dissolves in the acid, and is regenerated on adding sodium hydroxide, then in general, the substance is an amine. Later, tests will be encountered for distinguishing between primary, secondary and tertiary amine groups.

The infrared spectra of primary and secondary amines show characteristic N—H stretching bands at 3500–3300 cm^{-1}. The bands are not as strong as O—H stretching bands, and because hydrogen bonding is not as strong in amines, the amine bands are not as broad. Primary amines give two bands and secondary amines give one. These groups also have N—H bending absorption in the 1650–1550 cm^{-1} region, but these are often not strong and are not very useful for identification. Tertiary amines show only weak C—N vibration in that region of the spectrum, where many other bands appear.

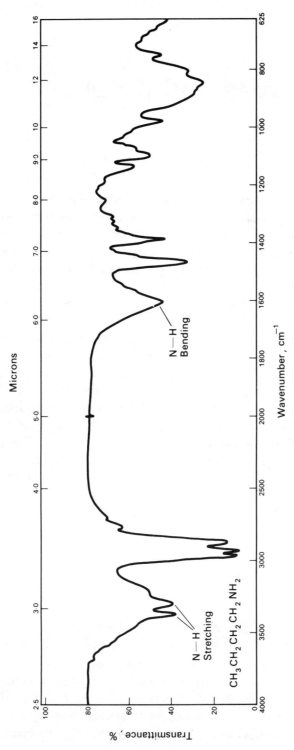

Fig. 31 Infrared spectrum of n-butylamine

8

Aldehydes and Ketones

Aldehydes and ketones present us with a versatility of reactions much greater than that of any group we have yet encountered. Nature has discovered this first and made use of their reactions in many and ingenious ways to carry out biological processes. Compounds of these classes are found in many types of material from the animal and vegetable kingdoms. For example, retinal, an aldehyde, plays a vital part in the vision process in the eye, and progesterone, a ketone, is the human pregnancy hormone.

Apart from specific reactions, the differences between aldehydes and ketones are more of degree than of type and so they are generally treated together. Simple aliphatic compounds of both groups have the general formula, $C_nH_{2n}O$, and both groups can be considered as oxidation products of alcohols, aldehydes being derived from primary alcohols and ketones from secondary alcohols. They both contain an oxygen atom joined by a double bond to carbon. The defining *difference* is that in aldehydes, the carbon joined to oxygen is also linked to hydrogen; in ketones, the carbon joined to oxygen carries no hydrogen.

$$\begin{array}{cc} \overset{\displaystyle H}{\underset{\displaystyle |}{}} & \overset{\displaystyle R'}{\underset{\displaystyle |}{}} \\ R-C=O & R-C=O \\ \text{aldehyde} & \text{ketone} \end{array}$$

Structure of Aldehydes and Ketones

The simplest aldehyde, formaldehyde, has the formula CH_2O, and if we give the atoms their usual valency, there is only one structure possible (I, below). This is confirmed by infrared and ultraviolet spectra as well as by the transformations that may be carried out on formaldehyde, such as its reduction to methanol.

Acetaldehyde C_2H_4O could have three possible structures, (II, III

142

or IV), but II can now be eliminated because this structure has been assigned to ethylene oxide (page 58), and since both its reduced and oxidized relatives (ethanol and acetic acid) have intact methyl groups, acetaldehyde should have a methyl group too. Acetaldehyde does not show typical alcohol reactions, so III is unlikely, though in certain reactions, we will find that acetaldehyde reacts as if it had structure III. On this kind of evidence, together with a great deal more, we assign acetaldehyde structure IV.

$$\begin{array}{ccccc} \text{H} & \quad & \text{H}\ \ \text{H} & \quad & \text{H}\ \ \text{H} & \quad & \text{H}\ \ \text{H}\\ | & & |\ \ \ | & & |\ \ \ | & & |\ \ \ |\\ \text{HC}{=}\text{O} & & \text{HC}-\text{CH} & & \text{HC}{=}\text{C}-\text{OH} & & \text{HC}-\text{C}{=}\text{O}\\ & & \diagdown\!\!\diagup & & & & |\\ & & \text{O} & & & & \text{H}\\[4pt] \text{I} & & \text{II} & & \text{III} & & \text{IV} \end{array}$$

$$\begin{array}{ccc} \text{H}\quad \text{O}\quad \text{H} & \qquad & \text{H}\ \ \text{H}\ \ \text{H}\\ |\quad\ \ \|\quad\ \ | & & |\ \ \ |\ \ \ |\\ \text{HC}-\text{C}-\text{CH} & & \text{HC}-\text{C}-\text{C}{=}\text{O}\\ |\qquad\quad | & & |\ \ \ |\\ \text{H}\qquad\quad \text{H} & & \text{H}\ \ \text{H}\\[4pt] \text{V} & & \text{VI} \end{array}$$

Acetone is the simplest known ketone, with formula C_3H_6O. It does not behave like an alcohol, but is similar to aldehydes in reactivity, which allows only two possible structures (V or VI). It cannot be oxidized to a carboxylic acid with the same number of carbon atoms, nor can it be reduced to n-propanol, which excludes structure VI. Therefore it must have structure V. Structure VI represents propionaldehyde.

Nomenclature

The common names of aldehydes are derived from the names of the acids to which they are related by simple oxidation-reduction, the name ending in -aldehyde. From *formic* acid we get *form*aldehyde; from *acetic* acid we get *acet*aldehyde. In the IUPAC system, aldehydes are named like the corresponding alkanes, but dropping the final -*e* and adding -*al*:

143

Formula	Trivial name	IUPAC name
HCHO*	formaldehyde	methanal
CH_3CHO	acetaldehyde	ethanal
CH_3CH_2CHO	propionaldehyde	propanal
$CH_3CH_2CH_2CHO$	butyraldehyde	butanal
$\overset{\displaystyle CH_3}{\overset{\displaystyle \vert}{CH_3CHCHO}}$	isobutyraldehyde	2-methylpropanal
$CH_3(CH_2)_3CHO$	valeraldehyde	pentanal
$\overset{\displaystyle CH_3}{\overset{\displaystyle \vert}{CH_3CHCH_2CHO}}$	isovaleraldehyde	3-methylbutanal

For complex molecules, the principal chain is the longest one containing the aldehyde group, and the carbon attached to oxygen is numbered 1. In the non-systematic names, Greek letters are occasionally used, the carbon *next* to the one bearing oxygen is called α. For example:

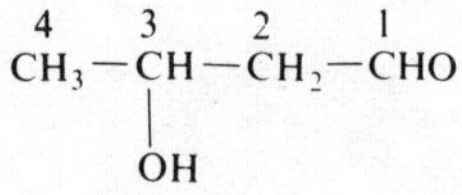

$$\overset{4}{CH_3}-\overset{3}{\underset{\underset{OH}{\vert}}{CH}}-\overset{2}{CH_2}-\overset{1}{CHO} \qquad\qquad \overset{\gamma}{CH_3}-\overset{\beta}{\underset{\underset{OH}{\vert}}{CH}}-\overset{\alpha}{CH_2}-CHO$$

3-hydroxybutanal   or 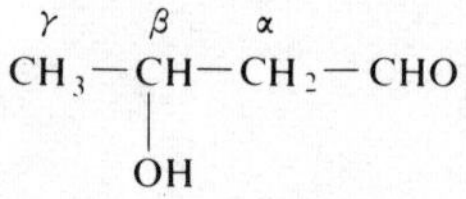   β-hydroxybutyraldehyde

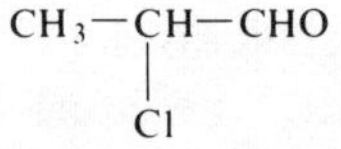

$$CH_3-\underset{\underset{Cl}{\vert}}{CH}-CHO$$
2-chloropropanal   or   α-chloropropionaldehyde

Ketones are commonly named like the ethers, giving the names of the two alkyl groups, followed by *ketone*, the simplest alkyl group coming first, and further substituents indicated by numbers or sometimes Greek letters, α, β, γ, etc.

The systematic IUPAC names for ketones are obtained by taking the whole alkyl chain to get the stem of the name and attaching the suffix *-one*, preceded by a number to indicate which atom is attached to oxygen. For example, acetone, which has three carbon atoms, is derived from propane; it is therefore called propanone; no number is necessary here, because there is only one possible

* The aldehyde group $-\overset{\displaystyle H}{\underset{}{C}}=O$ can be abbreviated to —CHO in print.

144

position for C=O giving a ketone. The numbering is chosen so as to give the lowest possible number to the ketone group.

The group C=O is referred to as a *carbonyl* group, whether it occurs in aldehydes, ketones, carboxylic acids, or esters and so on.

Formula	Trivial name	IUPAC name
$CH_3 - \overset{\displaystyle O}{\overset{\|}{C}} - CH_3$	acetone *or* dimethyl ketone	propanone
$CH_3 - \overset{\displaystyle O}{\overset{\|}{C}} - CH_2CH_3$	methyl ethyl ketone	butanone
$CH_3 - \overset{\displaystyle O}{\overset{\|}{C}} - (CH_2)_2CH_3$	methyl propyl ketone	pentan-2-one
$CH_3CH_2 - \overset{\displaystyle O}{\overset{\|}{C}} - CH_2CH_3$	diethyl ketone	hexan-3-one
$CH_3 - \overset{\displaystyle O}{\overset{\|}{C}} - \overset{\overset{\displaystyle CH_3}{\|}}{CH}\ CH_3$	methyl isopropyl ketone	3-methylbutanone

$$\overset{1}{C}H_2 - \overset{2}{C}H_2 - \overset{3}{\overset{\displaystyle O}{\overset{\|}{C}}} - \overset{4}{C}H_2 - \overset{5}{C}H - \overset{6}{C}H_3 \quad \text{or} \quad \overset{\beta}{C}H_2 - \overset{\alpha}{C}H_2 - \overset{\displaystyle O}{\overset{\|}{C}} - \overset{\alpha'}{C}H_2 - \overset{\beta'}{C}H - \overset{\gamma'}{C}H_3$$
$$\ \ |\qquad\qquad\qquad\qquad |\qquad\qquad\qquad\ |\qquad\qquad\qquad\qquad\quad |$$
$$Cl\qquad\qquad\qquad\qquad Cl\qquad\qquad\qquad Cl\qquad\qquad\qquad\qquad Cl$$

1,5-dichlorohexan-3-one   or   ß-chloroethyl-ß'-chloropropyl ketone

Physical Properties

Formaldehyde is a gas with a strong, irritating odour, and acetaldehyde is a low boiling liquid, also with an unpleasant odour. Only these two aldehydes are completely miscible with water. Higher aldehydes are oily liquids with more pleasant odours.

The lower ketones (C_3–C_{11}) are pleasant smelling liquids. Only acetone is completely miscible with water. Higher ketones ($>C_{11}$) are colourless solids with very faint but pleasant odours.

Table 13. Aldehydes

Name	Formula	m.p. °C	b.p. °C	Solubility in H_2O (g/100 g at 20°)
Formaldehyde	HCHO	−92	+21	miscible
Acetaldehyde	CH_3CHO	−121	20	miscible
Propionaldehyde	CH_3CH_2CHO	−81	49	16
Butyraldehyde	$CH_3(CH_2)_2CHO$	−99	76	3·6
Isobutyraldehyde	$(CH_3)_2CHCHO$	−65	61	10·0
Valeraldehyde	$CH_3(CH_2)_3CHO$	−91	103	slight
Isovaleraldehyde	$(CH_3)_2CHCH_2CHO$	−51	92	slight
Hexanal	$CH_3(CH_2)_4CHO$	—	130	—
Octanal	$CH_3(CH_2)_6CHO$	—	170	—
Octadecanal	$CH_3(CH_2)_{16}CHO$	38	—	—

Table 14. Ketones

Name	Formula	m.p. °C	b.p. °C	Solubility in H_2O (g/100 g at 20°)
Acetone	CH_3COCH_3	−94	56	miscible
Methyl ethyl ketone	$CH_3COCH_2CH_3$	−86	80	27·3
Diethyl ketone	$CH_3CH_2COCH_2CH_3$	−42	103	5·1
Methyl propyl ketone	$CH_3CO(CH_2)_2CH_3$	−78	102	6·0
Methyl isopropyl ketone	$CH_3COCH(CH_3)_2$	−92	95	6·5
Methyl butyl ketone	$CH_3CO(CH_2)_3CH_3$	−57	127	slight
Dipropyl ketone	$CH_3(CH_2)_2CO(CH_2)_2CH_3$	−32	143	—
Dibutyl ketone	$CH_3(CH_2)_3CO(CH_2)_3CH_3$	−6	190	—
Diamyl ketone	$CH_3(CH_2)_4CO(CH_2)_4CH_3$	14	226	—
Dihexyl ketone	$CH_3(CH_2)_5CO(CH_2)_5CH_3$	33	264	—

Preparation of Aldehydes and Ketones

A. *Methods useful for both classes*

1. *Oxidation of Alcohols.* Primary alcohols give aldehydes on mild oxidation, but because the aldehyde is more susceptible to oxidation than the alcohol, it must be removed from the mixture as it is formed. Fortunately simple aldehydes have lower boiling points than the corresponding alcohols, so the aldehyde is generally distilled from the solution. The best example is the formation of acetaldehyde (b.p. 20°). Ethanol (b.p. 78°) is dropped slowly into a solution of potassium dichromate in dilute sulphuric acid maintained at 50°:

$$CH_3CH_2OH \xrightarrow[H\oplus]{K_2Cr_2O_7} CH_3CHO$$

With higher aldehydes it becomes more difficult to separate the aldehyde as it is formed because the difference in boiling points becomes less.

Secondary alcohols give ketones on oxidation with the same reagents as for aldehydes, but because the ketones are more resistant to further oxidation than aldehydes, isolation of the product is less difficult.

Manganese dioxide suspended in acetone or hexane can also be used as an oxidizing agent for primary and secondary alcohols, and it has the advantage that the reagent does not attack double bonds, so, for example, an unsaturated secondary alcohol can be converted to an unsaturated ketone.

2. *Dehydrogenation.* On an industrial scale, the vapour phase dehydrogenation of alcohols is a good source of aldehydes and ketones. The catalyst used is copper or copper chromite at a temperature of 200° to 300°, or silver gauze at 450°:

$$CH_3CH_2OH \xrightarrow{Cu} CH_3CHO + H_2$$

$$CH_3-\underset{\underset{OH}{|}}{CH}-CH_3 \xrightarrow{Cu} CH_3COCH_3 + H_2$$

By controlled addition of air to the mixture, the hydrogen is oxidized to water:

$$RCH_2OH + \tfrac{1}{2}O_2 \longrightarrow RCHO + H_2O$$

3. *From Calcium Salts of Carboxylic Acids.* Dry distillation or pyrolysis of the calcium (or barium) salt of an acid gives calcium (or barium) carbonate and a symmetrical ketone:

$$\underset{RC}{\overset{O}{\overset{||}{}}}-OCaO-\underset{CR}{\overset{O}{\overset{||}{}}} \longrightarrow \underset{RCR}{\overset{O}{\overset{||}{}}} + CaCO_3$$

Yields for this reaction are poor, as in most organic reactions in which an infusible solid is heated.

If the salt is mixed with calcium formate, then some aldehyde is produced as well:

$$(R-\underset{}{\overset{O}{\overset{||}{C}}}-O)_2Ca + (H\overset{O}{\overset{||}{C}}O)_2Ca \longrightarrow R-CHO + CaCO_3$$

A variation of the method is to pass the vapour of the acids over a metal oxide catalyst (aluminium or thorium oxide): the same aldehyde or ketone products are formed, together with carbon dioxide.

4. *Grignard Reaction.* A Grignard reagent will react with an ester forming a ketone, but the ketone is more reactive towards the Grignard reagent than the starting material, so that a mixture of products is formed:

$$RMgX + R'—\overset{\overset{\displaystyle O}{\|}}{C}—OR'' \longrightarrow \left[R'—\overset{\overset{\displaystyle R}{|}}{\underset{\underset{\displaystyle OMgX}{|}}{C}}—OR'' \right] \longrightarrow R'—\overset{\overset{\displaystyle O}{\|}}{C}—R + R''OMgX$$

$$\downarrow RMgX$$

$$R'—\overset{\overset{\displaystyle R}{|}}{\underset{\underset{\displaystyle R}{|}}{C}}—OMgX$$

However, under special conditions Grignard reagents can be used to obtain both aldehydes and ketones. Ethyl orthoformate (prepared from chloroform, and sodium ethoxide) can be used successfully to give an aldehyde:

$$\overset{\overset{\displaystyle OEt}{|}}{\underset{\underset{\displaystyle OEt}{|}}{HC}}—OEt \xrightarrow{RMgX} \overset{\overset{\displaystyle R}{|}}{\underset{\underset{\displaystyle OEt}{|}}{HC}}—OEt \xrightarrow[H^{\oplus}]{H_2O} \overset{\overset{\displaystyle H}{|}}{RC}{=}O + 2EtOH$$

Higher orthoesters would give ketones, but these starting materials are not readily available. Ketones can be prepared by the addition of the Grignard reagent to a nitrile (or alkyl cyanide):

$$RC{\equiv}N + R'MgX \longrightarrow \left[\overset{\displaystyle RC{=}N—MgX}{\underset{\underset{\displaystyle R'}{|}}{}} \right] \xrightarrow{H_2O} \overset{\overset{\displaystyle O}{\|}}{RCR'} + NH_3 + MgXOH$$

Alternatively, the Grignard reagent can be converted to the less reactive zinc or cadmium alkyls, and these will react with an acid chloride, but not a ketone. The ketone can thus be obtained in good yield:

$$2\,RMgBr + CdBr_2 \longrightarrow RCdR + 2\,MgBr_2$$

$$RCdR + 2\,R'\overset{\overset{\displaystyle O}{\|}}{C}Cl \longrightarrow 2\,R\overset{\overset{\displaystyle O}{\|}}{C}R' + CdCl_2$$

5. *Ozonolysis provides a way of obtaining aldehydes and ketones from alkenes, the products depending upon the number of alkyl*

groups attached to the double bond. The unstable ozonide is reduced (e.g. catalytically) to give the carbonyl compound.

$$CH_3CH{=}CHCH_3 \xrightarrow{O_3} CH_3CH\underset{O-O}{\overset{O}{\diagdown\diagup}}CHCH_3 \xrightarrow{H_2} 2CH_3CHO + H_2O$$

$$\underset{CH_3}{\overset{CH_3}{|}}C{=}CHCH_3 \xrightarrow{O_3} \underset{CH_3}{\overset{CH_3}{|}}C\underset{O-O}{\overset{O}{\diagdown\diagup}}CHCH_3 \xrightarrow{H_2} \underset{CH_3}{\overset{CH_3}{|}}C{=}O + CH_3CHO + H_2O$$

6. *Hydrolysis of Acetylenes.* Acetylene itself, in dilute sulphuric acid with mercuric sulphate adds on water to form acetaldehyde (page 91). Substituted acetylenes react in the same way to form ketones.

$$HC{\equiv}CH + H_2O \xrightarrow[H_2SO_4]{HgSO_4} CH_3CHO$$

$$RC{\equiv}CH + H_2O \xrightarrow[H_2SO_4]{HgSO_4} RC\overset{O}{\overset{\|}{-}}CH_3$$

B. Methods for aldehydes only

7. *Rosenmund Reduction.* When an acid chloride is reduced with hydrogen and a metal catalyst, usually palladium, though Rosenmund originally used nickel, good yields of aldehydes are obtained. A small amount of quinoline or sulphur is added to 'poison' the catalyst so that it does not further reduce the aldehyde to alcohol. In practice, a vigorous stream of hydrogen is blown through the mixture to carry away the HCl, which can then be absorbed in water and titrated with alkali to see how far the reaction has gone.

$$RCOCl + H_2 \xrightarrow{Pd} RCHO + HCl$$

8. *Methods for reducing carboxylic acid derivatives to aldehydes must avoid conditions under which the aldehyde itself is reduced.* Lithium aluminium hydride normally reduces acid derivatives to alcohols, but two less reactive derivatives of LiAlH$_4$ can be used to stop the reduction at the aldehyde stage.

Lithium diethoxyaluminium hydride LiAlH$_2$(OEt)$_2$, made by adding the correct quantity of ethanol to LiAlH$_4$, reduces tertiary amides to aldehydes. The amides are easily made from the corresponding acid (see Chapter 9):

$$RC\overset{O}{\overset{\|}{N}}(CH_3)_2 + LiAlH_2(OEt)_2 \longrightarrow RC\overset{H}{\overset{|}{=}}O + (CH_3)_2NH$$

Lithium tri-t-butoxyaluminium hydride, made from LiAlH$_4$ and t-butanol, can reduce acid chlorides to aldehydes at low temperatures.

$$\underset{\text{RCCl}}{\overset{\overset{\displaystyle O}{\|}}{}} + \text{LiAlH[OC(CH}_3)_3]_3 \xrightarrow{-78°} \underset{\text{RC}=\text{O}}{\overset{\overset{\displaystyle H}{|}}{}}$$

9. *The Oxo Process*. This has already been mentioned in the chapters on alkenes and alcohols. An alkene, carbon monoxide and hydrogen are passed over catalysts at high temperature and pressure; mixtures of isomers are produced, but with great preponderance of the straight chain product.

$$\text{RCH}=\text{CH}_2 + \text{CO} + \text{H}_2 \longrightarrow \text{RCH}_2\text{CH}_2\text{CHO} + \underset{\overset{|}{\text{CHO}}}{\text{RCHCH}_3}$$

C. Methods for ketones only

10. *By Ester Condensation*. This is a very useful method for preparing ketones. Essentially it consists of condensing together two molecules of an ester, such as ethyl acetate, with a strong base, usually sodium ethoxide, as catalyst, to give a β-keto ester. This is hydrolysed with dilute alkali to a β-keto acid, which is unstable, and loses carbon dioxide on heating, and a ketone is formed. The mechanism of the reaction is discussed in Chapter 9.

$$\underset{\text{CH}_3\text{COEt}}{\overset{\overset{\displaystyle O}{\|}}{}} + \underset{\text{CH}_3\text{COEt}}{\overset{\overset{\displaystyle O}{\|}}{}} \xrightarrow{\text{NaOEt}} \underset{\text{CH}_3\text{CCH}_2\text{COEt}}{\overset{\overset{\displaystyle O}{\|}\quad\overset{\displaystyle O}{\|}}{}} + \text{EtOH}$$

$$\underset{\text{CH}_3\text{CCH}_2\text{COEt}}{\overset{\overset{\displaystyle O}{\|}\ \overset{\displaystyle O}{\|}}{}} + \text{H}_2\text{O} \xrightarrow{\text{NaOH}} \underset{\text{CH}_3\text{CCH}_2\text{COH}}{\overset{\overset{\displaystyle O}{\|}\ \overset{\displaystyle O}{\|}}{}} \longrightarrow \underset{\text{CH}_3\text{CCH}_3}{\overset{\overset{\displaystyle O}{\|}}{}} + \text{CO}_2$$

11. *Oppenauer Oxidation*. Oxidation of secondary alcohols has already been mentioned; the Oppenauer method uses aluminium t-butoxide in acetone as reagent. The secondary alcohol is oxidized to a ketone and the acetone is reduced to isopropanol. Since the reagent does not affect double bonds, unsaturated ketones can be obtained by this method. The mechanism of this reaction is described below under the Meerwein–Ponndorf reaction. Oppenaur oxidation and Meerwein–Ponndorf reduction are two aspects of the same equilibrium reaction. Here the reaction is driven forward by the large excess of acetone.

$$\underset{\overset{|}{\text{OH}}}{\text{R}-\text{CH}-\text{R}} + \underset{\text{CH}_3\text{CCH}_3}{\overset{\overset{\displaystyle O}{\|}}{}} \xrightarrow{\text{Al}(-\text{O}-\overset{\overset{\displaystyle \text{CH}_3}{|}}{\underset{\underset{\displaystyle \text{CH}_3}{|}}{\text{C}}}\text{CH}_3)_3} \underset{\text{R}-\text{C}-\text{R}}{\overset{\overset{\displaystyle O}{\|}}{}} + \underset{\overset{|}{\text{OH}}}{\text{CH}_3\text{CHCH}_3}$$

150

Reactions of aldehydes and ketones

Aldehydes and ketones undergo more varied and interesting reactions than any group of organic compounds yet encountered in this book. Before discussing these reactions, it is helpful to look again at the reason for this. The carbonyl bond is not a simple sharing of electrons; oxygen is a more electron-negative element than carbon, i.e. oxygen has a slightly greater attraction for the valency electrons, which leaves a deficiency of electron charge around the carbon atom, as indicated below. When a charged (or potentially charged) reagent approaches the group, the oxygen atom draws the electrons away more and more; the extreme case is depicted below where $A^{\oplus}$ represents an electrophilic reagent and $B^{\ominus}$ a nucleophilic reagent. So we see that $A^{\oplus}$ will become attached to carbonyl oxygen and $B^{\ominus}$ will link with carbonyl carbon. The attack of the nucleophile $B^{\ominus}$ on carbon is usually the first step of such reactions:

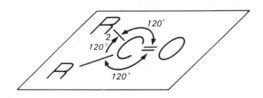

Considering the three-dimensional structure of the carbonyl group (Fig. 32) it is seen that the two groups (R_1 and R_2) attached to the carbonyl, and the carbon and oxygen atoms all lie in a plane.

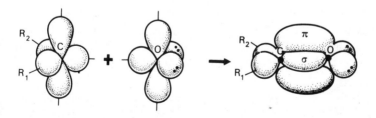

Fig. 32   Arrangement of the carbonyl group bonds

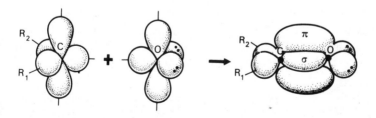

Fig. 33   Orbital picture of the formation of a carbonyl bond

Fig. 34   The polarization of π-electron charge towards the oxygen atom in a carbonyl group

In the language of atomic orbitals, a carbonyl bond is formed by combining carbon and oxygen atoms, both with three sp^2 hybrid orbitals and one p orbital. In oxygen two of the sp^2 hybrid orbitals are filled with unshared pairs of electrons. Overlap of the sp^2 and p orbitals leads to the formation of one σ bond and one π bond (Fig. 33). In order to represent the greater electron density on oxygen, the π bond can be pictured as in Fig. 34. This emphasizes that electrons in π orbitals are more easily displaced than σ electrons. It is because the carbonyl group contains a π bond that the electron density is less evenly distributed between carbon and oxygen than, for example, in a C—O single bond in ethers. This uneven sharing of electron density accounts for many of the characteristic reactions of carbonyl groups.

1. *Oxidation.* Aldehydes are easily oxidized to acids, so easily in fact, that they become partly oxidized on contact with air, through the intermediate formation of peroxides. Besides the common oxidizing agents, such as permanganate and dichromate, salts of heavy metals, such as silver and gold, oxidize aldehydes to acids. Ketones are stable to all but the most vigorous oxidizing agents (such as hot nitric acid), which fragment them into acids of lower molecular weight, and so the two groups can be distinguished by their behaviour towards mild oxidizing agents. One of these is Tollen's reagent, an ammoniacal silver nitrate solution (containing silver ions in solution as $Ag(NH_3)_2^+$), which is reduced by aldehydes, giving a 'silver mirror' on the walls of the test tube. Another is Fehling's solution, which is an alkaline solution containing cupric ions (held in solution as the tartrate complex): the solution is deep blue and on reaction with aldehydes, red cuprous oxide is precipitated. Both reagents are specific for aldehydes; ketones, alcohols or double bonds are not affected.

$$RCHO \; + \; 2Ag(NH_3)_2OH \longrightarrow RCOOH \; + \; 4NH_3 \; + \; 2Ag \; + \; H_2O$$

Tollen's reagent

$$RCHO + 2Cu^{\oplus\oplus} + 5NaOH \longrightarrow RCOONa + Cu_2O + 4Na^{\oplus} + 3H_2O$$

$\underbrace{\phantom{2Cu^{\oplus\oplus}}}_{\text{Fehling's solution}}$

2. *Reduction.* Both aldehydes and ketones are reduced catalytically with hydrogen to alcohols. The catalyst is usually platinum or Raney nickel (page 107).

If the carbonyl compound also contains a double bond, this will also be reduced, because carbonyl groups are not as readily hydrogenated as double bonds. Therefore an unsaturated carbonyl compound will give a saturated alcohol by catalytic hydrogenation:

$$CH_2\!=\!CH\!-\!CHO + H_2 \xrightarrow{\text{Pt}} CH_3CH_2CH_2OH$$

The Meerwein–Ponndorf reduction, which is simply the **Oppenauer** oxidation in reverse, reduces aldehydes and ketones with aluminium isopropoxide in isopropanol solution. Exchange takes place between the isopropoxy and carbonyl groups; acetone and the required alcohol are formed, and the equilibrium is displaced by continually distilling away the acetone:

$$R_2C\!=\!O + Al(OC_3H_7)_3 \longrightarrow Al(OCHR_2)_3 + CH_3\overset{\displaystyle O}{\overset{\|}{C}}CH_3$$

$$\downarrow H^{\oplus}$$

$$R_2CHOH$$

The reaction probably occurs through a cyclic exchange of electrons, starting with donation of charge from the electron-rich oxygen of the carbonyl compound to electro-positive aluminium:

The Meerwein–Ponndorf method has been largely replaced by the newer hydride reagents, such as lithium aluminium hydride and sodium borohydride which provide rapid and complete reduction to alcohols. These reagents do not affect double bonds.

$$4R_2CO + NaBH_4 + 3H_2O \longrightarrow 4R_2CHOH + NaH_2BO_3$$

A hydrogen atom becomes transferred to the carbonyl carbon, and oxygen becomes bound to boron or aluminium in much the same way as in the Meerwein–Ponndorf–Oppenauer reaction. Notice that

153

the hydrogen atom migrates with its electrons (behaving as if it were $H:^{\ominus}$).

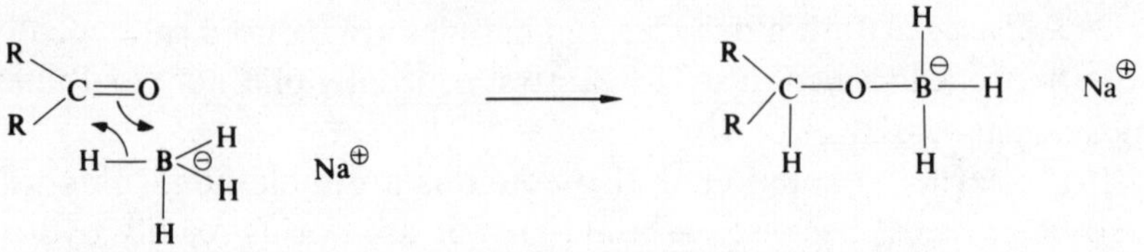

Complete reduction of $C=O$ to CH_2 is also possible by several methods. The first, Clemmensen reduction, uses amalgamated zinc and hydrochloric acid. The yield is variable:

$$R_2C = O + [H] \xrightarrow[HCl]{Zn/Hg} R_2CH_2 + H_2O$$

An alternative method is Wolff–Kishner reduction, sometimes known as the Huang–Minlon method, since it is Huang–Minlon's modification of the original which is used today. The aldehyde or ketone is heated with hydrazine to give the hydrazone (see paragraph 5 below) and this is then heated at 190° with alkali in ethylene glycol to give the hydrocarbon. This method usually gives better results for aliphatic compounds than the Clemmensen reduction.

$$R_2C = O + NH_2{-}NH_2 \longrightarrow R_2C{=}N{-}NH_2 \xrightarrow[190°]{KOH} R_2CH_2 + N_2$$

3. *Bisulphite Compounds.* Both aldehydes and ketones reversibly add a number of reagents to the $C=O$ bond under mild conditions. The bisulphite addition compounds are produced by shaking the carbonyl compound with a saturated solution of sodium bisulphite in water. The bisulphite addition compound has the structure of an hydroxy-sulphonic acid salt.

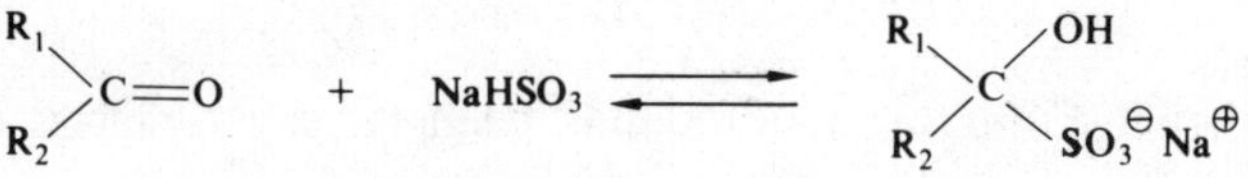

This is an example of the large group of nucleophilic additions to a carbonyl group:

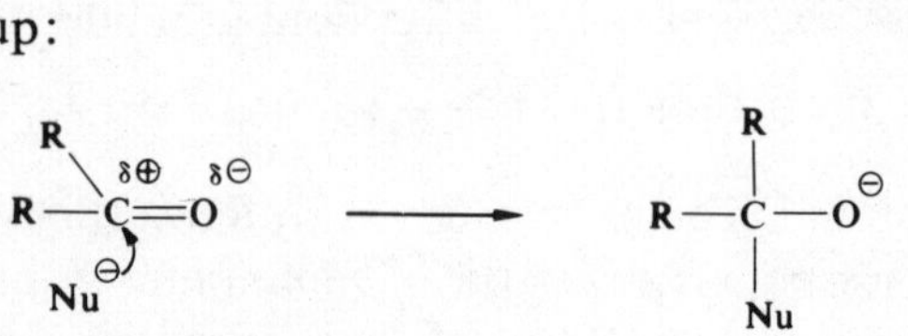

Here Nu represents any of a large number of nucleophilic reagents (e.g. $CN^\ominus$, NH_3, ROH, $H^\ominus$ of $BH_4^\ominus$). In this particular case the nucleophile is $HSO_3^\ominus$:

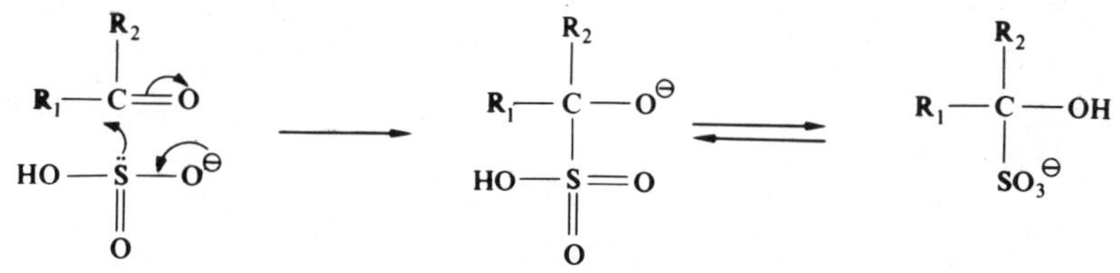

The ease of formation of the bisulphite compound depends upon the groups R_1 and R_2. For aldehydes (R_1=alkyl; R_2=H), the addition takes place easily; methyl ketones (R_1=alkyl; R_2=CH_3) also react in this way, but for ketones where both R_1 and R_2 are bulky alkyl groups, the carbonyl group is so 'hindered' that reaction does not occur to any appreciable extent. The salts are only slightly soluble in aqueous solution, and are precipitated as colourless crystals which can be filtered off. The bisulphite compound can be decomposed with dilute acid, or sodium carbonate, regenerating the carbonyl compound. By this method, any aldehydes or methyl ketones can be separated from a mixture of water-insoluble organic compounds.

4. *Cyanhydrins.* Hydrogen cyanide forms addition compounds called cyanhydrins with aldehydes and some ketones. As with bisulphite compounds, bulky alkyl groups hinder the formation of the compound. Either liquid hydrogen cyanide or aqueous sodium cyanide and dilute acid can be used, or sodium cyanide can be added to the bisulphite compound.

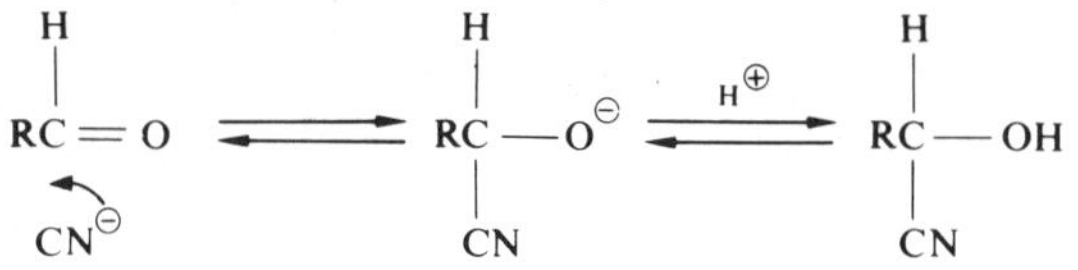

A

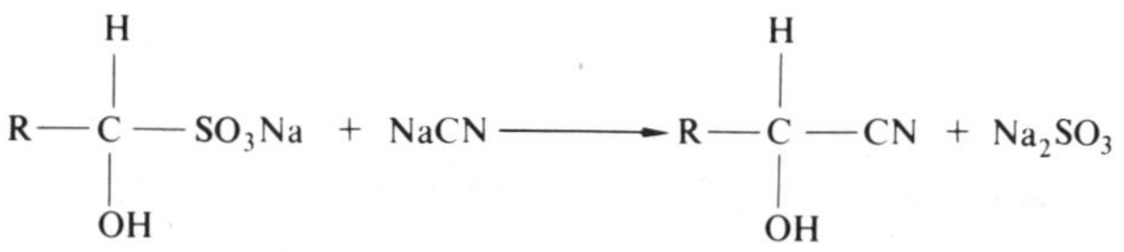

This was probably the very first organic reactions studied from the point of view of mechanism. Lapworth in Manchester in 1903 studied the effects of various catalysts on cyanhydrin formation. He found the reaction had to be slightly alkaline to go well (the $CN^\ominus$ ion is necessary). It was he that proposed that the reaction began with attachment of cyanide to carbonyl carbon to give the ion A (above), and was able to prepare a number of salts of this ion.

The aldehyde or ketone can be recovered by treating the cyanhydrin with moist silver oxide. Cyanhydrins can also be regarded as hydroxynitriles (page 206), and by acid-catalysed hydrolysis can give hydroxyacids:

$$\underset{\underset{OH}{|}}{\overset{\overset{H}{|}}{RC}}-CN + H_2O \xrightarrow{H^{\oplus}} \underset{\underset{OH}{|}}{RCH}-COOH + NH_3$$

5. *Nitrogen Derivatives.* Several groups of nitrogen compounds related to ammonia form products with aldehydes and ketones by nucleophilic attack at the carbon atom in weakly acid conditions. In most cases these products are crystalline, are easily purified and have sharp melting points, and so they are frequently used to confirm the identity of a carbonyl compound. The original compound can often be recovered from the derivative by hydrolysis with dilute acid.

(a) Hydroxylamine gives an oxime:

$$R_2C=O + NH_2OH \longrightarrow R_2C=N-OH$$
an oxime

(b) Hydrazine gives a hydrazone:

$$R_2C=O + NH_2-NH_2 \longrightarrow R_2C=N-NH_2$$
a hydrazone

(c) Phenylhydrazine gives a phenylhydrazone. The phenylhydrazones are generally more crystalline and more insoluble than hydrazones, which makes them better derivatives for low molecular weight carbonyl compounds:

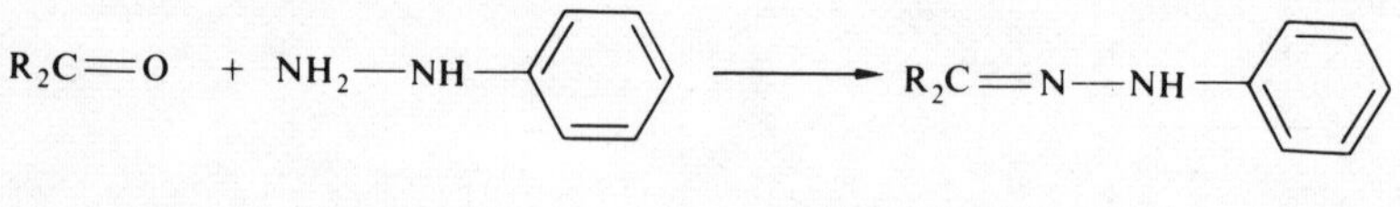

$$R_2C=O + NH_2-NH-\langle\rangle \longrightarrow R_2C=N-NH-\langle\rangle$$
a phenylhydrazone

(d) Semicarbazide gives a semicarbazone:

$$R_2C=O + NH_2-\overset{\overset{O}{\|}}{C}-NH-NH_2 \longrightarrow R_2C=N-NH-\overset{\overset{O}{\|}}{C}-NH_2$$
a semicarbazone

156

The probable sequence of addition of the nitrogen compound (represented as $R'NH_2$) followed by addition and rearrangement of protons, loss of water and a proton, is shown below for a typical aldehyde:

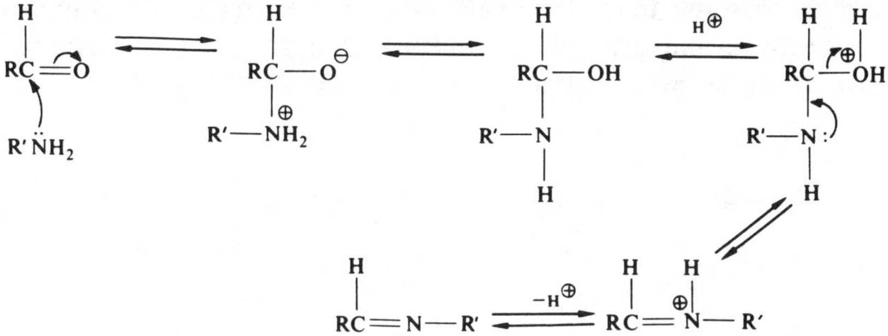

$$
\underset{\substack{|\\R'\ddot{N}H_2}}{\overset{\substack{H\\|}}{RC}}{=}O \;\rightleftharpoons\; \underset{\substack{|\oplus\\R'-NH_2}}{\overset{\substack{H\\|}}{RC}}{-}\overset{\ominus}{O} \;\rightleftharpoons\; \underset{\substack{|\\R'-N\\|\\H}}{\overset{\substack{H\\|}}{RC}}{-}OH \;\xrightarrow{H^\oplus}\; \underset{\substack{|\\R'-N:\\|\\H}}{\overset{\substack{H\quad H\\|\quad\oplus|}}{RC}}{-}OH
$$

$$
\underset{}{\overset{\substack{H\\|}}{RC}}{=}N{-}R' \;\underset{-H^\oplus}{\rightleftharpoons}\; \overset{\substack{H\quad H\\|\quad\oplus|}}{RC}{=}N{-}R'
$$

Note that an acid is required to protonate the OH groups, to form an incipient water molecule which is easily lost; hence catalysis by acid is found to occur, but if too much acid is used, the nitrogen compound is converted to the protonated form $R'N^\oplus H_3$ which cannot react with the carbonyl group. Either too little or too much acid will not catalyse the reaction.

In a similar manner aldehydes add dry ammonia to give unstable addition products and there is evidence that acetaldehyde is hydrated in aqueous solution. In these cases the OH group is not eliminated and the reaction is easily reversed:

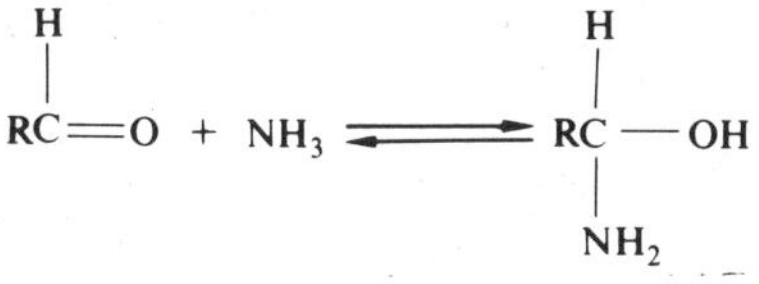

$$
\overset{\substack{H\\|}}{RC}{=}O + NH_3 \;\rightleftharpoons\; \underset{\substack{|\\NH_2}}{\overset{\substack{H\\|}}{RC}}{-}OH
$$

aldehyde ammonia

$$
\overset{\substack{H\\|}}{CH_3C}{=}O + H_2O \;\rightleftharpoons\; \underset{\substack{|\\OH}}{\overset{\substack{H\\|}}{CH_3C}}{-}OH
$$

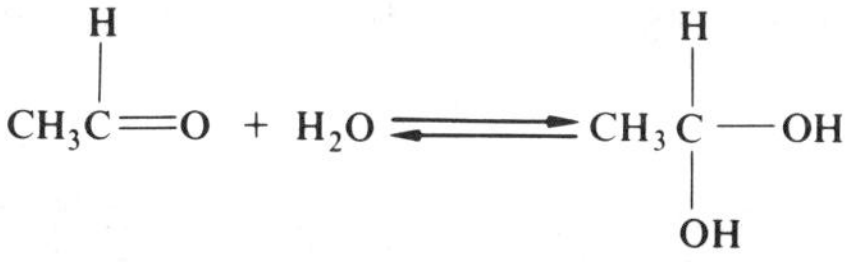

6. *Addition of Alcohols.* If an alcohol is added to an aldehyde, some heat is produced, showing that the two components react; the product is a hemi-acetal. Such compounds are unstable and are decomposed again on heating.

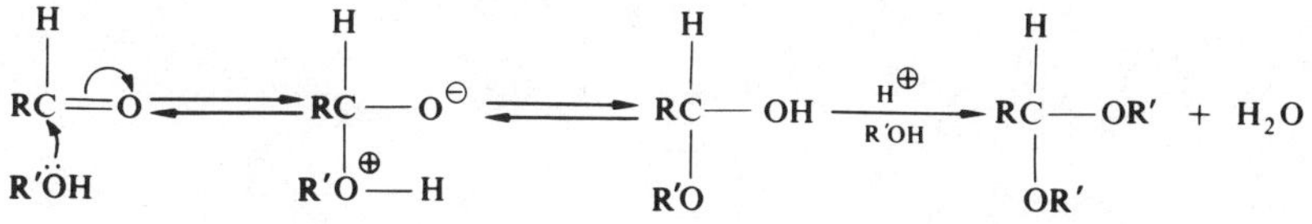

$$
\underset{\substack{|\\R'\ddot{O}H}}{\overset{\substack{H\\|}}{RC}}{=}O \rightleftharpoons \underset{\substack{|\oplus\\R'O-H}}{\overset{\substack{H\\|}}{RC}}{-}\overset{\ominus}{O} \rightleftharpoons \underset{\substack{|\\R'O}}{\overset{\substack{H\\|}}{RC}}{-}OH \xrightarrow[R'OH]{H^\oplus} \underset{\substack{|\\OR'}}{\overset{\substack{H\\|}}{RC}}{-}OR' + H_2O
$$

a hemi–acetal \qquad an acetal

If both reactants are dry and an anhydrous acid, such as hydrogen chloride gas, is added to the mixture, an acetal is formed, by the reaction of a further molecule of alcohol with the hemi-acetal. The acetals are stable to oxidation and alkali, but are readily decomposed to aldehyde and alcohol by aqueous acid. In this way, aldehyde groups can be 'protected' and later regenerated:

$$
\begin{array}{ccc}
\overset{\text{H}}{\underset{\text{R}'\text{O}}{\text{R}\overset{|}{\underset{|}{\text{C}}}\text{—OH}}} & \xrightarrow{\text{H}^{\oplus}} & \overset{\text{H}}{\underset{\text{R}'\text{O:}}{\text{R}\overset{|}{\underset{|}{\text{C}}}\text{—}\overset{\oplus}{\text{O}}\text{H}}} \rightleftharpoons \overset{\text{H}}{\underset{\overset{\oplus}{\text{RO}}}{\text{R}\overset{|}{\underset{||}{\text{C}}}}}
\end{array}
$$

$$
\overset{\text{H}}{\underset{\overset{\oplus}{\text{R}'\text{O}}}{\text{R}\overset{|}{\underset{||}{\text{C}}}}} + \text{R}'\overset{..}{\text{O}}\text{H} \rightleftharpoons \overset{\text{H}}{\underset{\text{R}'\text{O}}{\text{R}\overset{|}{\underset{|}{\text{C}}}\text{—}\overset{\oplus}{\text{O}}\text{R}}} \xrightarrow{-\text{H}^{\oplus}} \overset{\text{H}}{\underset{\text{R}'\text{O}}{\text{R}\overset{|}{\underset{|}{\text{C}}}\text{—OR}'}}
$$

an acetal

The ease of reaction depends upon the group R; the more bulky this group, the less readily is the acetal formed.

Ketones, with two R groups, do not form the corresponding ketals under these conditions: more forcing conditions are necessary. Usually this is accomplished by using ethyl orthoformate (page 195):

$$
\underset{\text{R}}{\overset{\text{R}}{>}}\text{C}=\text{O} + \overset{\text{OEt}}{\underset{\text{OEt}}{\text{HC}\overset{|}{\underset{|}{\text{—OEt}}}}} \longrightarrow \underset{\text{R}}{\overset{\text{R}}{>}}\text{C}\overset{\text{OEt}}{\underset{\text{OEt}}{<}}\overset{}{\text{OEt}} + \overset{\text{O}}{\overset{||}{\text{HCOEt}}}
$$

With thioalcohols (Chapter 10) both aldehydes and ketones react to give thioacetals and thioketals respectively:

$$
\overset{\text{H}}{\underset{}{\text{RC}}}=\text{O} + 2\text{R}'\text{SH} \longrightarrow \text{H}_2\text{O} + \underset{\text{SR}'}{\text{R}\text{—}\overset{\text{H}}{\overset{|}{\text{C}}}\overset{|}{\text{—SR}'}}
$$

a thioacetal

$$
\underset{\text{R}}{\overset{\text{R}}{>}}\text{C}=\text{O} + 2\text{R}'\text{SH} \longrightarrow \text{H}_2\text{O} + \underset{\text{R}}{\overset{\text{R}}{>}}\text{C}\overset{\text{SR}'}{\underset{\text{SR}'}{<}}
$$

a thioketal

7. *Grignard Reaction.* Carbonyl compounds react readily with Grignard reagents. This reaction also belongs to the general class of nucleophilic additions, for, as described earlier, the Grignard reagent is partly polarized and the R group behaves as a strongly nucleophilic reagent, attacking the electropositive carbon atom. The type of alcohol formed depends on the starting material:

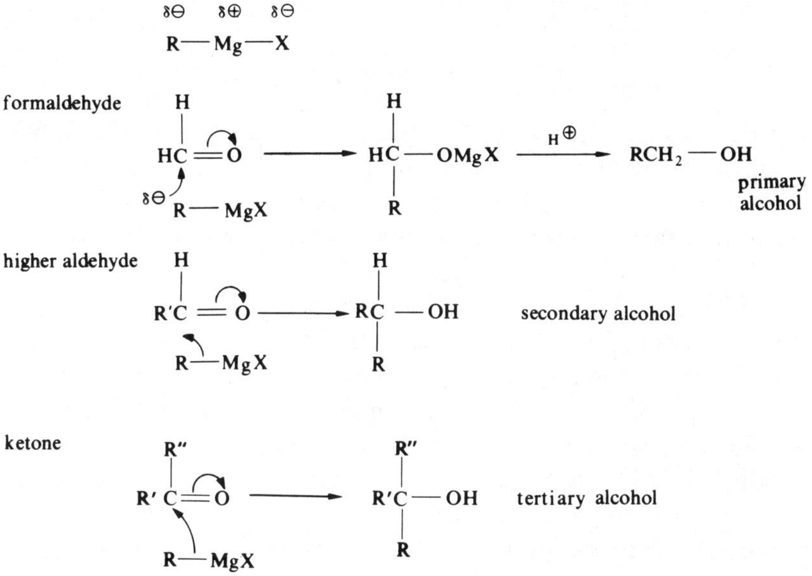

$$\overset{\delta\ominus}{R}\!-\!\overset{\delta\oplus}{Mg}\!-\!\overset{\delta\ominus}{X}$$

formaldehyde

$$\underset{\overset{\delta\ominus}{R\!-\!MgX}}{\overset{H}{\underset{|}{HC}}\!\!=\!\!O} \longrightarrow \overset{H}{\underset{\underset{R}{|}}{HC}}\!-\!OMgX \xrightarrow{H^{\oplus}} RCH_2\!-\!OH$$

primary alcohol

higher aldehyde

$$\underset{R\!-\!MgX}{\overset{H}{\underset{|}{R'C}}\!\!=\!\!O} \longrightarrow \overset{H}{\underset{\underset{R}{|}}{RC}}\!-\!OH \qquad \text{secondary alcohol}$$

ketone

$$\underset{R\!-\!MgX}{\overset{R''}{\underset{|}{R'C}}\!\!=\!\!O} \longrightarrow \overset{R''}{\underset{\underset{R}{|}}{R'C}}\!-\!OH \qquad \text{tertiary alcohol}$$

A special case is the addition to carbon dioxide to give a carboxylic acid:

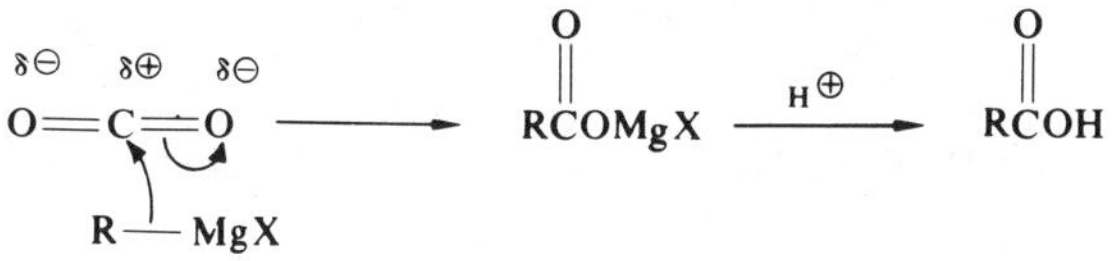

$$\underset{R\!-\!MgX}{\overset{\delta\ominus}{O}\!\!=\!\!\overset{\delta\oplus}{C}\!\!=\!\!\overset{\delta\ominus}{O}} \longrightarrow \overset{O}{\overset{\|}{RCOMgX}} \xrightarrow{H^{\oplus}} \overset{O}{\overset{\|}{RCOH}}$$

The Grignard reagent might be expected to attack the second $C\!=\!O$ bond, but by pouring the reagent onto excess of solid carbon dioxide ($-80°C$) the second reaction is virtually eliminated and a good yield of carboxylic acid is obtained.

8. *Geminal Dihalides.* Aldehydes and ketones and geminal dihalides are interconvertible. The action of phosphorus pentachloride on aldehydes and ketones gives geminal dichlorides:

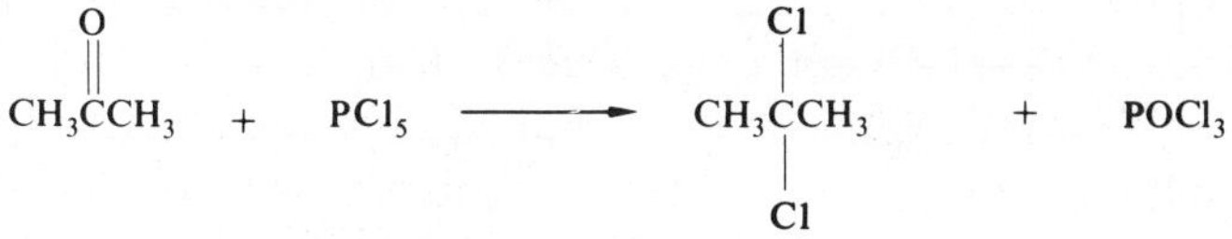

$$\overset{O}{\overset{\|}{CH_3CCH_3}} + PCl_5 \longrightarrow \overset{Cl}{\underset{\underset{Cl}{|}}{CH_3CCH_3}} + POCl_3$$

159

Phosphorus pentabromide converts aldehydes to dibromides but with ketones causes substitution of Br for H in the alkyl groups. Treatment of the dihalides with water converts them to the carbonyl compounds.

9. *Halogenation.* The action of halogen on aldehydes and ketones gives the corresponding α-halo-aldehyde or -ketone:

$$\underset{CH_3\overset{\displaystyle O}{\overset{\|}{C}}CH_3}{} + Br_2 \longrightarrow \underset{CH_3\overset{\displaystyle O}{\overset{\|}{C}}CH_2Br}{} + HBr$$

Aldehydes and ketones exhibit the property of *tautomerism*. When a substance exists in two interconvertible forms which are in equilibrium with each other, it is said to exhibit tautomerism (Greek tauto=the same, meros=parts) and the two forms are called tautomers. The most common origin of this kind of isomerism is when hydrogen can be attached at two alternative positions in the molecule. It is known from physical measurements that simple aldehydes and ketones contain a small amount of molecules of a second form, which is an unsaturated hydroxy compound (an alk*enol*), hence this is called the *enol* form. The familiar carbonyl form considered so far is called the *keto* form. Acetone, for example, consists of an equilibrium mixture of the two forms shown, with the equilibrium well to the left (the proportion of enol in this case is less than 0·001 %):

$$\underset{\text{keto}}{CH_3-\overset{\displaystyle O}{\overset{\|}{C}}-CH_3} \rightleftharpoons \underset{\text{enol}}{CH_3-\overset{\displaystyle OH}{\overset{|}{C}}=CH_2}$$

The enol form of acetaldehyde (Fig. III, page 143) is also known as vinyl alcohol, and can be 'trapped' in the form of its esters and ethers.

$$CH_3-\overset{\displaystyle H}{\overset{|}{C}}=O \rightleftharpoons CH_2=\overset{\displaystyle H}{\overset{|}{C}}-OH$$

Lapworth studied the halogenation of acetone and showed that the rate was the same for chlorine, bromine or iodine. He concluded that the slow step which governs the rate of the reaction is the change from the keto to the enol form. Halogens then add rapidly to the double bond of the enol form, from which by elimination of a proton (giving HX), the halo-aldehyde or ketone is formed. The enol form is steadily removed from the keto-enol equilibrium reaction until the carbonyl compound is completely halogenated. If the resulting compound can enolize again, the reaction need not stop here but can continue until all the α-hydrogen atoms have been replaced:

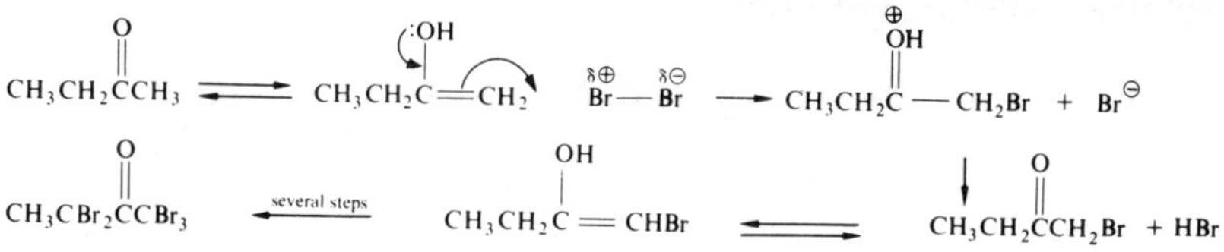

$$CH_3CH_2CCH_3 \rightleftharpoons CH_3CH_2C = CH_2 \quad Br-Br \longrightarrow CH_3CH_2C-CH_2Br + Br^\ominus$$

$$CH_3CBr_2CCBr_3 \xleftarrow{\text{several steps}} CH_3CH_2C = CHBr \rightleftharpoons CH_3CH_2CCH_2Br + HBr$$

The chlorination of acetaldehyde to give trichloroacetaldehyde (chloral, page 115) proceeds in the same way.

10. *The Iodoform Reaction.* A special case of the halogenation reaction is used to identify methyl ketones. Reaction of a methyl ketone with iodine in alkaline solution gives, as an intermediate step, a tri-iodoketone:

$$RCCH_3 \xrightarrow{3I_2} RCCI_3$$

With three electron-withdrawing iodine atoms attached to carbon, the group becomes susceptible to attack by hydroxyl ions, the C—C bond is broken and iodoform, a water-insoluble, heavy yellow solid (m.p. 119°) is formed:

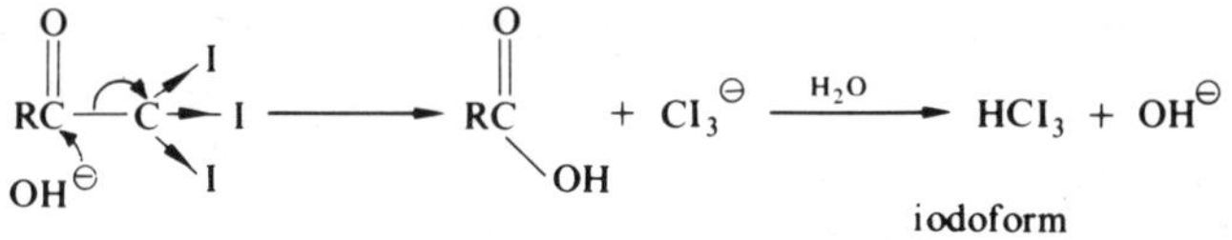

$$RC-C-I \longrightarrow RC + CI_3^\ominus \xrightarrow{H_2O} HCI_3 + OH^\ominus$$

iodoform

Iodine in alkaline solution contains hypoiodite ions $OI^\ominus$, a mild oxidizing agent, so that ethanol and groups such as —CHOHCH$_3$ which can be easily oxidized to methyl ketones also give iodoform in this test. Note that the other product of the reaction is a carboxylic acid with one less carbon atom than the ketone.

Similar reactions (haloform reactions) occur with hypochlorite and hypobromite, giving chloroform and bromoform respectively.

11. *Aldol Condensation.* Two molecules of an aldehyde react under mildly alkaline conditions to give an *ald*ehyde-alcoh*ol*. In the case of acetaldehyde the product is called *aldol*, and this is used as a class name for such condensation compounds:

$$CH_3CHO + CH_3CHO \xrightarrow{OH^\ominus} CH_3CH-CH_2CHO$$
$$| \atop OH$$

aldol

Either by heating or spontaneously, such compounds lose water to give α,β-unsaturated aldehydes. Acetaldehyde is used in this way as a commercial source of n-butanol:

$$2CH_3CHO \longrightarrow \underset{\underset{OH}{|}}{CH_3CHCH_2CHO} \longrightarrow CH_3CH\!=\!CHCHO$$

$$\overset{H_2/Pt}{\diagdown}$$

$$CH_3CH_2CH_2CH_2OH$$

If strong alkali and heat is applied, condensation goes on until a black resin is produced:

$$\underset{\underset{OH}{|}}{CH_3CHCH_2CHO} \longrightarrow \underset{crotonaldehyde}{CH_3CH\!=\!CHCHO} \xrightarrow[OH^\ominus]{CH_3CHO} \underset{further\ condensation}{CH_3CH\!=\!CH\ CH\!=\!CH\!-\!CHO} \ etc.$$

Both the condensation and elimination of water can be explained by the reactivity of α-hydrogen atoms in carbonyl compounds. In the presence of hydroxl ions, the following change can take place:

$$HO^\ominus\ H\!-\!CH_2\!-\!C\diagup^{O}_{\diagdown H} \longrightarrow H_2O \ + \left[CH_2\!=\!C\diagup^{O^\ominus}_{\diagdown H} \longleftrightarrow \overset{\ominus}{C}H_2\!-\!C\diagup^{O}_{\diagdown H} \right]$$

$$\text{I} \qquad\qquad\qquad \text{II}$$

At first glance this is rather surprising and contrary to what has been learned about the reactivity of the C—H bond. Here two special factors are operating; first the polarized carbonyl group draws away electron charge (making H rather more $\delta+$ than in a hydrocarbon). Though this effect falls off rapidly with distance it is still strongly felt at the α-carbon atom. Secondly, the enolate ion (I and II) is resonance stabilized.

When two valence forms can be written for a molecule or ion, in which the *atoms* occupy the same places and only the distribution of the electrons is changed, according to *resonance theory* the character of the molecule or ion is not that of either form (I or II) but something that combines the character of both, called a *resonance hybrid*. Note that the two forms are linked with a double-headed arrow, to indicate the two forms contributing to the average appearance of the hybrid. This is not the same as tautomerism (such as discussed earlier for the keto-enol equilibrium) where the two forms have independent existence and in favourable circumstances can be isolated. One of the consequences of resonance is that the hybrid has more stability than either of the contributing forms would be expected to have, and the more contributing forms one can write, the more stable the hybrid. Further examples of resonance (e.g. the carboxylate ion) will be met later.

We can attempt to picture the resonance hybrid thus:

$$CH_2 = C \overset{O}{\underset{H}{<}}$$

Molecular orbitals provide an alternative to the resonance theory. By this concept, after joining the CH_2, C and O atoms by σ bonds, there are four π electrons to be accommodated on three atoms. Such three-centre four-electron bonds have a definite stability equivalent to half a π-bond plus a negative charge distributed over the three atoms of the bond; the net result is the same as the resonance picture.

Once the enolate ion has been formed, it can attack another molecule of acetaldehyde with formation of aldol:

$$CH_3C\overset{O}{\underset{H}{<}} \quad \overset{\ominus}{C}H_2C\overset{O}{\underset{H}{<}} \longrightarrow CH_3\overset{O^\ominus}{\underset{H}{C}}-CH_2C\overset{O}{\underset{H}{<}} \xrightarrow{H^\oplus} CH_3\overset{OH}{\underset{H}{C}}-CH_2C\overset{O}{\underset{H}{<}}$$

As usual, protonation (the last stage) is here a rapid and easy step. The dehydration of aldol can be catalysed by acid or base:

$$CH_3\overset{}{\underset{OH}{C}}H-CH_2C\overset{O}{\underset{H}{<}} \xrightarrow{OH^\ominus} CH_3\overset{}{\underset{OH}{C}}H-\overset{\ominus}{C}\overset{}{\underset{H}{C}}C\overset{O}{\underset{H}{<}} \longrightarrow CH_3CH=CHC\overset{O}{\underset{H}{<}} + OH^\ominus$$

Note that it can only be the α-hydrogen atom which is removed in the aldol condensation:

$$RCH_2C\overset{O}{\underset{H}{<}} + RCH_2C\overset{O}{\underset{H}{<}} \longrightarrow RCH_2\overset{}{\underset{OH}{C}}H-\overset{R}{\underset{}{C}}HC\overset{O}{\underset{H}{<}}$$

$$\downarrow$$

$$RCH_2CH=\overset{R}{\underset{}{C}}-C\overset{O}{\underset{H}{<}}$$

Since condensation depends upon an enolate ion approaching a carbonyl group, so the more bulky the alkyl group linked to carbonyl the more hindered will be the reaction. Therefore simple aldehydes (one alkyl group) condense more readily than simple ketones (with two alkyl groups). From acetone, diacetone alcohol is first formed. The reaction is catalysed by acid or base. If dry hydrogen chloride is used as catalyst, this first product is dehydrated to mesityl oxide (so named because of an early misunderstanding of its structure), which then condenses with more acetone to give the unsaturated ketone phorone. The acid-catalysed condensation and dehydration can be described as follows:

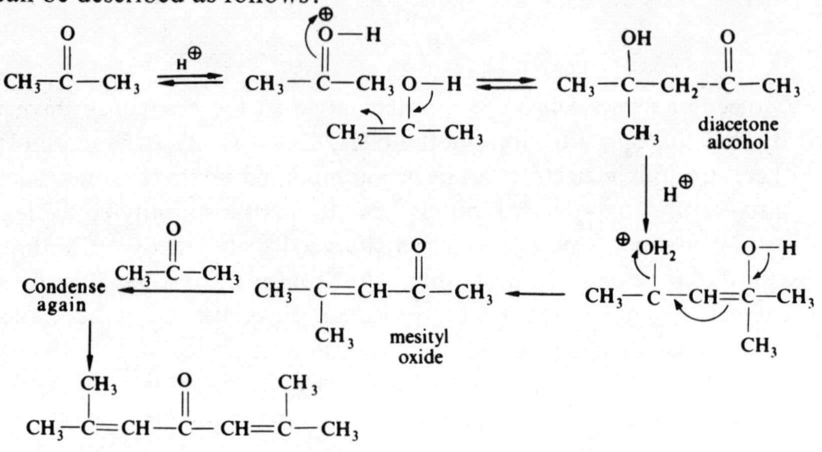

phorone

Notice that it is necessary to have at least one α-hydrogen atom for condensation to occur. This hydrogen atom is activated by its nearness to the carbonyl group. If two different aldehydes are mixed, all possible combinations will take place, but if a ketone and an aldehyde are mixed, the more reactive aldehyde C=O will tend to react preferentially with the enolate ion of the ketone, giving an hydroxy-ketone or an unsaturated ketone as well as some self-condensation product of the aldehyde.

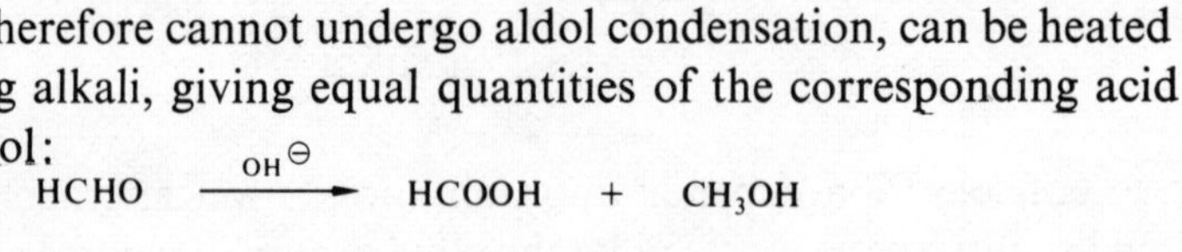

Many further variations of the aldol condensation are known.

12. *Cannizzaro Reaction*. Aldehydes which have no α-hydrogen and therefore cannot undergo aldol condensation, can be heated with strong alkali, giving equal quantities of the corresponding acid and alcohol:

$$HCHO \xrightarrow{OH^{\ominus}} HCOOH + CH_3OH$$

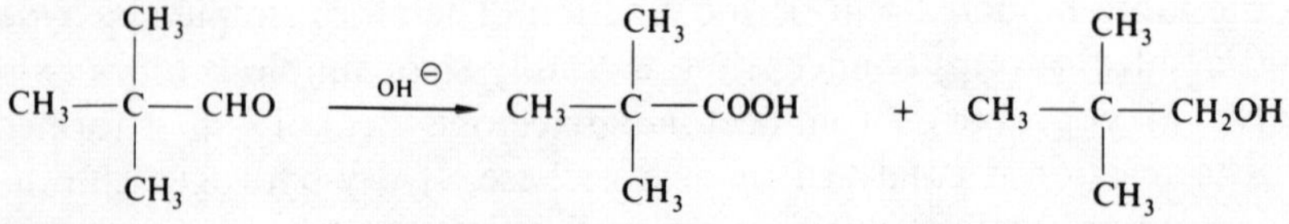

pivalic aldehyde   pivalic acid   neopentyl alcohol

164

13. *Wittig Reaction.* As already mentioned in Chapter 3, aldehydes and ketones react with phosphoranes to give alkenes. The phosphorane is prepared from triphenyl phosphine and an alkyl halide, followed by strong alkali treatment of the intermediate phosphonium halide:

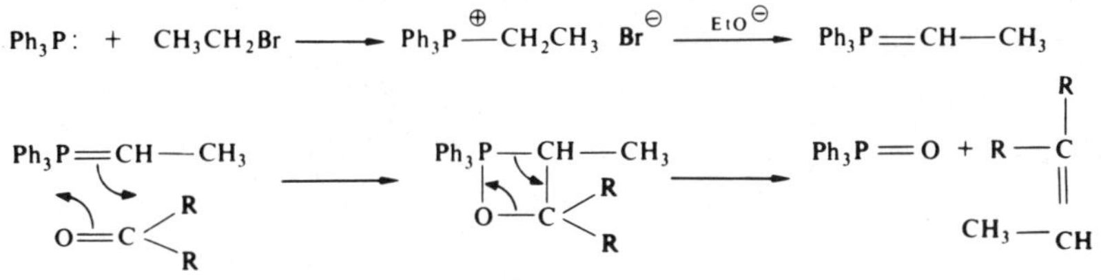

$$Ph_3P: \ + \ CH_3CH_2Br \ \longrightarrow \ Ph_3\overset{\oplus}{P}\!-\!CH_2CH_3 \ \ Br^{\ominus} \ \xrightarrow{EtO^{\ominus}} \ Ph_3P\!=\!CH\!-\!CH_3$$

$$Ph_3P\!=\!CH\!-\!CH_3 \qquad \longrightarrow \qquad Ph_3P\!-\!CH\!-\!CH_3 \qquad \longrightarrow \qquad Ph_3P\!=\!O \ + \ R\!-\!\overset{R}{\underset{\parallel}{\underset{CH_3-CH}{C}}}$$

Ph = phenyl group

14. *Polymerization.* The low molecular weight aldehydes are very easily polymerized in a variety of ways; the products obtained are described with the individual aldehydes.

Formaldehyde

Formaldehyde is prepared industrially by the vapour phase oxidation of methanol or methane. It is a colourless gas with a very strong irritating odour. In the liquid state (b.p. $-21°$) it tends to polymerize very quickly, but an aqueous solution of formaldehyde consists largely of the hydrate, which does not polymerize, so commercial formaldehyde is condensed into water and the resulting solution, called formalin, containing a little by-product methanol, is quite stable.

It is used industrially in the manufacture of resins, plastics, leather and dyes, and also as a disinfectant. Bakelite resin is a thermosetting polymer of formaldehyde and phenol. Formalin has powerful antiseptic properties and is commonly used for preserving biological specimens.

If an aqueous solution of formaldehyde is concentrated, the formaldehyde polymerizes to paraformaldehyde, a white insoluble powder. The material is a linear polymer.

$$n \ \ H\overset{|}{\underset{}{C}}\!=\!O \ \longrightarrow \ HOCH_2\!-\!O\!-\!CH_2\!-\!O\!-\!CH_2\!-\!O \cdots CH_2\!-\!O\!-\!CH_2OH$$
paraformaldehyde

165

A similar white solid, called polyoxymethylene, of higher molecular weight is formed using concentrated sulphuric acid as a catalyst. Both these polymers are easily decomposed by heat, back to monomer.

Formaldehyde gas at room temperature polymerizes to a cyclic solid trimer, trioxane or metaformaldehyde. This substance is soluble in water and organic solvents.

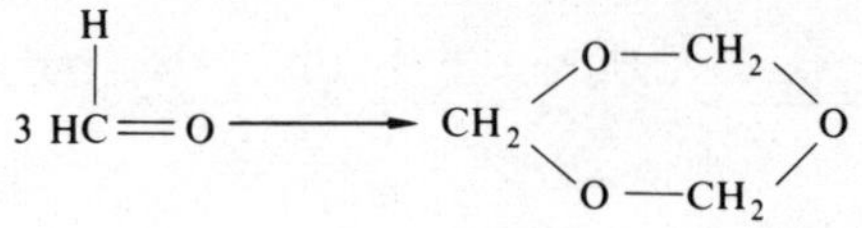

$$3\ HC{=}O \longrightarrow$$

trioxane

In recent years a solid linear polymer of formaldehyde, called 'Delrin' has been produced commercially. It is an extremely tough plastic which can be moulded and machined. It has the same poly-acetal structure as paraformaldehyde, but is not decomposed by heating. The polymer is stabilized by converting the ends of the chains to unreactive derivatives which prevent the start of the depolymerization reaction.

A different kind of polymer is formed by the action of weak alkalis, such as calcium hydroxide on formalin. A series of aldol condensations takes place, giving a syrupy mixture of sugars called acrose:

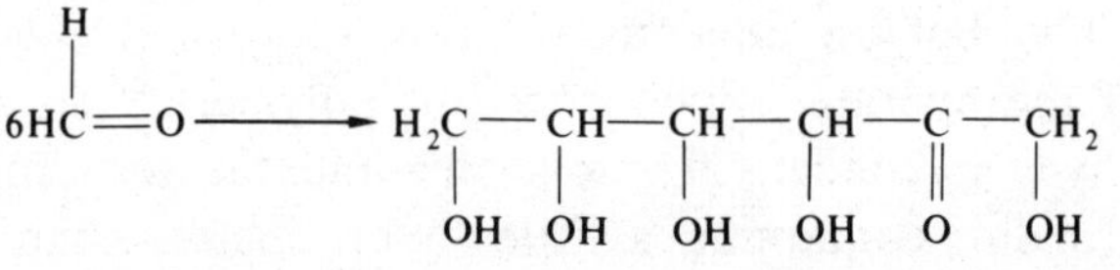

$$6HC{=}O \longrightarrow H_2C - CH - CH - CH - C - CH_2$$

acrose

Formaldehyde and ammonia do not form an aldehyde-ammonia, but give a white solid, hexamethylene tetramine or hexamine.

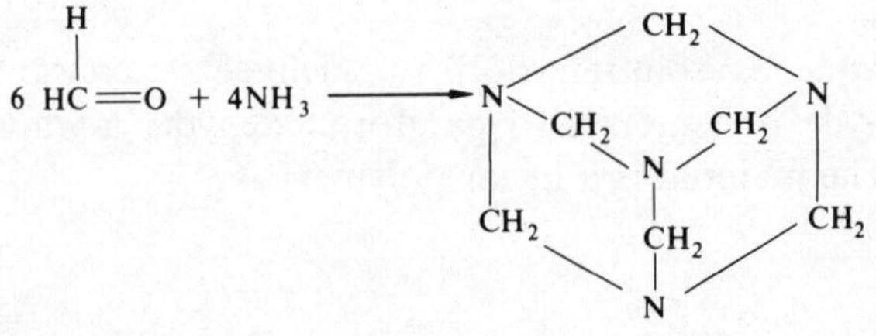

$$6\ HC{=}O + 4NH_3 \longrightarrow$$

hexamine

Hexamine is used medically as a urinary disinfectant. It burns with a quiet blue flame and is used as a source of heat in camping stoves. Nitration gives a powerful plastic explosive known as cyclonite or RDX.

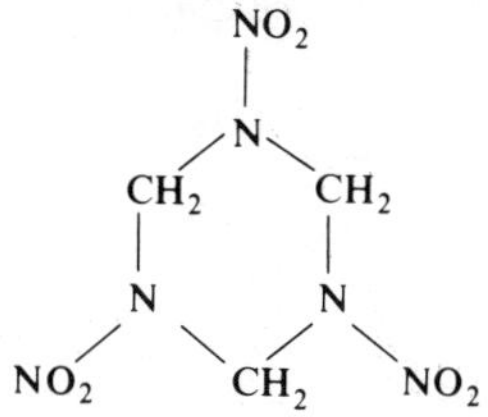

RDX

Formaldehyde and acetaldehyde react together in a combined aldol and Cannizzaro reaction to give pentaerythritol, a tetrahydroxy compound. The tetranitrate of this, known as PETN, is also a powerful explosive.

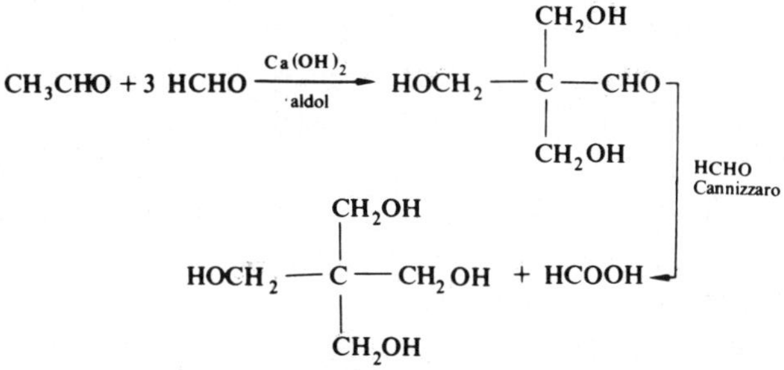

pentaerythritol

Acetaldehyde

Acetaldehyde is a low-boiling liquid (b.p. 21°) with a pungent smell, but not as unpleasant as formaldehyde. It is produced industrially by vapour phase oxidation of ethanol or by hydration of acetylene.

It also has a strong tendency to polymerize. Paraldehyde is a cyclic trimer, like trioxane. It is formed by adding a drop of concentrated sulphuric acid to acetaldehyde and can be decomposed again by distilling with a little dilute sulphuric acid. It is a neutral liquid and is used as a sedative in mental diseases.

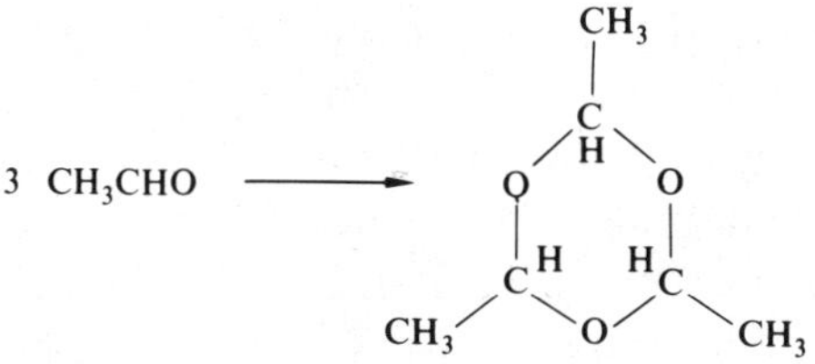

paraldehyde

167

At a lower temperature (0°C) sulphuric acid causes acetaldehyde to polymerize to a solid, metaldehyde, which is a cyclic tetramer. Like paraldehyde, it can be depolymerized with dilute acid. It is rather poisonous and is used in the garden to kill slugs.

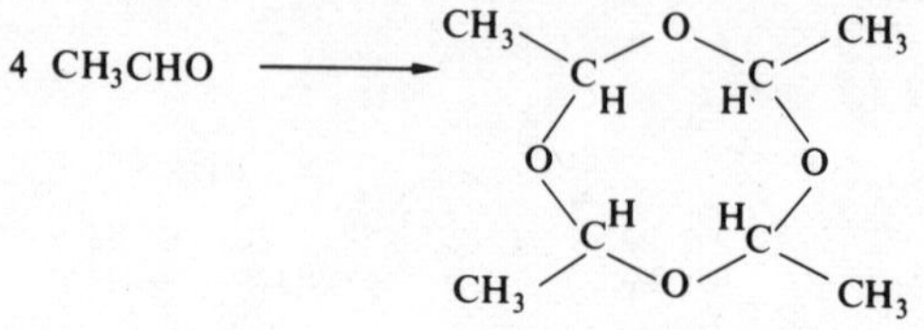

$$4 \ CH_3CHO \longrightarrow$$

metaldehyde

Acetone

Acetone is by far the most important ketone. It is a pleasant smelling liquid (b.p. 56°). It is prepared industrially from isopropanol or propylene by oxidation, from acetic acid over a calcium oxide catalyst or as a fermentation by-product. It is used chiefly as a solvent. It does not polymerize like the lower aldehydes. The polymerization of acetone with various catalysts has been reported, but so far no practical product has emerged. A stable polymer of acetone would be an exciting discovery as well as having considerable industrial possibilities.

Acetone is decomposed at 700° on a metal filament to give ketene, a very reactive compound (see page 202).

$$\overset{\text{O}}{\underset{\|}{CH_3CCH_3}} \longrightarrow CH_2{=}C{=}O + CH_4$$

ketene

Identification of aldehydes and ketones

Aldehydes and ketones are best identified by infrared spectroscopy followed by chemical tests to obtain further information. All compounds containing the C=O linkage (including carboxylic acids, esters, amides, etc.) show a strong and easily recognized absorption in the infrared region between 1800 cm^{-1} and 1600 cm^{-1} approximately. Simple saturated aldehydes have this absorption near 1725 cm^{-1}, and also a band at 2700 cm^{-1} for the stretching vibration of the C—H bond next to the carbonyl group. Aliphatic ketones have a C=O stretching absorption near 1710 cm^{-1}. The position can be affected by a number of factors, such as double bonds, hydroxyl groups or halogen atoms attached near the carbonyl group, but these positional shifts occur in predictable ways and the accurate determination of the band position can give very useful information in complex compounds.

Carbonyl groups in general can be recognized by the formation of a 2,4-dinitrophenylhydrazone (precipitated as a yellow or orange solid) when a small amount of the unknown compound is treated with 2,4-dinitrophenylhydrazine. The test is not completely reliable, for example a carbonyl group flanked by bulky alkyl groups may not react, and a few ketones form liquid 2,4-dinitrophenylhydrazones.

Alternatively, the formation of a solid derivative with phenyl-hydrazine or hydroxylamine can be used as a test. Extensive lists of the melting points of derivatives of all the common aldehydes and ketones are available, so often from the melting points of one or two such derivatives, a compound can be identified.

Provided other functional groups are absent, aldehydes are easily distinguished from ketones by their sensitivity to mild oxidizing agents, e.g. Tollen's or Fehling's test. But in the presence of other functional groups (e.g. —OH) such tests can be misleading. Other tests, such as formation of the bisulphite compounds and the iodoform test are used to give further or confirmatory evidence.

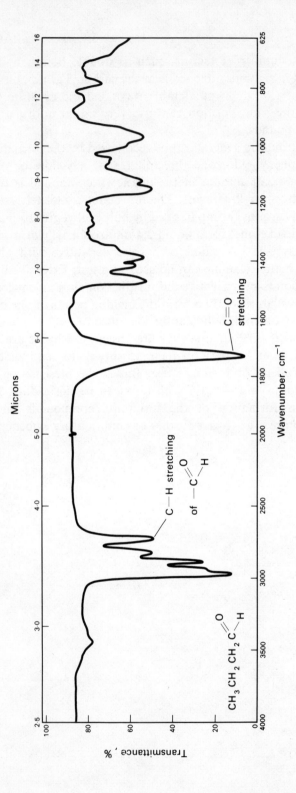

Fig. 35   Infrared spectrum of n-butyraldehyde

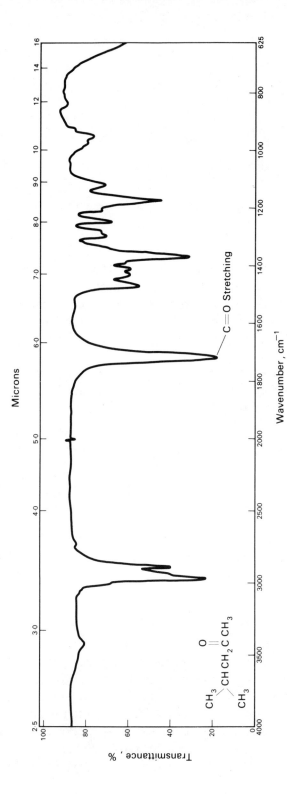

Fig. 36 Infrared spectrum of methyl isobutyl ketone

9

Carboxylic Acids and their Derivatives

The name carboxyl is derived from *carb*onyl and hydr*oxyl* because the carboxyl group contains both these groups sharing the same carbon atom. The properties of the carboxyl group are not simply those of a carbonyl and hydroxyl combined; the two groups interact in an interesting way which gives carboxylic acids their own distinctive properties.

We find that in the carboxyl group the polarity of the carbonyl group is diminished compared with aldehydes and ketones, because of electron donation from hydroxyl:

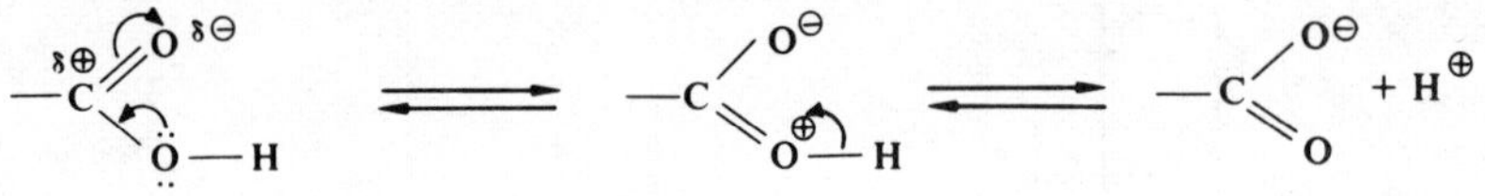

so that the carbon atom of carboxyl carries less positive charge than in aldehydes and ketones, and is therefore not so readily attacked by nucleophilic reagents. This is known as a *neutralized system*. One of us was the first to suggest that the carboxylic acids, amides and similar substances contained a neutralized system in which an electron donor and an electron acceptor co-operate according to simple rules.

The drawing away of electronic charge from the hydroxyl group leaves it more electropositive than in an alcohol. That is, the proton is more readily given up to the solvent or other proton-accepting substance; in other words a carboxyl group is a stronger acid than an alcohol. Not only is the proton given up more readily, but the carboxylate ion once formed is stabilized by spreading the charge over both oxygen atoms (cf. enolate ions, p. 162). We can write two resonance forms for the carboxylate ion, A and B; the resonance hybrid can be depicted as C, with $1\frac{1}{2}$ bonds between C and O:

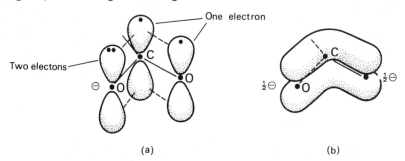

X-Ray measurements on crystals have confirmed that the C—O bond lengths in carboxylate ions are equal (e.g. $1 \cdot 27$ Å in the acetate ion). The alkoxide ion $RO^{\ominus}$ has no resonance possibilities to stabilize it, so the acid strengths of carboxylic acids are several powers of ten greater than those of alcohols.

In the language of molecular orbitals, the carbon of carboxylate is in an sp^2 hybrid state, with one p electron. It is bound to each oxygen atom by a σ bond, and there is overlap of the carbon p orbital with p orbitals of both oxygen atoms. As described on page 163 for the enolate ion, this leads to a stable three-centre four-electron bond (Fig. 37) with a negative charge.

Two electons —

One electron

Fig. 37 Formation of a three-centre four-electron bond in the carboxylate ion (a) The three p orbitals which combine to form the molecular orbital (b)

Acidic strength is usually represented by the dissociation constant K_a, which is defined for the dissociation equation:

$$RCOOH \rightleftharpoons H^{\oplus} + RCOO^{\ominus}$$

by the expression:

$$K_a = \frac{[RCOO^{\ominus}][H^{\oplus}]}{[RCOOH]}$$

where the quantities in square brackets represent the concentration of those species in the equilibrium mixture.

The greater the dissociation into ions, the greater is K_a, and the stronger is the acid. An atom or group which attracts electrons will draw away some of the negative charge on a carboxylate ion. The

173

charge will be distributed over a larger number of atoms and the ion will be stabilized. In 1926, one of us stated as one of the consequences of his electronic theory of organic reactions that groups which draw charge away from the carboxylate ions will tend to make the corresponding carboxylic acid a stronger acid. For example, substituting halogen atoms into acetic acid will make the resultant acids stronger than acetic acid:

$$CH_3-C\underset{OH}{\overset{O}{\diagdown}} \qquad Br \rightarrow CH_2 \rightarrow C\underset{OH}{\overset{O}{\diagdown}} \qquad Cl \rightarrow CH_2 \rightarrow C\underset{OH}{\overset{O}{\diagdown}} \qquad F \rightarrow CH_2 \rightarrow C\underset{OH}{\overset{O}{\diagdown}}$$

K_a 1·75 x 10^{-5} K_a 125 x 10^{-5} K_a 136 x 10^{-5} K_a 260 x 10^{-5}

Three strongly electronegative fluorine atoms make trifluoracetic acid a completely ionized acid in aqueous solution, like the strong mineral acids.

The measurement of acidic strength of carboxylic acids has been one of the most useful tools for measuring and studying the inductive effect. For example, the inductive effect can be shown to fall off rapidly with increasing distance between the electronegative group and carboxyl:

$$CH_3-CH_2-CH \leftarrow COOH \qquad\qquad CH_3-CH-CH_2-COOH$$
$$\downarrow \qquad\qquad\qquad\qquad\qquad\qquad \downarrow$$
$$Cl \qquad\qquad\qquad\qquad\qquad\qquad Cl$$

K_a 140 x 10^{-5} K_a 9·0 x 10^{-5}

$$CH_2-CH_2-CH_2-COOH \qquad\qquad CH_3-CH_2-CH_2-COOH$$
$$\downarrow$$
$$Cl \qquad\qquad\qquad\qquad\qquad\qquad K_a\ 1·5\ x\ 10^{-5}$$

K_a 3·0 x 10^{-5}

In general, the inductive effect falls to a negligible value after passing through a chain of more than three saturated carbon atoms. By comparing the strength of the following acids we find confirmation that a methyl group is a weak electron donor:

$$H-COOH \qquad\qquad CH_3 \leftarrow COOH$$

K_a 17·7 x 10^{-5} K_a 1·75 x 10^{-5}

$$CH_3 \rightarrow CH_2 \rightarrow COOH \qquad\qquad CH_3 \overset{CH_3}{\underset{CH_3}{\overset{\uparrow}{\underset{\downarrow}{\rightarrow C}}}} \leftarrow COOH$$

K_a 1·3 x 10^{-5}

K_a 0·97 x 10^{-5}

Hydrogen Bonding

Like alcohols, acids can form hydrogen bonds to water or other acid molecules. As a result, the lower members of the series are water-soluble. In general, the solubility properties of carboxylic acids resemble those of alcohols. In the pure liquid state, acids are found to exist as dimers, linked through hydrogen bonds; indeed the vapour of acetic acid is still dimeric just above its boiling point. This accounts for the surprisingly high boiling points of the carboxylic acids (Table 15).

	CH_3COOH	$CH_3CH_2CH_2OH$	$CH_3CH_2CH_2CH_3$
mol. wt.	60	60	58
b.p. °c	118	97	−0·5

If we take the effective molecular weight of dimeric acetic acid (120) its boiling point (118°C) is close to that of n-octane (mol. wt. 116, b.p. 126°).

Table 15. Normal carboxylic acids

Common name	IUPAC Name	Formula	m.p. °C	b.p. °C	K_a
Formic	Methanoic	HCOOH	8·5	100·5	$1·77 \times 10^{-4}$
Acetic	Ethanoic	CH_3COOH	17	118	$1·75 \times 10^{-5}$
Propionic	Propanoic	CH_3CH_2COOH	−22	141	$1·3 \times 10^{-5}$
Butyric	Butanoic	$CH_3(CH_2)_2COOH$	−5	163	$1·5 \times 10^{-5}$
Valeric	Pentanoic	$CH_3(CH_2)_3COOH$	−35	187	$1·6 \times 10^{-5}$
Caproic	Hexanoic	$CH_3(CH_2)_4COOH$	−1·5	205	$1·4 \times 10^{-5}$
Oenanthic	Heptanoic	$CH_3(CH_2)_5COOH$	−10	223	$1·3 \times 10^{-5}$
Caprylic	Octanoic	$CH_3(CH_2)_6COOH$	16	237	$1·4 \times 10^{-5}$
Pelargonic	Nonanoic	$CH_3(CH_2)_7COOH$	13	254	$1·1 \times 10^{-5}$
Capric	Decanoic	$CH_3(CH_2)_8COOH$	32	268	
Undecylic	Undecanoic	$CH_3(CH_2)_9COOH$	30	213*	
Lauric	Dodecanoic	$CH_3(CH_2)_{10}COOH$	48	225*	
Myristic	Tetradecanoic	$CH_3(CH_2)_{12}COOH$	58	250*	
Palmitic	Hexadecanoic	$CH_3(CH_2)_{14}COOH$	64	269*	
Stearic	Octadecanoic	$CH_3(CH_2)_{16}COOH$	69	287*	

Nomenclature

Many of the aliphatic carboxylic acids have trivial names, dating from the early days of organic chemistry, and denoting the source from which the acid was obtained. For example, formic from *Formica* (L. ants), acetic from *acetum* (L. vinegar), butyric from *butyrum* (L.

* At 100 mm pressure.

butter), caproic from *caper* (L. goat), palmitic from *Palma* (L. oil palm), stearic from *stear* (Gk. fat). As a group, they are known as fatty acids because many occur as esters in natural fats and oils.

The IUPAC names are formed by dropping the final *-e* from the name of the corresponding alkane and adding the suffix *-oic* plus *acid*. For example, the simple C_5 acid is called pentanoic acid.

Many branched chain acids also have trivial names, and many can be named as derivatives of acetic acid. The systematic names are derived from the longest chain present, with numbered and named branches, as for alkanes. The carbon atom of carboxyl is numbered 1. Some examples are given in Table 16.

The numbering of other substituents follows the rules used in Chapter 8 (Greek letters for trivial names, numbers for systematic names). Note that whereas numbering starts from the carboxyl carbon atom, the Greek letter system starts from the adjacent carbon atom:

$$\overset{5}{CH_3}-\overset{4}{CH}-\overset{3}{CH_2}-\overset{2}{CH}-\overset{1}{COOH} \quad or \quad \overset{\delta}{CH_3}-\overset{\gamma}{CH}-\overset{\beta}{CH_2}-\overset{\alpha}{CH}-COOH$$

2-chloro-4-hydroxypentanoic acid   or   α-chloro-γ-hydroxyvaleric acid

Preparation of Acids

1. *Oxidation of a primary alcohol or aldehyde of the same number of carbon atoms.* The usual reagent is sodium or potassium dichromate in sulphuric acid. The acid is very resistant to further oxidation. Industrially, air oxidation is employed:

$$3 RCH_2OH + 2Na_2Cr_2O_7 + 8H_2SO_4 \rightarrow RCOOH + 2Na_2SO_4 + 2Cr_2(SO_4)_3 + 11H_2O$$

Oxidation of alkenes also gives acids. The oxidation of oleic acid (an unsaturated acid) with permanganate or dichromate gives two acids, a mono- and a di-carboxylic acid:

$$CH_3(CH_2)_7 CH=CH(CH_2)_7COOH \xrightarrow{KMnO_4} CH_3(CH_2)_7COOH + HOOC(CH_2)_7COOH$$

oleic acid   nonanoic acid   azelaic acid or nonanedoic acid

2. *Hydrolysis of Cyanides.* Alkyl cyanides, which can be prepared from alkyl halides (page 35) are hydrolysed to acids by either strong acid or base. Because of this relationship to acids, alkyl cyanides are also called acid nitriles. This sequence produces an acid with one more carbon atom than the starting halide.

$$CH_3Br \xrightarrow{KCN} CH_3CN \xrightarrow[H^{\oplus}]{H_2O} CH_3COOH + NH_3$$

methyl cyanide or acetonitrile

Table 16.   Branched carboxylic acids

Formula	Trivial name	Systematic name	b.p. (°C)
CH$_3$ $\diagdown$ CHCOOH $\diagup$ CH$_3$	isobutyric *or* dimethylacetic	2-methylpropanoic	154
CH$_3$ ∣ CH$_3$CCOOH ∣ CH$_3$	pivalic *or* trimethylacetic	2,2-dimethylpropanoic	164
CH$_3$ ∣ CH$_3$CHCH$_2$COOH	isovaleric *or* isopropylacetic	3-methylbutanoic	175
CH$_3$ ∣ CH$_3$CH$_2$CHCOOH	α-methylbutyric *or* methylethylacetic	2-methylbutanoic	175

3. Grignard Reaction. Another way of producing an acid from an alkyl halide with one less carbon atom is first to make the Grignard reagent and to treat this with solid carbon dioxide.

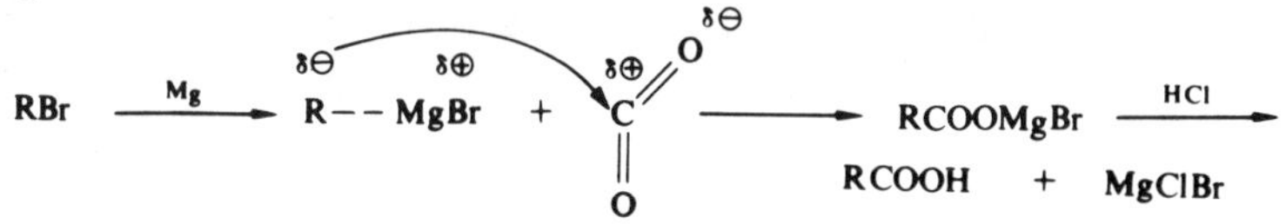

$$RBr \xrightarrow{\text{Mg}} R \text{--} MgBr + \overset{\delta\oplus}{C}\overset{\delta\ominus}{\diagup}\overset{O}{\diagdown} \longrightarrow RCOOMgBr \xrightarrow{\text{HCl}}$$
$$RCOOH + MgClBr$$

A Grignard reagent will normally react with a carboxylic acid; how-ever, at the low temperature of the reaction (about —80°C) and by pouring the Grignard reagent onto a large amount of carbon dioxide, reaction between reagent and product is avoided. The method is not used for low molecular weight acids, which are readily available in other ways, but it is very useful for less common acids, and for secondary or tertiary halides, it works better than the nitrile method, because potassium cyanide is a rather alkaline re-agent, and tends to cause alkene formation; for example, t-butyl chloride and potassium cyanide give chiefly isobutene.

Aliphatic and Alicyclic Compounds

4. *Hydrolysis of Esters.* Many acids occur in nature as their esters of glycerol or other alcohols. The free acid can be obtained by boiling the ester with a dilute acid or alkali:

$$\text{RCOOR}' + \text{H}_2\text{O} \xrightarrow{\text{H}^{\oplus} \text{ or } \text{H}^{\ominus}} \text{RCOOH} + \text{R}'\text{OH}$$

$$
\begin{array}{ccc}
\text{CH}_2\text{—OCOR} & & \text{CH}_2\text{OH} \\
| & & | \\
\text{CH—OCOR} \quad + \quad 3\text{H}_2\text{O} \longrightarrow & & \text{CHOH} \quad + \quad 3\text{RCOOH} \\
| & & | \\
\text{CH}_2\text{—OCOR} & & \text{CH}_2\text{OH}
\end{array}
$$

a triglyceride   glycerol

5. *Malonic Ester Synthesis.* An alkyl halide will react with the sodium derivative of diethyl malonate to give a substituted malonic ester. This can be hydrolysed and decarboxylated to give an acid with two more carbon atoms than the starting halide:

$$\text{R—X} + \text{Na}^{\oplus} \; ^{\ominus}\text{CH} \overset{\text{COOEt}}{\underset{\text{COOEt}}{\big\langle}} \longrightarrow \text{R—CH} \overset{\text{COOEt}}{\underset{\text{COOEt}}{\big\langle}} \xrightarrow[\text{H}_2\text{O}]{\text{H}^{\oplus}} \text{RCH} \overset{\text{COOH}}{\underset{\text{COOH}}{\big\langle}}$$

$$\downarrow \text{heat}$$

$$\text{RCH}_2\text{COOH} + \text{CO}_2$$

6. *Carbonylation.* The reaction of carbon monoxide with an alcohol under a pressure of 500 atmospheres and using a boron trifluoride catalyst gives acids. A newer variation of this method is to heat an alkene with carbon monoxide and steam under pressure with phosphoric acid. In both these reactions, rearrangement of the alkyl group can occur, so a mixture of isomeric branched chain acids is produced:

$$\text{ROH} + \text{CO} \xrightarrow[125\text{ - }180°]{\text{BF}_3} \text{R}'\text{COOH}$$

$$\text{RCH}{=}\text{CH} + \text{CO} + \text{H}_2\text{O} \xrightarrow[300-400°]{\text{H}_3\text{PO}_4} \text{R}'\text{COOH}$$

Sodium formate is prepared by the carbonylation reaction with sodium hydroxide:

$$\text{NaOH} + \text{CO} \xrightarrow{200°} \text{HCOONa}$$

178

Two other special methods of preparation of acids already encountered are the haloform reaction from methyl ketones (page 161) and the Canizzaro reaction from aldehydes that are unable to undergo aldol condensation (page 164).

Physical Properties of Acids

The lower members of the acid series are bitter-tasting water-soluble liquids. With increasing molecular weight (from butyric acid upward) they become sparingly soluble in water and more oily in appearance. The C_3—C_{10} acids have very unpleasant odours (butyric acid is the cause of the odour of rancid butter). With still higher molecular weight they are waxy solids, insoluble in water and odourless.

The melting points of the acids plotted against molecular weight lie on two smooth curves, one for the acids with an even number of carbon atoms and one for the odd-numbered acids. This is attributed to the way the molecules pack together in the crystal. The molecules with an even number of carbon atoms pack more closely and therefore have higher melting points.

Reactions of Acids

The reactions of acids fall into two groups, first, reactions that lead to the alteration of the carboxyl group, and secondly, reactions of the methylene group next to the carboxyl group.

1. *Salts.* Reactive metals dissolve in the acids, liberating hydrogen and forming the metal salts. The salts are also formed by reaction of the acids with hydroxides, and, because carboxylic acids are stronger acids than carbonic acid, with carbonates and bicarbonates.

$$RCOOH + Na \longrightarrow RCOO^{\ominus} \ Na^{\oplus} + \tfrac{1}{2}H_2$$

$$RCOOH + NaOH \longrightarrow RCOO^{\ominus} Na^{\oplus} + H_2O$$

$$RCOOH + NaHCO_3 \longrightarrow RCOO^{\ominus} Na^{\oplus} + CO_2 + H_2O$$

The third reaction is a simple and convenient test for carboxylic acids, since only they (and sulphonic acids, Chapter 10) of the common functional groups will dissolve in bicarbonate solution liberating carbon dioxide.

As a general rule the alkali metal salts of carboxylic acids are soluble in water, but the calcium and barium salts are insoluble.

2. *Formation of Acyl Derivatives.* That is, formation of esters, acid halides, acid anhydrides and amides, where the OH group of the

acid is replaced by OR′, X, OOCR′, and NR₂′ respectively. The RCO radical common to these acid derivatives is known as an *acyl* group.

(a) Esters. An acid and an alcohol react together very slowly to form an ester. The reaction comes slowly to equilibrium, and in the case of ethanol and acetic acid, the equilibrium mixture consists of 66% ester and water and 33% acid and alcohol:

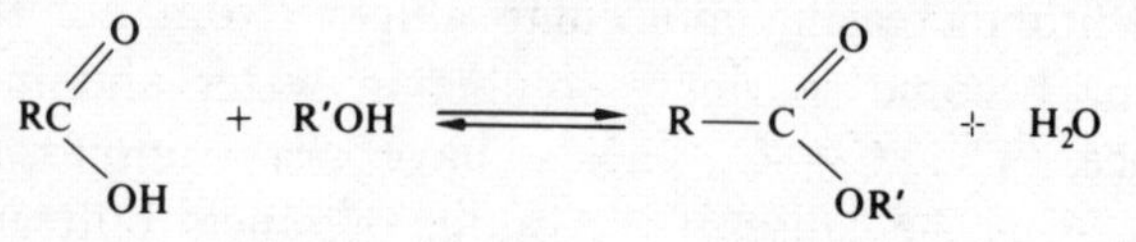

$$
RC\overset{O}{\underset{OH}{\diagup}} + R'OH \rightleftharpoons R-C\overset{O}{\underset{OR'}{\diagup}} + H_2O
$$

The rate of the reaction is found to depend on the concentration of both acid and alcohol.

In order to discover whether the C—O bond of the acid or alcohol is broken in forming the ester, experiments were carried out with alcohols containing ^{18}O in the hydroxyl group, in an early example (1938) of the use of isotopic labelling to study organic reactions. It was found that when methanol labelled with ^{18}O in the hydroxyl group was converted to an ester, the heavy isotope was found in the ester. Therefore the C—O bond of the acid is broken:

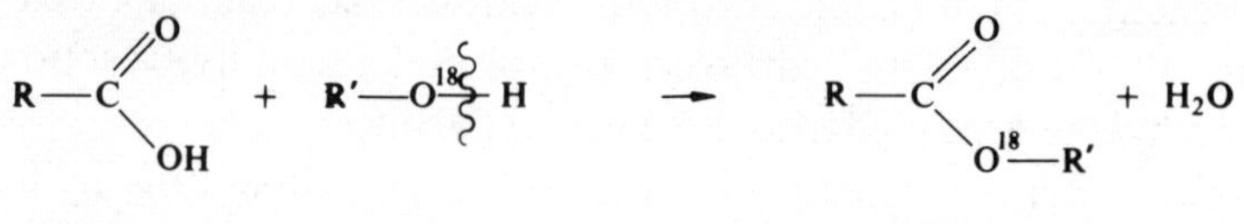

$$
R-C\overset{O}{\underset{OH}{\diagup}} + R'-O^{18}-H \longrightarrow R-C\overset{O}{\underset{O^{18}-R'}{\diagup}} + H_2O
$$

The kinetics and mechanism of the esterification reaction have been studied in great detail. The first stage is the attack of alcohol (a nucleophile) on the carboxyl carbon atom (just like the attack of a nucleophile on an aldehyde or ketone) but here, the positive nature of the carbon atom is much weaker than in ketones, which accounts for the reaction being a very slow one:

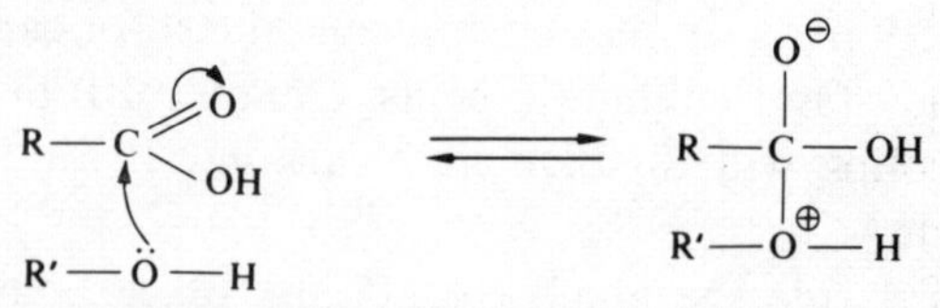

$$
R-C\overset{O}{\underset{OH}{\diagup}} \quad \rightleftharpoons \quad R-C\overset{O^{\ominus}}{\underset{O^{\oplus}-H}{\mid}}-OH
$$
$$
R'-\overset{..}{O}-H \qquad\qquad R'-O^{\oplus}-H
$$

By a rearrangement of protons we can eliminate a molecule of water and form an ester:

180

$$R-C \overset{O^{\ominus}}{\underset{OH}{|}} \quad R'-O \quad H \rightleftharpoons R-C \overset{O}{\underset{R'O}{||}} + H_2O$$

The net result is the replacement of —OH by —OR′ from the alcohol. All of the stages are reversible and we can see that attack of H_2O on the ester can lead back to the acid and alcohol:

$$R-C \overset{O}{\underset{H-\ddot{O}-H}{|}} OR' \rightleftharpoons R-C \overset{O^{\ominus}}{\underset{H-O-H}{|}} OR' \rightleftharpoons R-C \overset{H}{\underset{H-O}{|}} OR' \rightleftharpoons R-C \overset{O}{\underset{HO}{||}} + R'OH$$

The rate of formation of an ester is increased by adding a strong acid. Protons from the strong acid protonate the oxygen of the carboxyl group, increasing the electropositive nature of the carbon atom and making attack by the nucleophilic alcohol easier:

$$RC \overset{O}{\underset{OH}{\diagdown}} \rightleftharpoons RC \overset{OH}{\underset{OH}{\diagdown}} \rightleftharpoons RC \overset{OH}{\underset{R'-O-H}{-}}OH \rightleftharpoons RC \overset{O-H}{\underset{R'-O \quad H}{-}}OH \rightleftharpoons RC \overset{OH}{\underset{R'O}{\diagup}} + H_2O$$

$$RC \overset{O}{\underset{OR'}{\diagdown}} + H^{\oplus}$$

The reaction can thus be broken down into five stages (i) the addition of $H^{\oplus}$ to the acid, (ii) the addition of alcohol to the electropositive carbon atom to form a tetrahedral intermediate (this is the most difficult step, and therefore governs the rate of the overall reaction), (iii) rearrangement of protons, (iv) loss of water to reform the less crowded trigonal carboxyl grouping and (v) removal of a proton. Only a catalytic amount of strong acid is needed, since for each proton added in the initial stage one is regenerated in the final stage.

Since all the stages are reversible, it is evident that an ester can also be hydrolysed to acid and alcohol with acid catalysts.

In formation of esters, the reaction is driven forward usually by using a large excess of one of the reagents (often the alcohol is the cheaper), and by removing water, for example by distilling it out as it is formed. Sulphuric acid acts both as an acid catalyst and a remover of water (by the formation of hydrates).

Primary and secondary alcohols react in this way but when tertiary alcohols are labelled with ^{18}O, the isotopic label is not found

181

in the ester. This is because the tertiary alcohol is a bulky group and the hydroxyl cannot as easily approach the carboxyl group; the inductive effects of three alkyl groups loosen the C—O bond, and a carbonium ion is formed:

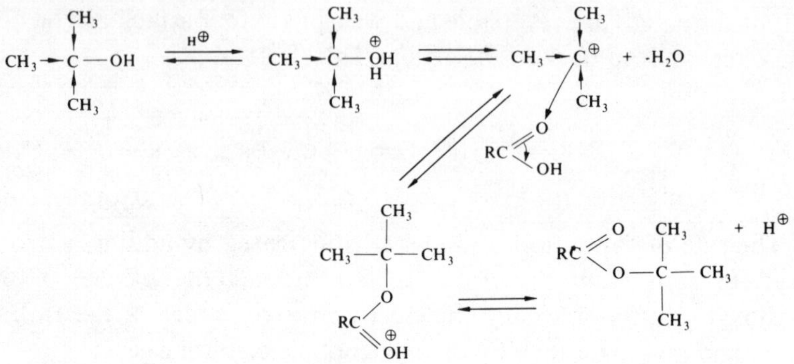

Sulphuric acid is not suitable as a catalyst for esterifying tertiary alcohols because it causes dehydration to an alkene; dry hydrogen chloride gas (about 1% by weight) dissolved in the alcohol is more suitable (this is sometimes known as the Fisher–Speier esterification method).

Not only proton acids, but Lewis acids such as boron trifluoride can catalyse esterification reactions. A solution of boron trifluoride in methanol can give good yields of methyl esters after only two minutes heating at 100° with a carboxylic acid. In this case the initial step is the formation of a complex between acid and catalyst:

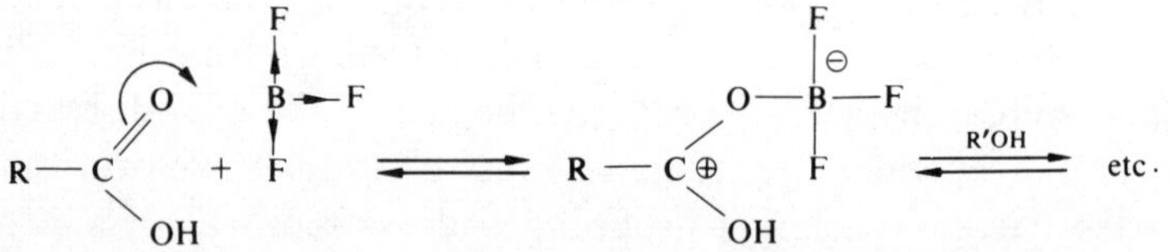

More evidence that attack of the alcohol at the carbon atom of carboxyl is a necessary part of the reaction is deduced from the observation that carboxylic acids with large and bulky groups near the acid group are very difficult to convert to esters. For example pivalic acid gives very low yields of ester by this method:

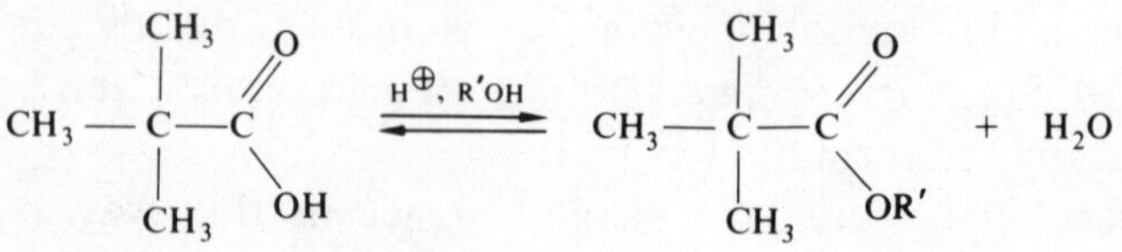

pivalic acid

(b) Acid Halides. The hydroxyl of an acid can be replaced by chloride by the action of thionyl chloride, phosphorus trichloride or phosphorus pentachloride:

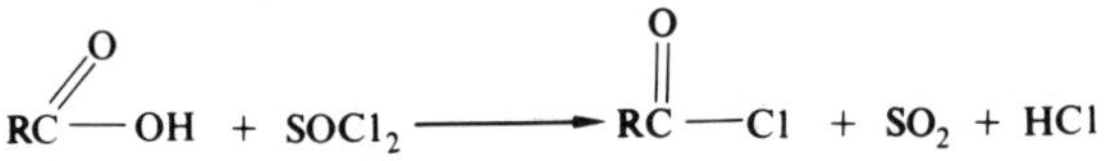

$$RC\overset{O}{\diagup}OH\ +\ SOCl_2\ \longrightarrow\ RC\overset{O}{\diagup}Cl\ +\ SO_2\ +\ HCl$$

We do not know the detailed mechanism of this reaction, but the scheme as follows is probable:

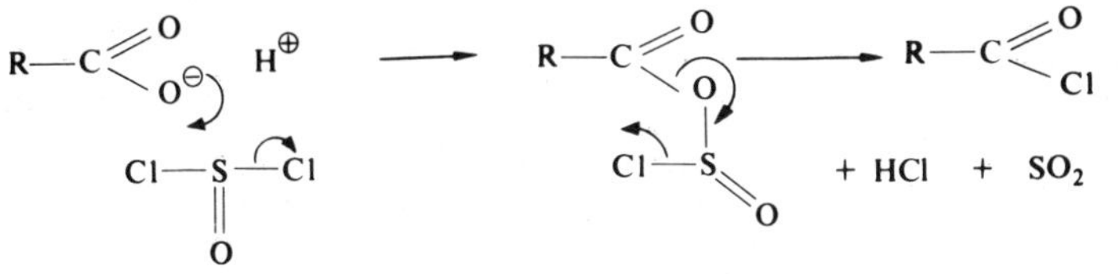

Acid fluorides, bromides and iodides can also be prepared but for ease of preparation and high reactivity, acid chlorides are generally preferred.

(c) Acid Anhydrides. The loss of a molecule of water between two molecules of an acid (e.g. using phosphorus pentoxide) gives an acid anhydride. Anhydrides can also be obtained conveniently from an acid chloride and the salt of an acid. The reaction occurs by attack of the anion of the acid on the acid chloride, similar to the attack of alcohol on the acid in the formation of an ester:

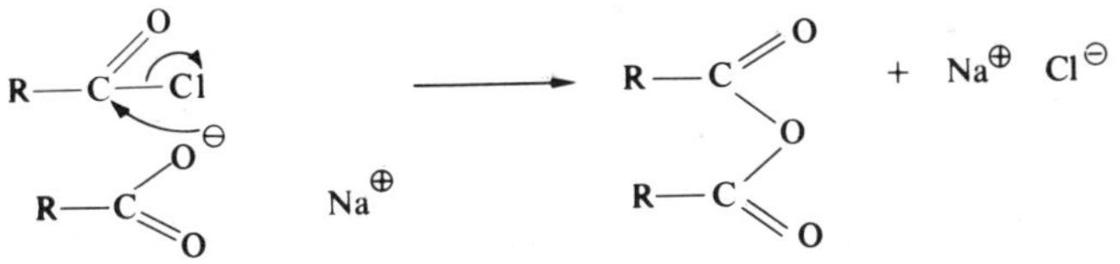

(d) Amides. An amide is derived from an acid by replacing the hydroxyl group by an amine group. Amides can be formed by heating the ammonium salt of an acid:

$$R-C\overset{O}{\diagdown}O^{\ominus}\quad NH_4^{\oplus}\ \longrightarrow\ RC\overset{O}{\diagup}NH_2\ +\ H_2O$$

They can also be formed by treating an ester or an acid chloride with ammonia (to give a simple amide) or an amine (to give a substituted amide). Both of these reactions resemble the formation of esters and anhydrides, in that they involve attack of a nucleophile (here the amine or ammonia) on the electropositive carbon atom and loss of a group (alcohol or chloride ion):

183

$$RC\overset{O}{\underset{\overset{\cdot\cdot}{N}H_3}{\diagdown}}\!\!\!\diagup^{O}_{OR} \longrightarrow RC\overset{O^{\ominus}}{\underset{\overset{\oplus}{N}H_3}{\diagdown}}\!\!-OR' \longrightarrow RC\overset{O^{\ominus}}{\underset{NH_2}{\diagdown}}\!\!\overset{\oplus}{\underset{H}{-OR'}} \longrightarrow RC\overset{O}{\underset{NH_2}{\diagdown}} + R'OH$$

$$RC\overset{O}{\underset{\overset{\cdot}{N}H_2R'}{\diagdown}}\!\!\!\diagup^{O}_{Cl} \longrightarrow RC\overset{O}{\underset{\overset{\oplus}{N}R'}{\diagdown}}\!\!\overset{H}{\underset{H}{}} \quad Cl^{\ominus}\!\!\longrightarrow RC\overset{O}{\underset{NHR'}{\diagdown}} + HCl$$

3. *Reduction*. The carboxylic acid group is comparatively resistant to reduction. High pressure hydrogenation with metal catalysts can produce primary alcohols. This method is used commercially to obtain high molecular weight alcohols from the acids of natural fats and oils:

$$RCOOH \xrightarrow[\text{cat.}]{H_2} RCH_2OH + H_2O$$

Lithium aluminium hydride reduces acids smoothly at room temperature to primary alcohols. The first step is formation of a mole of hydrogen and a complex salt of the acid and the aluminium hydride ion:

$$RC\overset{O}{\underset{OH}{\diagdown}} + Li^{\oplus} \quad H\!-\!\underset{\underset{H}{|}}{\overset{\overset{H}{|}}{Al}}^{\ominus}\!\!-H \longrightarrow RC\overset{O}{\underset{O-\underset{H}{\overset{H}{Al}}^{\ominus}\!\!-H}{\diagdown}} \quad Li^{\oplus} + H_2$$

This is followed by successive transfer of $H^{\ominus}$ to the carboxylate ion (the details are not known) until a complex aluminium salt of the alcóhol is formed, which is decomposed by aqueous hydrochloric acid. The balanced equation is as follows:

$$4RCOOH + 3LiAlH_4 \longrightarrow (RCH_2O)_4Al^{\ominus}\ Li^{\oplus} + 4H_2 + 2LiAlO_2$$
$$\downarrow HCl$$
$$\longrightarrow 4RCH_2OH + AlCl_3 + LiCl$$

Acid chlorides (via the Rosenmund method) and esters are more easily reduced than the acids.

4. *Decarboxylation*. This may be accomplished in several ways. If a sodium or potassium salt of a carboxylic acid in aqueous solution

is electrolysed, a paraffin of higher molecular weight is produced. (Kolbe electrolysis, page 12):

$$2RCOO^{\ominus} \; Na^{\oplus} + 2H_2O \longrightarrow R—R + 2CO_2 + 2KOH + H_2$$

The essential steps of the reaction are the loss of an electron from the carboxylate ion to give a carboxyl free radical, which then loses carbon dioxide to give an alkyl free radical, two of which combine to give an alkane:

$$RC\begin{smallmatrix}O\\O^{\ominus}\end{smallmatrix} \xrightarrow{-e} RC\begin{smallmatrix}O\\O\cdot\end{smallmatrix} \longrightarrow R\cdot + CO_2$$

$$\downarrow \cdot R$$

$$R—R$$

Heating an alkali salt with soda lime gives an alkane and a number of by-products. Heating the calcium or barium salt produces a ketone (page 147) and if the silver salt of an acid is treated with bromine, in an inert solvent, an alkyl bromide is formed (the Hunsdiecker reaction):

$$RCOOAg + Br_2 \longrightarrow AgBr + RBr + CO_2$$

5. *Halogenation.* The α-CH_2 group of carboxylic acids can be halogenated in much the same way that aldehydes and ketones are substituted by halogen. As pointed out at the beginning of this chapter, the activating effect of the carbonyl group is largely neutralized by the hydroxyl part in carboxylic acids, so that the amount of enol form present is very small, and substitution takes place very slowly:

$$R—CH_2—C\begin{smallmatrix}O\\OH\end{smallmatrix} \rightleftharpoons R—CH{=}C\begin{smallmatrix}OH\\OH\end{smallmatrix} \xrightarrow{Br_2} RCH\underset{Br}{|}—C\begin{smallmatrix}O\\OH\end{smallmatrix} + HBr$$

By adding a little red phosphorus the reaction proceeds much more rapidly and mono- or di-halo-acids (or in the case of acetic acid, trihalo-acetic acids) can be formed. Apparently a little phosphorus trihalide is first formed, which converts some of the acid into the acid halide; this enolizes much more readily than the acid, because the halide atom is electronegative. For example with bromine and acetic acid:

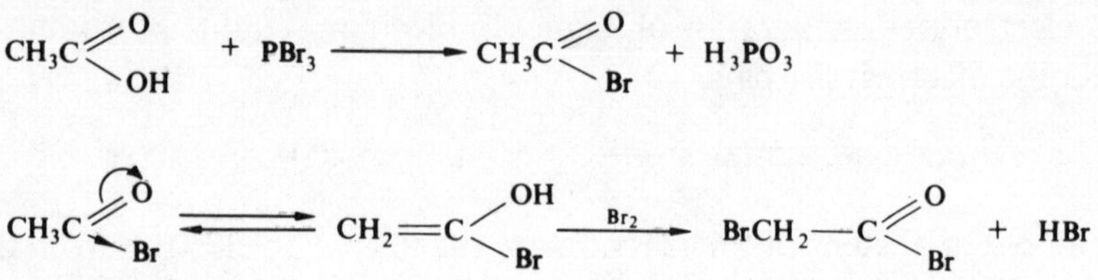

$$CH_3C\!\!\diagdown^{\displaystyle O}_{\displaystyle OH} \;+\; PBr_3 \longrightarrow CH_3C\!\!\diagdown^{\displaystyle O}_{\displaystyle Br} \;+\; H_3PO_3$$

$$CH_3C\!\!\diagdown^{\displaystyle O}_{\displaystyle Br} \rightleftharpoons CH_2\!\!=\!\!C\!\!\diagdown^{\displaystyle OH}_{\displaystyle Br} \xrightarrow{Br_2} BrCH_2\!\!-\!\!C\!\!\diagdown^{\displaystyle O}_{\displaystyle Br} \;+\; HBr$$

The bromo-acid bromide then can exchange with another molecule of acid:

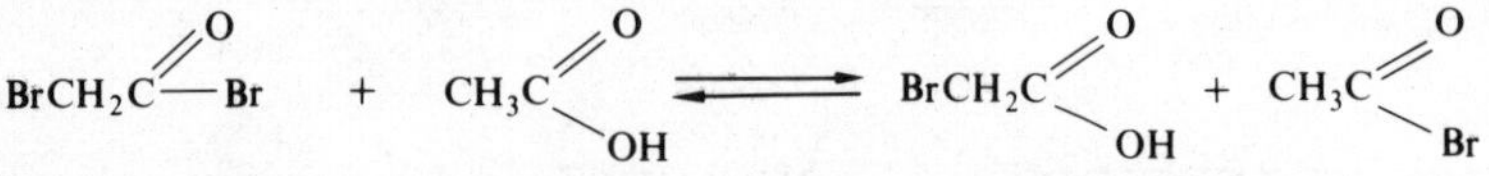

$$BrCH_2C\!\!-\!\!Br \;+\; CH_3C\!\!\diagdown^{\displaystyle O}_{\displaystyle OH} \rightleftharpoons BrCH_2C\!\!\diagdown^{\displaystyle O}_{\displaystyle OH} \;+\; CH_3C\!\!\diagdown^{\displaystyle O}_{\displaystyle Br}$$

If sufficient halogen is added, all available α-hydrogen atoms can be replaced by halogen:

$$CH_3COOH \xrightarrow[P]{Cl_2} ClCH_2COOH \xrightarrow{Cl_2} Cl_2CHCOOH \xrightarrow{Cl_2} Cl_3CCOOH$$

| chloroacetic acid | dichloroacetic acid | trichloroacetic acid |

But only α-hydrogen atoms are replaceable:

$$CH_3CH_2CH_2COOH \xrightarrow[P]{\text{excess } Br_2} CH_3CH_2CBr_2COOH$$

The reaction is known by the joint names of Hell, Volhard and Zelinsky, each of whom discovered this reaction independently between 1881 and 1887.

Formic Acid

Because it has no alkyl groups, formic acid stands apart from other acids in its behaviour. The name comes from the red ant (*Formica rufa*) from which the acid was obtained by distillation in the seventeenth century. Formic acid also occurs in stinging nettles and, in small quantities, in numerous living tissues.

Industrially, formic acid is produced by air oxidation of methane, methanol or formaldehyde, and by synthesis from carbon monoxide. Producer gas ($CO+N_2$) is mixed with sodium hydroxide under pressure, or synthesis gas ($CO+H_2O$) is used with sodium ethoxide catalyst:

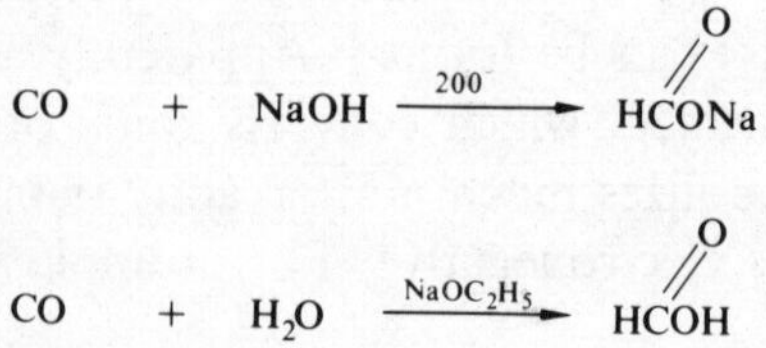

$$CO \;+\; NaOH \xrightarrow{200} HCONa$$

$$CO \;+\; H_2O \xrightarrow{NaOC_2H_5} HCOH$$

186

In addition to the general methods of preparing acids already described, formic acid can be made by heating oxalic acid. The yield is better if the oxalic acid is mixed with glycerol; an intermediate oxalic ester is formed which decomposes smoothly:

$$\underset{\text{oxalic acid}}{HOC - COH} \longrightarrow HCOOH + CO_2$$

Pure formic acid is a colourless liquid, strongly corrosive, and with a pungent smell. Its boiling point ($100 \cdot 5°$) is so close to that of water that it cannot be obtained anhydrous by distillation. The anhydrous acid is obtained by mixing concentrated sulphuric acid and sodium formate and distilling out the formic acid. The conditions must be maintained so that formic acid is always in excess, to dilute the sulphuric acid and prevent decomposition (see below).

Formic acid is the only simple carboxylic acid that is susceptible to oxidation. It reduces Tollen's reagent (in this respect it resembles the aldehydes) and many other mild oxidizing agents.

$$HCOOH \xrightarrow{[O]} H_2O + CO_2$$

It is easily decomposed in a number of ways:

(a) by heating under pressure at $160°$:

$$HCOOH \longrightarrow H_2 + CO_2$$

(b) by heating with concentrated sulphuric acid. This is used as a convenient source of carbon monoxide:

$$HCOOH \xrightarrow{H_2SO_4} H_2O + CO$$

(c) by heating formate salts with alkali:

$$HCOONa + NaOH \longrightarrow Na_2CO_3 + H_2O$$

(d) by heating the calcium salt, formaldehyde is formed:

$$(HCOO)_2Ca \longrightarrow HCHO + CaCO_3$$

Heating the alkali metal salt of formic acid produces an alkali metal oxalate:

$$\text{HCOONa} \longrightarrow \underset{\text{sodium oxalate}}{\text{NaO}\overset{\text{O}}{\overset{\|}{\text{C}}}-\overset{\text{O}}{\overset{\|}{\text{C}}}\text{ONa}} + \text{H}_2$$

Formic esters are readily formed by heating the acid (without a catalyst, because, as a strong acid, its esterification is self-catalysed) with alcohols. Formic anhydride and chloride are unknown. (A mixture of carbon monoxide and hydrogen chloride behaves like formyl chloride in many reactions however.)

The acid is used as a sterilizer, like formaldehyde, and in many industrial processes requiring a strong acid, where its cheapness gives it an advantage over sulphuric or other inorganic acids.

Acetic Acid

Acetic acid in aqueous solution, as sour wine or vinegar, has been known since ancient times. It occurs widely in nature in various forms, and many micro-organisms are able to produce the free acid by the destruction of organic material. Industrially, it is prepared by the oxidation of acetaldehyde (from acetylene) with air, by the reaction of methanol and carbon monoxide at high temperature and pressure, and by fermentation, from wine, malt or sawdust and other wood waste. When pure it is a pungent, corrosive liquid; in cold weather it solidifies to a clear, glassy mass (m.p. 16·6°), so anhydrous acetic acid is often referred to as *glacial* acetic acid. It is miscible with water. It is stable to oxidizing agents, and can be used as a solvent for oxidation reactions (e.g. with chromium trioxide). It is used industrially as a solvent, in the preparation of solvent esters and flavours and in plastics (e.g. vinyl acetate) but the greatest proportion of it is converted to acetic anhydride for the production of cellulose acetate from wood.

Preparation of Esters

1. The preparation of esters from carboxylic acids and alcohols has already been described. The method works best for primary alcohols and unhindered acids.

2. *From Acid Chlorides or Anhydrides.* Acid chlorides and anhydrides react more rapidly than carboxylic acids with alcohols to give esters. The method works well for primary, secondary and tertiary alcohols.

A protonated hydroxyl group is lost in the penultimate stage* of acid-catalysed esterification:

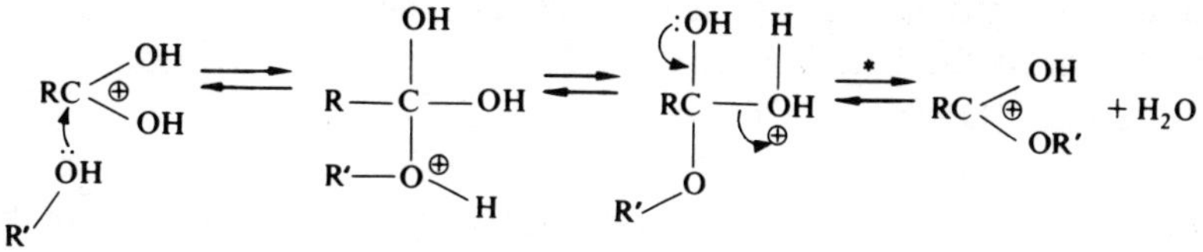

$$RC\underset{OH}{\overset{OH}{<}} \rightleftharpoons R-\underset{\underset{H}{\overset{|}{O^{\oplus}}}}{\overset{OH}{\underset{|}{C}}}-OH \rightleftharpoons RC\underset{\underset{R'}{O}}{\overset{:OH\ H}{\underset{\oplus}{-OH}}} \rightarrow RC\underset{OR'}{\overset{OH}{<}} + H_2O$$

In reaction of an acid chloride with alcohol, firstly the carbon atom is more electronegative because of the withdrawal effect of the chlorine atom, and secondly, the chlorine atom is lost without further rearrangements, as a chloride ion:

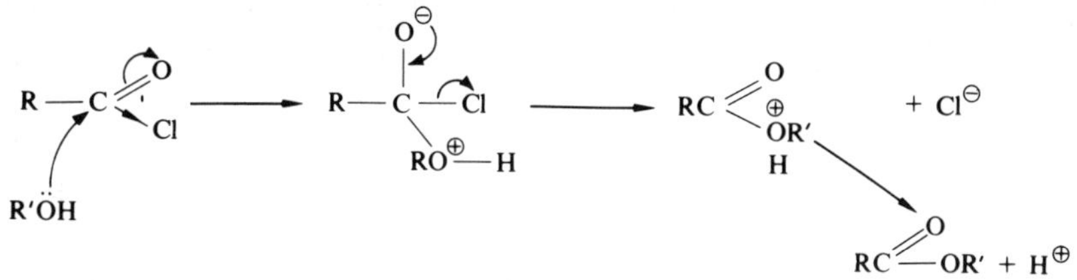

$$R-C\underset{Cl}{\overset{O}{<}} \longrightarrow R-\underset{RO^{\oplus}-H}{\overset{O^{\ominus}}{\underset{|}{C}-Cl}} \longrightarrow RC\underset{\underset{H}{OR'}}{\overset{O}{<}}_{\oplus} + Cl^{\ominus}$$

(R'ÖH)

$$\searrow RC\overset{O}{\diagup}-OR' + H^{\oplus}$$

The stability of the chloride ion makes it a good *leaving group* and helps to make the reaction energetically favourable. The reaction is not reversible.

The reaction of an anhydride is very similar, here the leaving group is a carboxylate ion, also relatively stable:

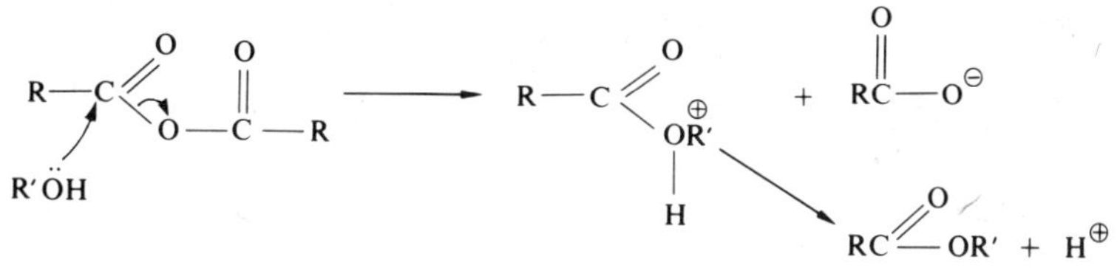

$$R-C\underset{O-C-R}{\overset{O\ \ \ O}{<}} \longrightarrow R-C\underset{\underset{H}{OR'}}{\overset{O}{<}}_{\oplus} + RC-O^{\ominus}$$

(R'ÖH)

$$\searrow RC\overset{O}{\diagup}-OR' + H^{\oplus}$$

Whereas the general reaction of a nucleophile with a ketone is one of addition:

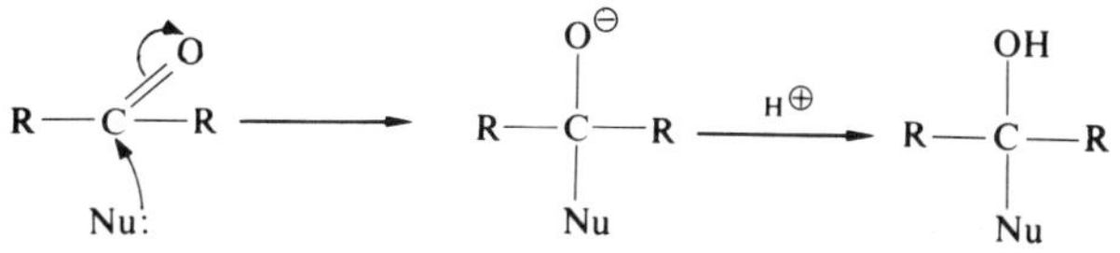

$$R-\underset{Nu:}{\overset{O}{\underset{|}{C}-R}} \longrightarrow R-\underset{Nu}{\overset{O^{\ominus}}{\underset{|}{C}-R}} \overset{H^{\oplus}}{\longrightarrow} R-\underset{Nu}{\overset{OH}{\underset{|}{C}-R}}$$

At a carboxylic acid derivative, it is one of displacement:

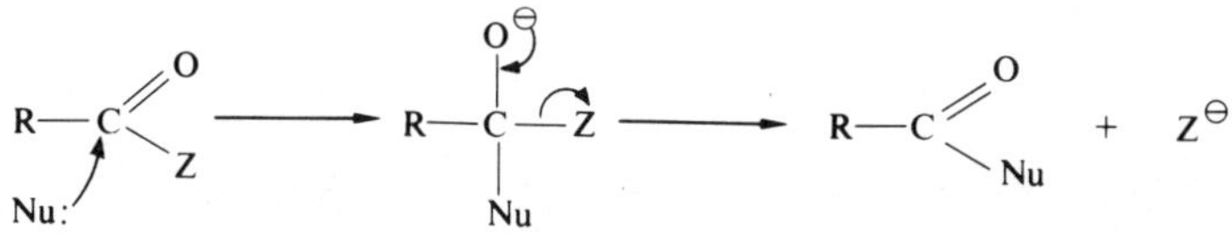

$$R-C\underset{\underset{Nu:}{Z}}{\overset{O}{<}} \longrightarrow R-\underset{Nu}{\overset{O^{\ominus}}{\underset{|}{C}-Z}} \longrightarrow R-C\underset{Nu}{\overset{O}{<}} + Z^{\ominus}$$

where Z is the leaving group; if we compare acids (Z=OH), anhydrides (RCOO), and acid chlorides (Cl), we can appreciate that the greater the stability of the leaving group, the more rapid and complete will be the reaction:

Stability: $OH^{\ominus} < RCOO^{\ominus} < Cl^{\ominus}$

3. *Carboxylate Salts and Alkyl Halides.* Primary halides that readily undergo S_N2 reactions can be converted to esters in high yields in this way:

$$RCOO^{\ominus} \ Na^{\oplus} + R'CH_2I \longrightarrow RCOOCH_2R' + NaI$$

Variations on this method are the use of a silver or triethylamine salt of the acid:

$$RCOOAg + C_2H_5I \longrightarrow RCOOC_2H_5 + AgI$$
$$RCOO^{\ominus}Et_3N^{\oplus}H + C_4H_9Br \longrightarrow RCOOC_4H_9 + Et_3N^{\oplus}HBr^{\ominus}$$

4. *Transesterification (Alcoholysis).* The ester of one alcohol may often be converted to another by refluxing it with an excess of the second alcohol containing a little dissolved sodium, or a strong acid as catalyst. If a low molecular weight alcohol is being replaced by one of higher molecular weight, distilling out the volatile alcohol forces the reaction in the forward direction:

$$RCOOCH_3 + C_4H_9OH \overset{NaOC_4H_9}{\rightleftharpoons} RCOOC_4H_9 + CH_3OH$$

The reaction falls into the same nucleophilic displacement group described above. The sodium gives rise to a catalytic amount of the alkoxide of the alcohol ($RO^{\ominus}$), a strong nucleophile:

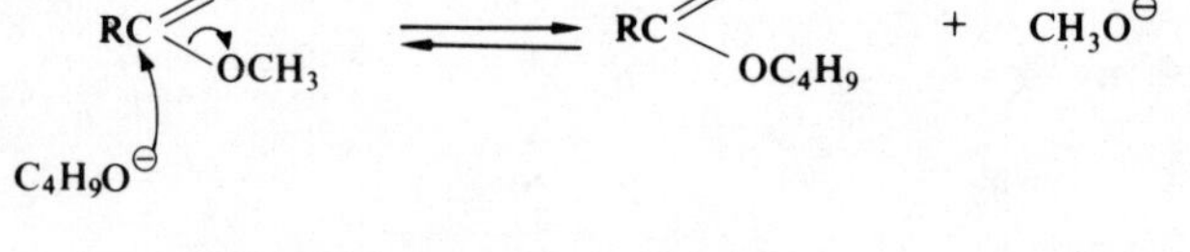

$$CH_3O^{\ominus} + C_4H_9OH \rightleftharpoons C_4H_9O^{\ominus} + CH_3OH$$

(excess)

Acid catalysed transesterification is similar to the acid catalysed reaction of an alcohol with a carboxylic acid.

5. *Diazomethane.* This reacts smoothly and completely with acids to give methyl esters without applying heat. Diazomethane is a

poisonous explosive yellow gas, and so must be handled with care, but it can be conveniently prepared in ether solution, which is added slowly to the carboxylic acid until no more nitrogen gas is evolved and the yellow colour of the diazomethane persists:

$$R-\overset{\overset{\displaystyle O}{\|}}{C}-OH \ + \ CH_2N_2 \ \longrightarrow \ R\overset{\overset{\displaystyle O}{\|}}{C}OCH_3 \ + \ N_2$$

diazomethane

A simple valency picture of diazomethane cannot be given; it can be shown as a resonance hydrid or as containing a three-centre, two-electron bond:

$$\underset{H}{\overset{H}{>}}C = \overset{\oplus}{N} = \overset{\ominus}{N} \quad \longleftrightarrow \quad \underset{H}{\overset{H}{>}}\overset{\ominus}{C} - \overset{\oplus}{N} = N \quad \text{or} \quad \underset{H}{\overset{H}{>}}C \cdots N \equiv N$$

Physical Properties

Esters are neutral pleasant-smelling liquids or low melting solids. The methyl and ethyl esters are always lower boiling than the corresponding acid (methyl formate is lower boiling than either of its components), because there is no opportunity in esters for hydrogen bond formation. For solid esters, methyl esters always have higher melting points than the ethyl esters. They are soluble in ether, benzene, acetone, etc., but only the lowest of the series are appreciably soluble in water, and none are miscible with water. The flavours of fruits are due largely to esters of simple acids and alcohols, and mixtures of synthetic esters are used as artificial flavouring in foods and in perfumery. Methyl butyrate, for example is used as an artificial rum flavour.

The biggest use of the lower esters is as solvents in paints and lacquers. The higher esters are used as plasticizers, to improve the flexibility of plastics. Some indication of use is given in Table 17, *f* indicates esters used in flavouring and *s* those used as solvents.

Reactions of Esters

1. *Hydrolysis.* It has already been pointed out that reaction of an acid and alcohol to give an ester is an equilibrium reaction. Each of the stages of the mechanism of the acid-catalysed esterification is reversible. In the presence of a large excess of water and a strong acid catalyst the esterification can be reversed:

191

Aliphatic and Alicyclic Compounds

Table 17.   Carboxylic esters

Compound	Formula	m.p. °C	b.p. °C	Use
Methyl formate	$HCOOCH_3$	−99	32	
Ethyl formate	$HCOOC_2H_5$	−81	54	f
Methyl acetate	CH_3COOCH_3	−99	57	
Ethyl acetate	$CH_3COOC_2H_5$	−84	77	f, s*
Propyl acetate	$CH_3COOC_3H_7$	−93	101	f, s
Butyl acetate	$CH_3COOC_4H_9$	−77	126	f, s
Amyl acetate	$CH_3COOC_5H_{11}$	−71	148	f
t-Butyl acetate	$CH_3COOC(CH_3)_3$	—	97	
Methyl propionate	$C_2H_5COOCH_3$	−88	80	
Ethyl propionate	$C_2H_5COOC_2H_5$	−74	99	
Ethyl butyrate	$C_3H_7COOC_2H_5$	−93	120	f
Ethyl hexanoate	$C_5H_{11}COOC_2H_5$	−68	166	
Ethyl octanoate	$C_7H_{15}COOC_2H_5$	−45	207	
Ethyl octadecanoate	$C_{17}H_{35}COOC_2H_5$	33·4	199†	

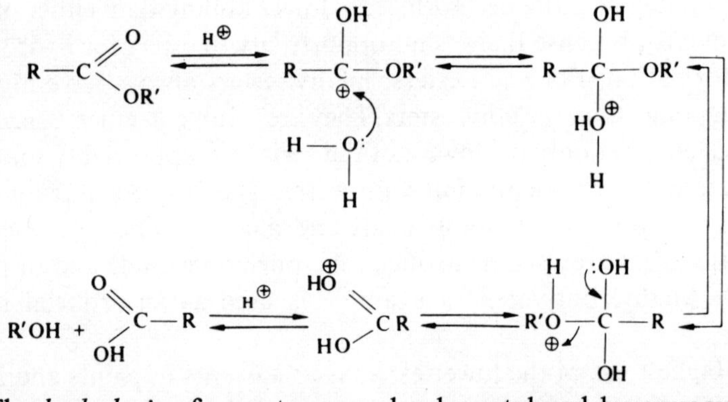

The *hydrolysis* of an ester can also be catalysed by aqueous base (the carboxylic acid, of course, is obtained as the salt of that base):

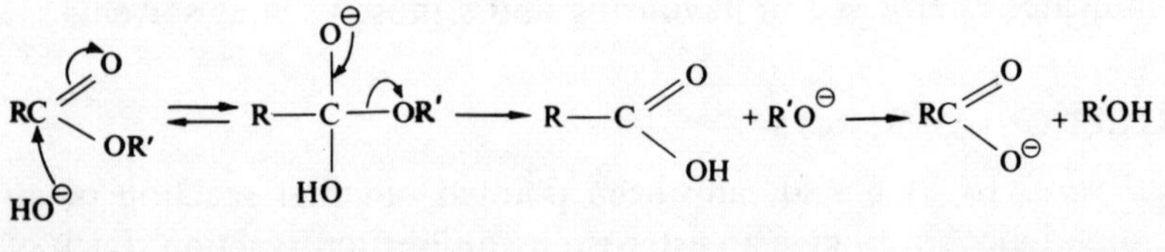

The reaction is reversible in the first stage but once the free acid is formed, it is converted to the salt and so the second stage of the

* f=flavouring; s=solvent.   † At 10 mm pressure.

192

reaction is not reversible. Therefore although hydrolysis of esters may be catalysed by aqueous acid or base, esterification is only *acid* catalysed.

Alkaline hydrolysis is similar in mechanism to the (reversible) transesterification reaction.

2. *Amides.* Esters react with ammonia to give amides. In many cases the ester is mixed with cold concentrated aqueous ammonia and the amide slowly crystallizes on standing. In more difficult cases, dry ammonia gas is dissolved in ethanol and mixed with the ester:

$$RCOOR' + NH_3 \longrightarrow RCONH_2 + R'OH$$

The reaction is another example of nucleophilic attack at the carbonyl carbon atom:

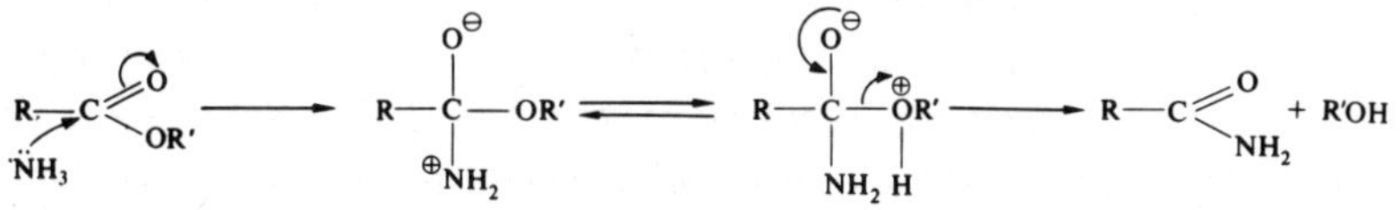

All amides of normal chain acids above formic acid are colourless crystalline solids and so amides are often used as derivatives to identify the acid portion of esters:

$$CH_3COOEt + NH_3 \longrightarrow CH_3CONH_2 + EtOH$$
$$\text{acetamide}$$

Substances related to ammonia, e.g. hydrazine, react similarly:

$$RCOOR' + NH_2-NH_2 \longrightarrow RCONH-NH_2 + R'OH$$
$$\text{an acid hydrazide}$$

3. *Grignard Reagents.* Esters react with Grignard reagents to give tertiary alcohols. The reaction cannot normally be stopped at the first stage, because the ketone formed has a more electropositive carbon atom than the ester and therefore reacts more rapidly with the reagent than the ester:

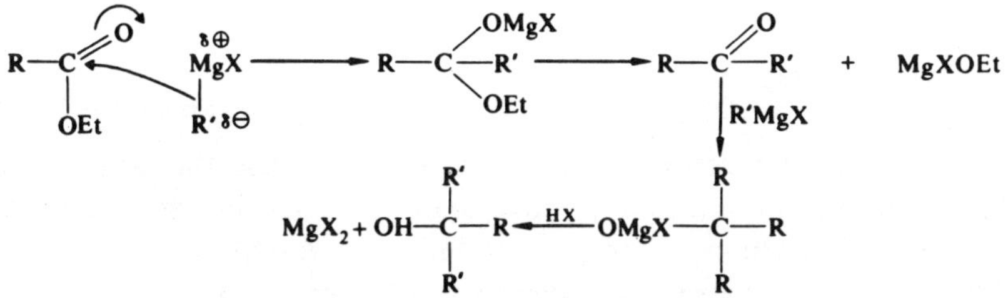

In cases where the ketone has bulky groups, steric hindrance prevents further reaction, and better yields of ketone are obtained.

4. *Reduction.* Esters are more susceptible to reduction than acids.

193

The reduction may be carried out catalytically over a copper chromite catalyst at high pressure (200–300 atmospheres). This method is used on a large scale to produce alcohols from vegetable oils for synthetic detergent manufacture:

$$CH_{11}H_{23}COOR + H_2 \xrightarrow[250°]{CuCrO_3} C_{11}H_{23}CH_2OH + ROH$$

lauric esters of glycerol   lauryl alcohol

Chemical reduction may be carried out by the Bouveault method (Chapter 5), or with lithium aluminium hydride:

$$RCOOR' + Na + EtOH \longrightarrow RCH_2OH + R'OH + NaOEt$$

$$RCOOR' \xrightarrow{LiAlH_4} RCH_2OH + R'OH$$

5. *Claisen Condensation.* Esters which possess an α-hydrogen atom undergo a condensation, similar to the aldol condensation of aldehydes and ketones. Because of the reduced reactivity of the carbonyl groups in esters compared with ketones, a strong base (e.g. sodium ethoxide or hydride or sodamide) is required. The mechanism is probably similar to that of the aldol condensation, the product is a β-keto-ester:

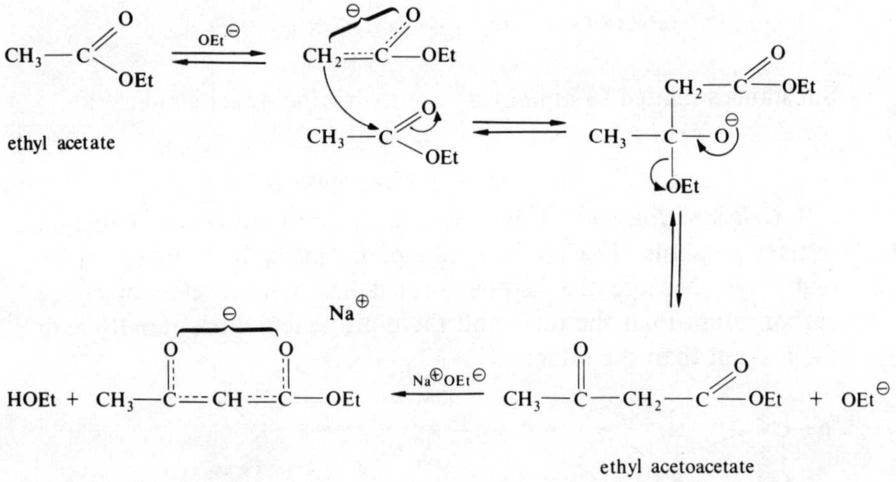

ethyl acetoacetate

The reversible reaction is driven forward because the acetoacetic ester formed is immediately converted into its salt by excess sodium ethoxide. The anion of this salt is stabilized by distributing the charge over two oxygen atoms and the central CH group. It is much more stable than the enolate ions of simple aldehydes, ketones or esters.

6. *The Acyloin Reaction.* When an ester is treated with metallic sodium in an *inert* solvent such as toluene, reduction and condensa-

tion occur to give as final product after acidification an α-hydroxy-ketone known as an acyloin:

$$
\underset{\overset{\displaystyle RC-OR'}{\underset{\displaystyle \| }{O}}}{\overset{\overset{\displaystyle O}{\|}}{RC-OR'}}
\quad\xrightarrow{\ Na\ }\quad
\underset{\displaystyle RC-O^{\ominus}Na^{\oplus}}{\overset{\displaystyle RC-O^{\ominus}Na^{\oplus}}{\big\|}}
\quad\xrightarrow{\overset{\oplus}{H}}\quad
\underset{\displaystyle RC=O}{\overset{\overset{\displaystyle H}{|}}{RC-OH}}
$$

The mechanism of this reaction is discussed in Chapter 11 where a special case of it is described.

Orthoesters

These are the esters of the hypothetical orthoacids. The most important orthoesters are those of orthoformic acid. These can be prepared from chloroform and sodium alkoxides, or from hydrogen cyanide, alcohol and hydrogen chloride.

$$
HCCl_3 \ + \ NaOEt \ \longrightarrow \ \underset{\displaystyle OEt}{\overset{\displaystyle OEt}{HC-OEt}}
$$

ethyl orthoformate

$$
HC\equiv N \ + \ CH_3OH \ \xrightarrow{\ HCl\ } \ HC\overset{\displaystyle OCH_3}{\underset{\displaystyle NH}{\diagdown}} \ \xrightarrow{\ CH_3OH\ } \ \underset{\displaystyle OCH_3}{\overset{\displaystyle OCH_3}{HC-OCH_3}} \ + \ NH_3
$$

Higher orthoformates can be prepared from trichloroalkanes or alkyl cyanides by the above methods.

Orthoesters are readily hydrolysed by acids but are resistant to alkali (in this way they behave like acetals). They have a strong tendency to lose two alkoxy groups to give a normal ester, as in the formation of ketals.

$$
\underset{CH_3}{\overset{CH_3}{\diagdown}}C=O \ + \ \underset{\displaystyle OEt}{\overset{\displaystyle OEt}{HC-OEt}} \ \xrightarrow{\ \overset{\oplus}{H}\ } \ \underset{CH_3}{\overset{CH_3}{\diagdown}}C\overset{\displaystyle OEt}{\underset{\displaystyle OEt}{\diagup}} \ + \ \overset{\overset{\displaystyle O}{\|}}{HCOEt}
$$

Oils, Fats and Waxes

Oils, fats and waxes are three important groups of naturally occurring esters of high molecular weight. The division between oils and fats is simply a practical one, dependent upon their physical state. Oils are liquid esters of glycerol and fats are solid esters of glycerol. The oils are liquid because they contain a larger proportion of unsaturated acids than fats, and unsaturated acids and esters have lower melting points than the saturated acids of the same chain length. The principal saturated acids present are lauric (C_{12}), myristic (C_{14}), palmitic (C_{16}) and stearic (C_{18}) acids, and the commonest unsaturated acid is oleic acid, though this by no means completes the list.

$$CH_3(CH_2)_{10}COOH \qquad CH_3(CH_2)_{16}COOH \qquad CH_3(CH_2)_7CH{=}CH(CH_2)_7COOH$$

lauric acid $\qquad\qquad$ stearic acid $\qquad\qquad\qquad$ oleic acid

The interesting points about these acids are that they all contain unbranched chains and they all contain an even number of carbon atoms. Only traces of branched chains or acids of 13, 15 or 17 carbon atoms are usually found in natural oils and fats.

Glycerol is a trihydroxy alcohol and in fats and oils all three hydroxyls can be esterified with a mixture of saturated and unsaturated acids in all possible combinations. For example, tristearin probably occurs to a very small extent naturally, whereas lauro-oleo-stearin would be a typical glyceride.

tristearin $\qquad\qquad\qquad\qquad\qquad$ lauro–oleo–stearin

The most important sources of fats and oils are butter, lard, soya beans, tallow, peanuts, cottonseed, sunflowers, coconut and olives, in that order. The mixture of acids can be released from the glycerides by steam, by acid or alkaline hydrolysis or by transesterification, using methanol, to give the methyl esters. The methyl esters can be separated into individual esters by very careful fractional distillation (or by chromatography, for very pure laboratory materials), but for most industrial purposes, the mixture of acids or esters is used.

The chief uses of the fats and oils are, first, in foods as butter, margarine, cooking oil and so on; in soap and in paints. Lesser uses are as lubricants and in cosmetics, such as creams, ointments, lipstick and nail varnish.

Milk fat contains a higher proportion of glycerides of short-chain acids than other fats; the lower melting glycerides of the short-chain acids give butter a softer consistency than other animal fats, and the smell of rancid butter is due to the free short-chain acids (e.g. butyric and hexanoic acid) produced by hydrolysis. Margarine is made from liquid vegetable oils, rich in unsaturated acids. The oil is partially hydrogenated under pressure, using a nickel catalyst to give a higher content of saturated acids and a similar consistency but different fatty acid content from butter.

Waxes are esters of high molecular weight acids with high molecular weight alcohols. A typical example is beeswax, which contains triacontanyl hexadecanoate (i.e. palmitate).

$$CH_3(CH_2)_{14} \overset{\displaystyle O}{\overset{\|}{C}}\!-\!O\!-\!(CH_2)_{29}CH_3$$

triacontanyl palmitate

Soaps

The metal salts of higher fatty acids are called soaps. Commercial soap consists chiefly of the potassium and sodium salts of lauric, myristic, palmitic and stearic acids, produced by alkaline or steam hydrolysis of oils or fats. Alkaline hydrolysis of esters is often known as saponification, since soaps are obtained from fats and oils by this reaction. After hydrolysis, sodium chloride is added to the solution, precipitating the soap. Glycerol is recovered from the aqueous solution by distillation.

A sodium soap consists of sodium ions and the ion of the long chain acid. The acid ion has two opposing features, the carboxylate ion which is soluble in water and insoluble in organic solvents and the long alkyl chain attached to it, which is insoluble in water and soluble in organic solvents, oils and greases. The soap dissolves in water to form a colloidal solution. The colloidal particles consist of clusters of alkyl chains with the carboxylate ions arranged at the surface of the particles, their negative charges being balanced by the positive charges of the sodium ions in the aqueous solution (Fig. 38). Part of the washing action of soaps is due to the dissolving of grease and dirt in the colloidal particles of soap, though the complete explanation is as yet unknown. Calcium and magnesium salts of long chain acids are insoluble in water, therefore when an alkali

197

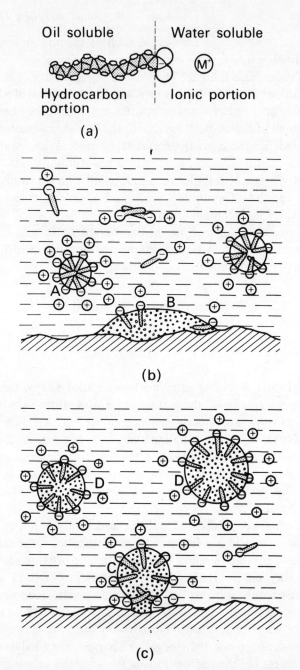

Oil soluble | Water soluble

Hydrocarbon portion | Ionic portion

(a)

(b)

(c)

Fig. 38 The cleansing action of soap (a) A typical soap molecule (b) Soap molecules in water tend to form aggregates called micelles A, the hydrocarbon part of the molecules dissolve in grease B and detach the grease from the surface C, and produce stabilized droplets in the aqueous phase D, as shown in (c).

metal soap is added to hard water, a scum or precipitate of the calcium soap is formed.

Modern synthetic detergents are commonly organic derivatives of sulphuric acid. The earliest of these were salts of alkylbenzene-sulphonic acids, for example:

$$CH_3-\underset{\underset{CH_3}{|}}{\overset{\overset{CH_3}{|}}{C}}-CH_2-\underset{\underset{CH_3}{|}}{\overset{\overset{CH_3}{|}}{C}}-\langle\bigcirc\rangle-\underset{\underset{O}{\|}}{\overset{\overset{O}{\|}}{S}}-O^{\ominus}\quad Na^{\oplus}$$

Such molecules possess the same character of a large oil-soluble molecule with an ionized water-soluble group attached, and behave as soaps. But the branched alkyl chains are not readily broken down (bio-degraded) by sewage bacteria, so that large amounts of detergent were released into rivers essentially unchanged, causing serious foaming problems, and killing fish. Newer detergents contain unbranched alkyl chains that are more easily destroyed in sewage. These are either n-alkylbenzenesulphonates or long chain mono-alkyl sulphates, made either from petroleum fractions or fatty alcohols obtained by hydrogenating vegetable oils, for example:

$$CH_3(CH_2)_{10}CH_2-O-\underset{\underset{O}{\|}}{\overset{\overset{O}{\|}}{S}}-O^{\ominus}\quad Na^{\oplus}$$

sodium lauryl sulphate

The calcium and magnesium salts of these sulphates and sulphonates are water-soluble, so synthetic detergents still possess their cleansing action in hard water.

Acid Halides

Acid halides and anhydrides are two very reactive groups of acid derivatives. Unlike esters and salts, acid halides do not occur in nature, but as laboratory chemicals are of prime importance.

The acid halides are so reactive because the electron withdrawing power of the halogen, together with the carbonyl group, leave the carbon atom to which they are both attached deficient in electron charge, and therefore very easily attacked by a variety of nucleophilic reagents:

199

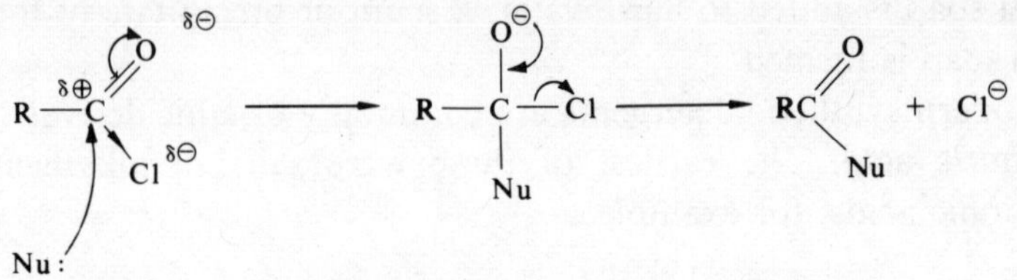

Acid halides are named by changing the final *-ic* of the corresponding acid to *-yl* and adding the name of the halogen; this applies both to trivial and Geneva names. For example:

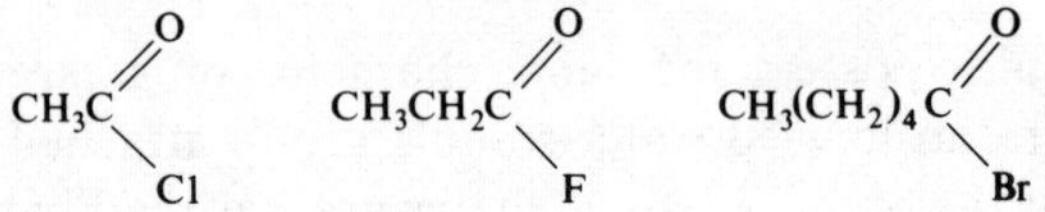

acetyl chloride   propionyl fluoride   hexanoyl bromide

Preparation of Acid Halides

The one general method of preparation is by the action of phosphorus or thionyl halides on acids. Thionyl chloride (b.p. 76°) is especially convenient for acids with higher boiling points, since any excess thionyl chloride can be easily distilled away at the end of the reaction. For acetyl chloride (b.p. 51°), which cannot easily be separated from thionyl chloride, phosphorus trichloride or pentachloride is used:

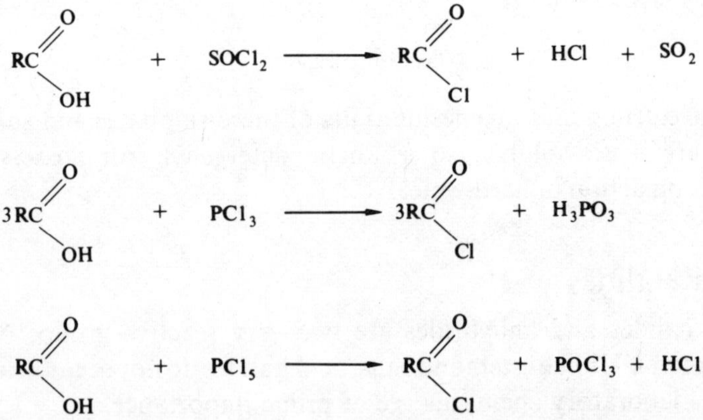

Acid bromides can be made from the acid and phosphorus tribromide or pentabromide, or from the acid chlorides and hydrogen bromide. Acid iodides are made by halogen exchange between acid chlorides and hydrogen iodide, calcium iodide or magnesium iodide.

200

Properties and Reactions of Acid Halides

The chlorides are colourless liquids; the lower members of the series have sharp odours and fume in air, due to reaction with water vapour, giving hydrogen chloride. The formyl halides are unstable; formyl fluoride is known, but formyl chloride decomposes spontaneously to carbon monoxide and hydrogen chloride. Among the halides of any one acid, reactivity decreases in the order $F > Cl > Br > I$ because in this series it is the electronegativity of the halogen atom that makes the compound reactive. This is the opposite of the order of reactivity in alkyl halides.

1. *With Nucleophiles.* Acid halides react rapidly with water, alcohols, ammonia, hydrazine, etc., to give respectively acids, amides and hydrazides plus hydrogen halide:

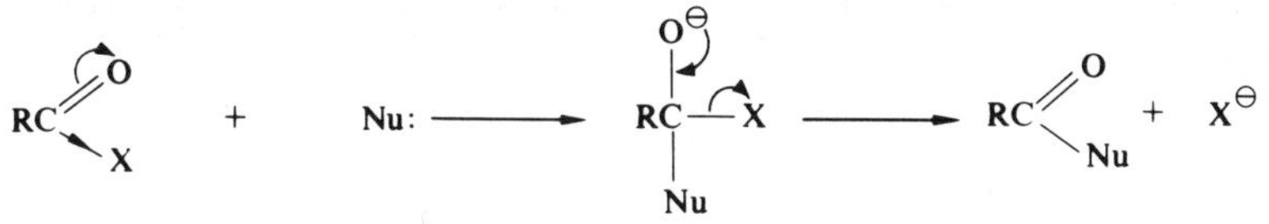

$$RC\overset{O}{\underset{X}{\diagup}} \quad + \quad Nu: \longrightarrow RC\overset{O^{\ominus}}{\underset{Nu}{-}}X \longrightarrow RC\overset{O}{\underset{Nu}{\diagup}} + \; X^{\ominus}$$

2. *Reduction.* Acid chlorides can be reduced to aldehydes (Rosenmund reduction, page 149) or alcohols with hydrogen and a metal catalyst:

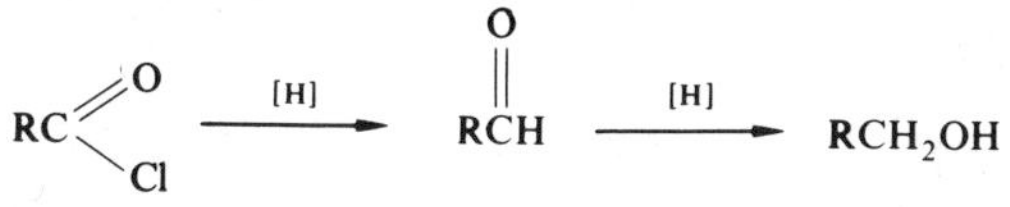

$$RC\overset{O}{\underset{Cl}{\diagup}} \xrightarrow{[H]} \overset{O}{\overset{\|}{RCH}} \xrightarrow{[H]} RCH_2OH$$

3. *Grignard Reactions.* Unlike esters, the chlorides can be used with one equivalent of a Grignard reagent to prepare ketones. The reaction can be stopped at the ketone stage because the acid chloride reacts more readily with the Grignard reagent than the resulting ketone does:

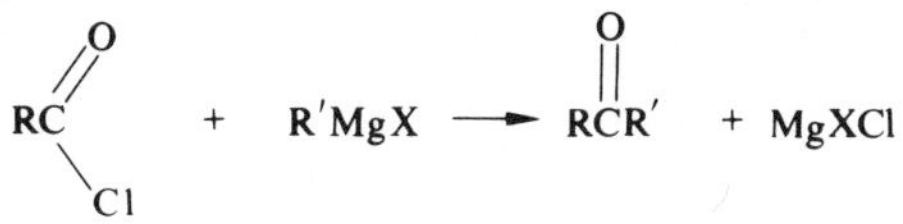

$$RC\overset{O}{\underset{Cl}{\diagup}} \quad + \quad R'MgX \longrightarrow \overset{O}{\overset{\|}{RCR'}} + MgXCl$$

The reaction can be improved by converting the Grignard reagent to an organo-cadmium compound by refluxing with cadmium chloride. Organo-magnesium compounds are more reactive than organo-cadmium compounds, which will react with an acid chloride but not a ketone:

$$2R'MgX \ + \ CdCl_2 \ \longrightarrow \ R'CdR' \ + \ MgCl_2 + MgX_2$$

$$2RCOCl \ + \ R'CdR' \ \longrightarrow \ 2RCOR' + CdCl_2$$

4. *Anhydrides.* Acid halides react with salts of carboxylic acids to give anhydrides:

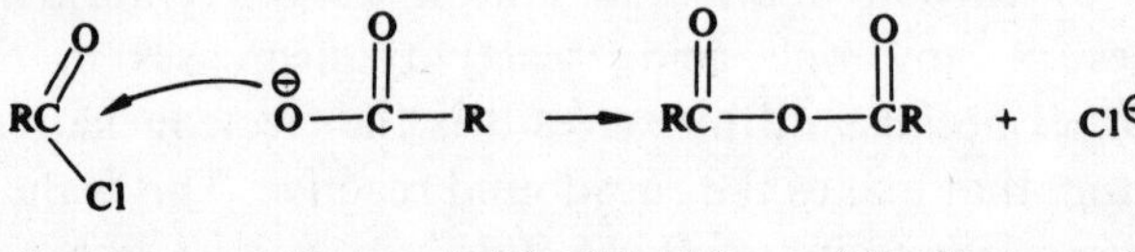

$$\underset{Cl}{\overset{O}{\underset{\|}{RC}}} \quad \overset{\ominus}{O} \underset{\|}{\overset{O}{-C}} - R \longrightarrow \underset{\|}{\overset{O}{RC}} - O - \underset{\|}{\overset{O}{CR}} \ + \ Cl^{\ominus}$$

5. *With Diazomethane.* Acid chlorides with diazomethane give diazoketones. The latter react with water, alcohols or ammonia in the presence of silver oxide to give an acid, ester or amide with one more carbon atom than the acid chloride. These reactions provide a method for extending a carboxylic acid by one carbon atom at a time (the Arndt–Eistert reaction):

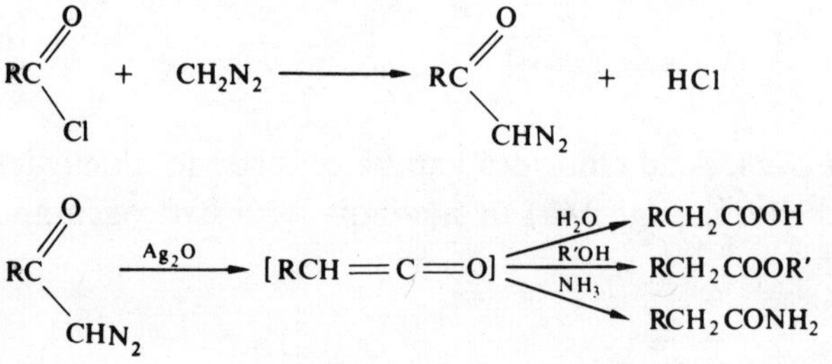

$$\underset{Cl}{\overset{O}{\underset{\|}{RC}}} \ + \ CH_2N_2 \ \longrightarrow \ \underset{CHN_2}{\overset{O}{\underset{\|}{RC}}} \ + \ HCl$$

$$\underset{CHN_2}{\overset{O}{\underset{\|}{RC}}} \ \xrightarrow{Ag_2O} \ [RCH = C = O] \underset{\underset{NH_3}{\xleftarrow{R'OH}}}{\overset{H_2O}{\xrightarrow{\hspace{1cm}}}} \begin{array}{l} RCH_2COOH \\ RCH_2COOR' \\ RCH_2CONH_2 \end{array}$$

The intermediate in brackets is a *ketene* (page 168). These are a very reactive group of intermediates which react rapidly with a number of reagents to give acid derivatives.

Acid Anhydrides

An acid anhydride consists of two acyl groups linked by an oxygen atom, or as the name indicates, two acids linked by the elimination of one molecule of water. Like ethers, they can be symmetrical or unsymmetrical. The systematic names are rather cumbersome and in most cases the trivial names are used.

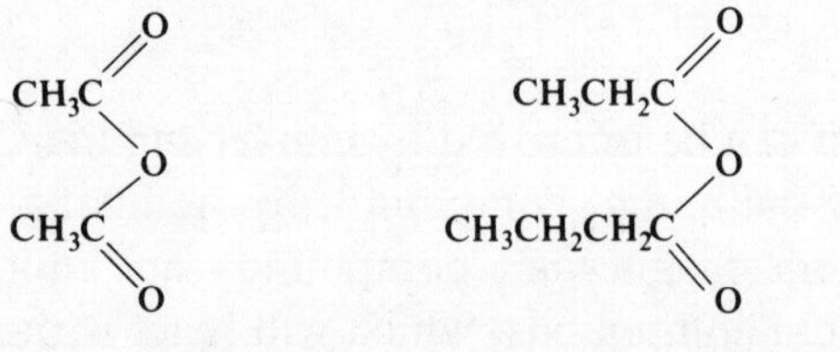

$$\underset{CH_3C}{\overset{O}{\nearrow}} \searrow_O \qquad \underset{CH_3CH_2C}{\overset{O}{\nearrow}} \searrow_O$$

acetic anhydride   propionic butyric anhydride

Preparation of Acid Anhydrides

The anhydrides can be prepared from the acids with phosphorus pentoxide, but in better yield from a salt and an acid chloride. The second method can give mixed anhydrides:

$$
\underset{\text{RCCl}}{\overset{O}{\overset{\|}{}}} \quad + \quad \underset{\text{R'CO}^{\ominus}}{\overset{O}{\overset{\|}{}}} \; \text{Na}^{\oplus} \quad \longrightarrow \quad \underset{\text{RC}-\text{O}-\text{CR'}}{\overset{O\qquad O}{\overset{\|\quad\ \|}{}}} \quad + \quad \text{NaCl}
$$

Another method is to add ketene (page 168 and above) to an acid. This method always gives a mixed anhydride of acetic acid:

$$
\text{RCOOH} \quad + \quad \text{CH}_2\!=\!\text{C}\!=\!\text{O} \quad \longrightarrow \quad \underset{\text{RC}-\text{O}-\text{CCH}_3}{\overset{O\qquad\ O}{\overset{\|\qquad\|}{}}}
$$

If a mixed anhydride is heated, it will rearrange, and some of the symmetrical lower anhydride and higher anhydride will be formed. If the temperature is high enough, the lower anhydride will distil out, and eventually all the mixed anhydride will be decomposed:

$$
\underset{\text{RC}-\text{O}-\text{CCH}_3}{\overset{O\qquad\ O}{\overset{\|\qquad\|}{}}} \ \rightleftharpoons \ \underset{\text{RC}-\text{O}-\text{CR}}{\overset{O\qquad O}{\overset{\|\qquad\|}{}}} + \underset{\text{CH}_3\text{C}-\text{O}-\text{CCH}_3}{\overset{O\qquad\quad O}{\overset{\|\qquad\quad\|}{}}}
$$

Properties and Reactions of Anhydrides

The anhydrides are high boiling liquids insoluble in water. Formic anhydride is unknown; mixed formic anhydrides with other acids have been prepared. Acetic anhydride is a colourless liquid with a sharp odour, it is slightly soluble in water and reacts slowly with cold water, and rapidly when the water is heated or when a drop of concentrated sulphuric acid is added.

The reactions of anhydrides parallel those of the acid chlorides, but anhydrides are a little less reactive:

(a) hydrolysis $(RCO)_2O + H_2O \rightarrow 2RCOOH$
(b) alcoholysis $(RCO)_2O + R'OH \rightarrow RCOOR' + RCOOH$
(c) ammonolysis $(RCO)_2O + R'NH \rightarrow RCONHR' + RCOOH$

Industrially, where a reactive acid derivative is required, anhydrides are used because they are cheaper than acid chlorides, e.g. in the manufacture of cellulose acetate.

Amides

Amides are derived from a carboxylic acid and an amine or ammonia, by elimination of a molecule of water. They represent another example of a neutralized system, for they do not show the typical reactivity of either a carbonyl group or an amine. The bond of the amide group is somewhere between the extremes of *a* and *b*.

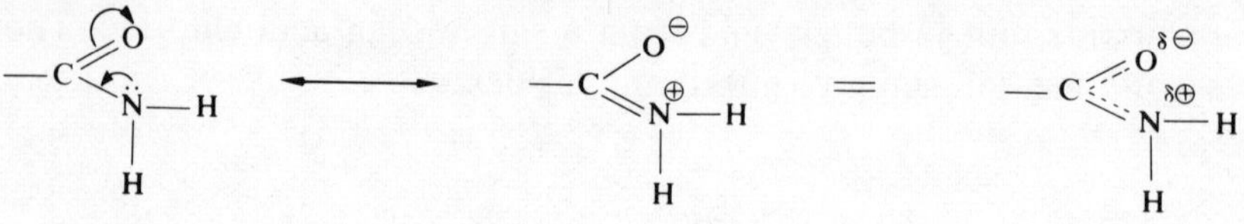

(a)   (b)

The partial double-bond character of the C—N bond means that the NH$_2$ groups cannot rotate about the C—N bond, and the unshared pair of electrons on nitrogen, responsible for its basic and nucleophilic properties, are not available for reactions as in amines. X-Ray analysis shows that the amide group is planar and the C—N bond length (1·32 Å) is intermediate between that of a C—N single bond (1·47 Å) and a C=N double bond (1·29 Å). Amides have higher melting and boiling points than corresponding acid chlorides, because they are associated by hydrogen bonding:

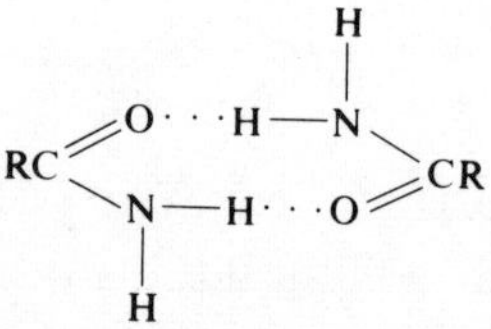

Amides are named systematically by dropping *-oic* from the acid name and replacing it by *-amide*, e.g. from hexanoic acid, we get hexanamide. Trivial names are formed similarly, e.g. acetamide, propionamide, butyramide.

Ammonia gives unsubstituted amides, primary amines give monosubstituted amides and secondary amines, disubstituted amides. To indicate which groups are attached to nitrogen, they are prefixed by *N*:

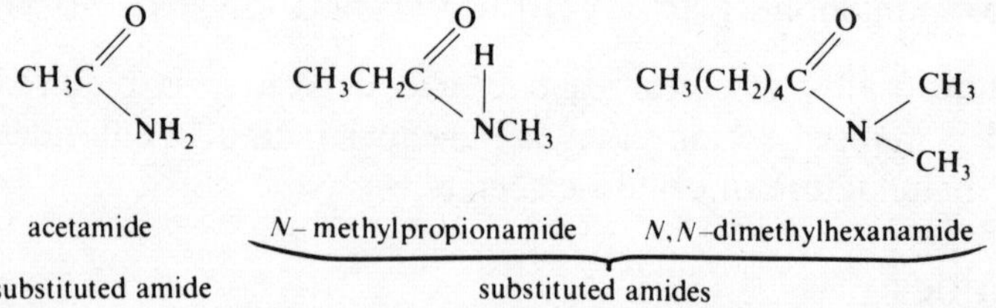

acetamide   *N*–methylpropionamide   *N,N*–dimethylhexanamide

unsubstituted amide   substituted amides

Preparation of Amides

1. *From Ammonium Salts.* When dry ammonium salts of carboxylic

acids are heated they lose water to give amides. This is the usual industrial preparation of amides:

$$RCOO^{\ominus} \ NH_4^{\oplus} \longrightarrow RCONH_2 + H_2O$$

2. *From Esters.* If an ester is treated with ammonia, an amide is formed. Some substituted amides can be formed in this way also. For example, an ester dissolved in a solution of methylamine slowly precipitates the methylamide:

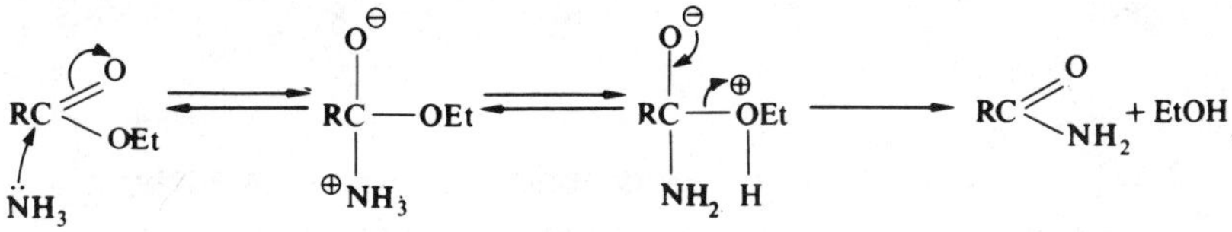

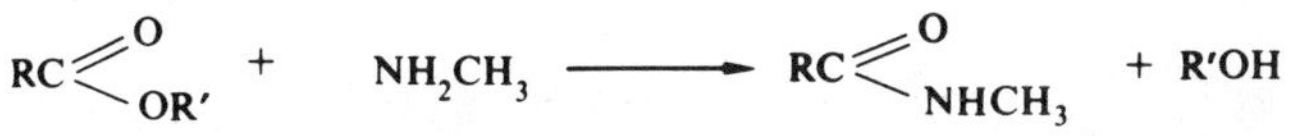

3. *From Acid Halides.* Acid halides react vigorously with ammonia or primary or secondary amines to give amides. This is the usual laboratory method for making amides:

$$RCOCl + 2R'NH_2 \longrightarrow RCONHR' + R'NH_3^{\oplus} \ Cl^{\ominus}$$

Note that one mole of amine is consumed in reacting with the hydrogen chloride produced in the reaction. Since most amides are nicely crystalline solids, they serve as useful derivatives of acids. A sample of an acid is converted to the acid chloride with thionyl chloride and this is added to concentrated aqueous ammonia. The solid amide is collected and its melting point determined and compared with a table of amide melting points to identify the acid.

4. *From Anhydrides.* The reaction is similar to method 3. Methods 2, 3 and 4 are all examples of ammonolysis (cf. hydrolysis) of acid derivatives. Substituted or simple amides can be made by each of these methods.

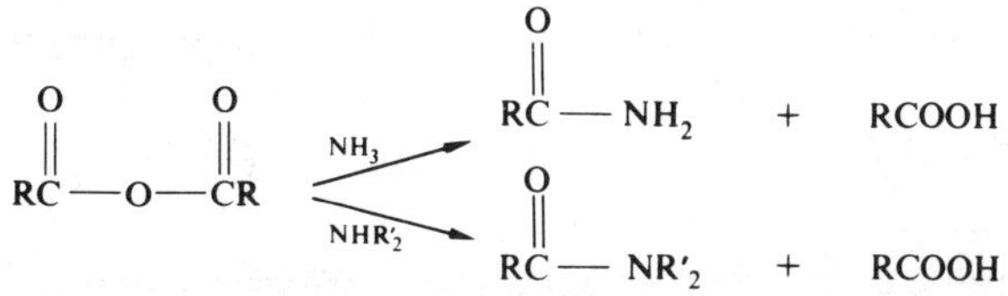

5. *From Nitriles.* Partial hydrolysis of a nitrile gives an unsubstituted amide. The hydrolysis is best carried out with concentrated sulphuric or hydrochloric acid in the cold, followed by pouring the reaction mixture into water:

$$RC\equiv N + H_2O \xrightarrow{H_2SO_4} RCON H_2$$

Reactions of Amides

1. *Hydrolysis.* Amides are only slightly attacked by water, many amides are conveniently recrystallized from water; acids hydrolyse them slowly, alkali more rapidly gives the amine and acid salt.

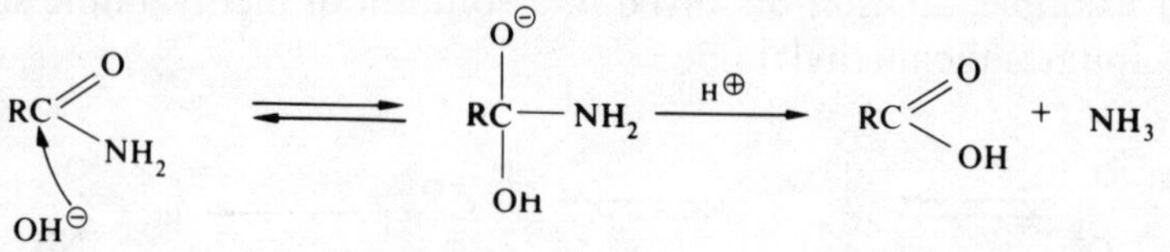

$$RC\overset{O}{\underset{NH_2}{\diagdown}} \overset{OH^{\ominus}}{\rightleftharpoons} RC\overset{O^{\ominus}}{\underset{OH}{-}}NH_2 \xrightarrow{H^{\oplus}} RC\overset{O}{\underset{OH}{\diagdown}} + NH_3$$

2. *Dehydration.* Unsubstituted amides are dehydrated by phosphorus pentoxide or acetic anhydride to give nitriles:

$$RC\overset{O}{\underset{NH_2}{\diagdown}} \xrightarrow{P_2O_5} RC\equiv N + H_2O$$

3. *Nitrous Acid.* Unsubstituted amides are converted to acids by nitrous acid. This is a more satisfactory way of recovering an acid from an amide than hydrolysis:

$$RCONH_2 + HNO_2 \longrightarrow RCOOH + N_2$$

4. *Reduction.* Lithium aluminium hydride reduces unsubstituted amides to amines:

$$RCONH_2 \xrightarrow{LiAlH_4} RCH_2NH_2$$

Lithium diethoxyaluminium hydride (page 149) gives aldehydes from substituted amides.

$$RC\overset{O}{\underset{NR'_2}{\diagup}} \xrightarrow{LiAl(OEt)_2H} RC\overset{H}{=}O + NHR'_2$$

Amides of long chain fatty acids are used in large quantities as foam stabilizers, i.e. 'frothing agents' in synthetic detergents, in making water-repellent cloth and as a thickening agent in paint and varnish.

Nitriles

Nitriles can be considered either as derivatives of acids or of alkanes, this is reflected in the two ways in which they are named, and the two general ways in which they can be prepared. The compound, CH_3CN, can be made from methyl bromide and an alkali cyanide, and is therefore called *methyl cyanide*, but it can also be prepared from acetamide by dehydration, and is therefore called *acetonitrile*. The latter system of naming is preferred.

$$H\!-\!C\!\equiv\!N \qquad CH_3CH_2C\!\equiv\!N \qquad CH_3(CH_2)_6C\!\equiv\!N$$

hydrogen cyanide   ethyl cyanide   heptyl cyanide
or formonitrile   or propionitrile   or octanonitrile

Hydrogen cyanide is one of those compounds which can be classed either as an inorganic or organic substance. As a nitrile, it forms the first member of the homologous series and can be considered a derivative of formic acid. Like other formic acid derivatives, in many reactions it differs from higher members of the series because it has no alkyl group.

Preparation of Nitriles

1. *From Alkyl Halides.* Potassium cyanide and primary alkyl halides in alcohol or dimethyl sulphoxide solution give nitriles (page 35):

$$RCH_2X + KCN \longrightarrow RCH_2C\equiv N$$

2. *From Amides.* Dehydration with phosphorus pentoxide or acetic anhydride gives the nitrile from unsubstituted amides. Higher molecular weight amides are dehydrated under milder conditions, simply by heating:

$$RCONH_2 \xrightarrow{P_2O_5} RC\equiv N + H_2O$$

3. *Industrially,* nitriles are made directly by passing an acid and ammonia over an alumina catalyst at 500°:

$$RCOOH + NH_3 \xrightarrow{Al_2O_3} RC\equiv N + 2H_2O$$

Hydrogen cyanide is made from carbon monoxide and ammonia by this method.

$$CO + NH_3 \xrightarrow{Al_2O_3} HC\equiv N + H_2O$$

Hydrogen cyanide is also made by oxidation of a mixture of methane and ammonia over platinum-iridium at 1000°:

$$CH_4 + NH_3 + 1\tfrac{1}{2}O_2 \xrightarrow{cat.} HC\equiv N + 3H_2O$$

Properties and Reactions of Nitriles

Hydrogen cyanide is a very poisonous gas (b.p. 20°C) with a distinct smell of almonds, though not everyone is able to detect the smell. Higher nitriles are less poisonous, colourless liquids or solids with pleasant odours.

1. *Hydrolysis*. Careful hydrolysis of nitriles, with cold concentrated acid gives unsubstituted amides. Refluxing the nitrile with alkali gives the acid and ammonia:

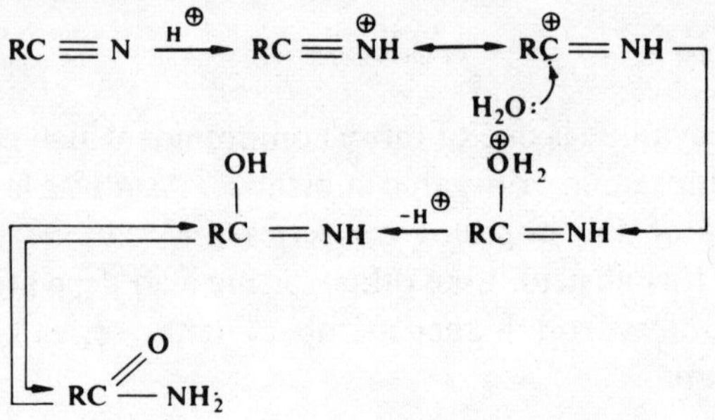

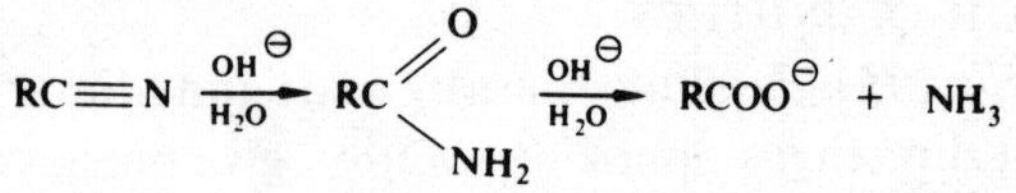

$$RC \equiv N \xrightarrow{H^{\oplus}} RC \equiv \overset{\oplus}{N}H \longleftrightarrow R\overset{\oplus}{C} = NH$$

$$H_2O:$$

$$\overset{OH}{\underset{|}{RC}} = NH \xleftarrow{-H^{\oplus}} \overset{\overset{\oplus}{OH_2}}{\underset{|}{RC}} = NH$$

$$RC \overset{O}{\diagup} NH_2$$

$$RC \equiv N \xrightarrow[H_2O]{OH^{\ominus}} RC \overset{O}{\underset{NH_2}{\diagup}} \xrightarrow[H_2O]{OH^{\ominus}} RCOO^{\ominus} + NH_3$$

The cyanhydrins obtained from aldehydes and ketones can be hydrolysed to hydroxy acids or amides by this method:

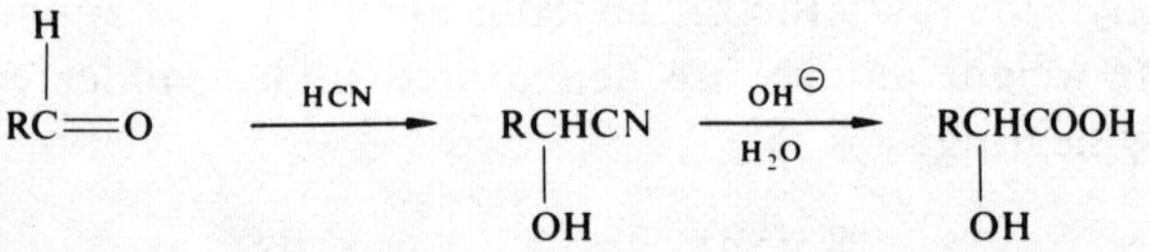

$$\overset{H}{\underset{}{RC}} = O \xrightarrow{HCN} \overset{}{\underset{OH}{RCHCN}} \xrightarrow[H_2O]{OH^{\ominus}} \overset{}{\underset{OH}{RCHCOOH}}$$

2. *Esterification*. If a nitrile is heated with an alcohol containing

Table 18. Some acid derivatives

Acid	Acid Chloride b.p. °C	Anhydride b.p. °C	Amide b.p. °C	Amide m.p. °C	Nitrile b.p. °C
Formic	—	—	210*	2·5	26
Acetic	51	140	222	82	82
Propionic	80	169	213	81	97
Butyric	101	198	216	115	118
Pentanoic	108	218	—	106	141
Hexanoic	151	255	255	100	163
Heptanoic	125	260	—	96	183
Octanoic	195	282	—	110	205
Decanoic	114†	—	—	108	244

* Decomposes on distillation.   † At 15 mm pressure.

sulphuric acid, an ester is formed (cf. the stepwise equation for hydrolysis above):

$$RC \equiv N + R'OH \xrightarrow{\ H^{\oplus}\ } RC \Big\langle \begin{matrix} {}^O \\ {}_{OR'} \end{matrix} + NH_3$$

3. *Reduction.* Reduction of nitriles with stannous chloride and hydrogen chloride gives an aldehyde, reduction with sodium in ethanol gives a primary amine:

$$RC \equiv N \xrightarrow{\ [H]\ } RCH_2NH_2$$

Identification of Acids and Derivatives

Several classes of compounds will dissolve in aqueous alkali but carboxylic acids are the only group (with a few exceptions, not yet encountered) containing only carbon, hydrogen and oxygen, which dissolve in aqueous bicarbonate solution with the evolution of bubbles of carbon dioxide. This simple and reliable test is recommended as the first to perform on a substance suspected of being a carboxylic acid.

The infrared spectra of acids are easily recognized. Because of the strong intermolecular hydrogen bonds in acids, the O—H stretching absorption is strong and broad, and at lower frequency than in alcohols. It is centred around 3000 cm^{-1} and overlies the narrow C—H band. The carbonyl group has a strong absorption near 1700 cm^{-1}. The presence of these two bands (Fig. 39) together indicate a carboxylic acid. The spectra of salts are quite different, with a broad absorption near 1600 cm^{-1} due to the ionized carboxyl group.

Individual acids are usually identified by conversion to an amide, an ester or other crystalline derivative and comparing the melting point of the derivative with those listed in tables.

All of the derivatives of carboxylic acids can in theory be recognized by hydrolysing them to the acid and identifying it and the other component. In practice, acid chlorides are such reactive compounds that they are seldom encountered except as intermediates. They can be recognized by their rapid hydrolysis to an acid or conversion to an amide with ammonia. They show strong C=O infrared absorption near 1800 cm^{-1} (Fig. 41).

Anhydrides also can be recognized by their reactivity and by *two* infrared C=O bands between 1850 and 1750 cm^{-1}, with a strong absorption due to C—O—C found near 1100 cm^{-1} in simple anhydrides (Fig. 41).

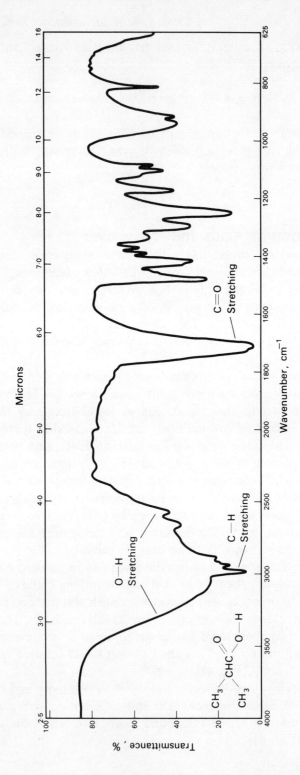

Fig. 39 Infrared spectrum of isobutyric acid

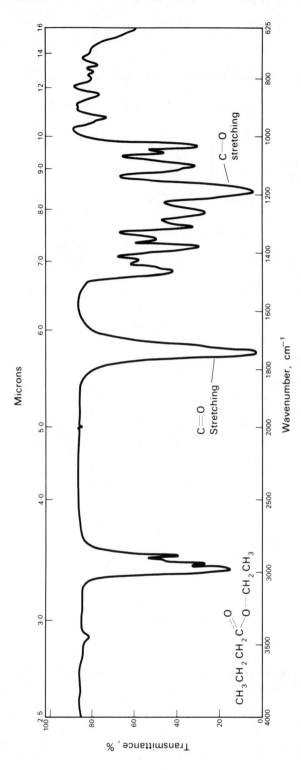

Fig. 40   Infrared spectrum of ethyl n-butyrate

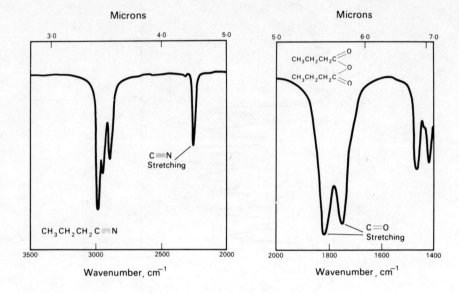

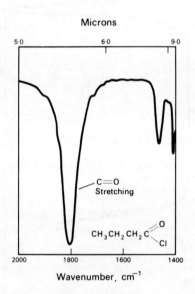

Fig. 41 Parts of the infrared spectra of butyryl chloride, butyric anhydride and butyronitrile

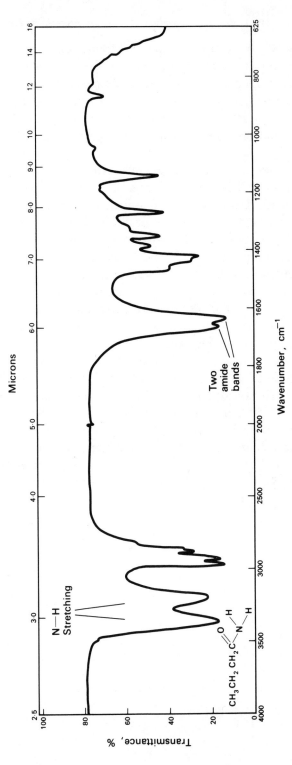

Fig. 42 Infrared spectrum of butyramide

Esters (Fig. 40) show infrared carbonyl frequencies very similar to those of ketones (1740 cm^{-1}) but also a strong C—O stretching absorption near 1200 cm^{-1} not seen in ketones, though this band cannot be used reliably in complex molecules with many absorptions in this region, but esters dissolve slowly in boiling aqueous alkali, and do not undergo the typical derivative-forming reactions of ketones.

Neutral compounds containing nitrogen which show strong absorption in the 1700–1600 cm^{-1} region of the infrared spectrum can be recognized as amides. The identification is best confirmed by hydrolysis to the acid and ammonia or amine. Unsubstituted amides show two fairly strong N—H absorptions in the region 3500–3400 cm^{-1} (Fig. 42), monosubstituted amides have one band in this region, and disubstituted amides of course have none. The pattern of carbonyl absorption also depends upon the number of groups attached to nitrogen, but also on the conditions under which the spectrum is obtained. Amides in solution, as a pure liquid or the solid made into a paste with paraffin oil (a mull) all show slightly different spectra because the physical state affects the amount of hydrogen bonding. All amides show a C=O band between 1700 and 1600 cm^{-1}. Unsubstituted (near 1600 cm^{-1}) and monosubstituted amides (near 1550 cm^{-1}) show additional strong bands due to complex vibrations, which are not shown by disubstituted amides.

Nitriles are recognized chiefly by their inertness and their C≡N absorption near 2250 cm^{-1} (Fig. 41), which is rather weak but in a region where few other groups (except other triple bonds, for example C≡C) absorb.

10

Alkyl Derivatives of Sulphur

From its position in the Periodic Table of elements we might expect sulphur to form alkyl compounds similar to those of oxygen. As a first approximation this is true, but with four important qualifications.

1. *The S—H bond* (348 kJ/mole) is weaker than the O—H bond (463 kJ/mole), so that S—H compounds are more acidic than O—H compounds, as hydrogen sulphide ($K_a = 6 \times 10^{-8}$) is more acidic than water ($K_a = 10^{-14}$).

2. *Sulphur* is only weakly electronegative, so that hydrogen bonding in sulphur compounds is almost negligible.

3. *Elements from the second row are reluctant to form π bonds.* For example, O_2 is a stable compound, but S_2 is not (sulphur prefers to form S_8 rings with S—S single bonds). As a result, the C=O group is stable but the C=S group is much less so.

4. Most important of all, second row elements (particularly Si, P and S) can expand their valencies by the use of empty *d* orbitals and so we can expect to find sulphur compounds with higher valencies.

Each of these points will be taken up in turn in this chapter.

Thiols

The sulphur analogues of the alcohols are called *thiols*. The prefix *thio-* is used in general in chemistry to indicate the presence of sulphur or where sulphur has replaced oxygen (Greek, *theion* = sulphur). The systematic names are formed just as for alcohols except that the final *e* of the alkane name is retained here for euphony (from ethane we get *ethanol*, but *ethanethiol*). Formerly thiols were known as mercaptans (from Latin, *mercurium captans*) because they react readily with mercuric oxide to form solid derivatives. The name mercaptan is still encountered and metal salts of thiols are sometimes called mercaptides.

Thiols occur in nature in small quantities, for example as traces in radishes, cabbage and broccoli. They generally have such strong and unpleasant odours that they cannot be tolerated in more than traces. The skunk uses butane-l-thiol to drive away its enemies. The same compound is added to natural gas in minute amounts to make gas leaks detectable by smell. Thiols also occur in petroleum and have to be removed during refining (page 22).

Preparation of Thiols

1. *From an Alkyl Halide.* The S_N2 reaction of an alkyl halide with bisulphide ion gives a thiol. The reaction does not work as well for halides that undergo S_N1 reaction, so the method is essentially one for making primary thiols.

$$CH_3CH_2Br + SH^{\ominus} \longrightarrow CH_3CH_2SH + Br^{\ominus}$$

ethanethiol

2. *From Alkenes.* By a reaction superficially analogous to the hydration of alkenes, hydrogen sulphide adds to double bonds. Hydration, however, is an ionic reaction, but this is probably a free radical reaction. This method can be used for secondary and tertiary thiols:

$$CH_3CH{=}CH_2 + H_2S \xrightarrow{H_2SO_4} CH_3CHCH_3$$
$$\underset{\displaystyle SH}{|}$$

propane-2-thiol

3. *From Grignard Reagents.* One must keep air away from Grignard reagents to prevent their reaction with oxygen:

$$RMgX + O \longrightarrow ROMgX$$

The similar reaction with sulphur can be used to prepare thiols:

$$RMgX + S \longrightarrow RSMgX \xrightarrow{HX} RSH + MgX_2$$

Properties of Thiols

Methanethiol is a gas, higher members of the series are liquids. Because hydrogen bonding is less effective than in alcohols, thiols are less soluble in water and have lower boiling points than comparable alcohols. As they are alkyl derivatives of hydrogen sulphide, thiols are relatively acidic compared with alcohols, and will dissolve in aqueous alkali. The unpleasant odour of lower thiols diminishes with increasing molecular weight.

Reactions of Thiols

Reactions of thiols fall into two groups (i) those where the thiol behaves like an alcohol and (ii) those where sulphur is oxidized to a higher valency state.

1. *Formation of Thioethers.* The alkali metal salts of thiols react with alkyl halides to form thioethers (cf. the Williamson synthesis of ethers, page 126):

$$RS^\ominus + R'Br \longrightarrow RSR' + Br^\ominus$$

2. *With acid chlorides, thiols form thioesters:*

$$\underset{\displaystyle \overset{O}{\|}}{R'CCl} + RSH \longrightarrow \underset{\displaystyle \overset{O}{\|}}{R'C}{-}SR + HCl$$

Reacting an acid with a thiol does not give a very good yield of thioester because the equilibrium of the following equation is well to the left:

$$\underset{\displaystyle \overset{O}{\|}}{R'COH} + RSH \rightleftharpoons \underset{\displaystyle \overset{O}{\|}}{R'CSR} + H_2O$$

3. *Metal Derivatives.* Alkali metals dissolve in thiols, liberating hydrogen and giving alkali mercaptides. These compounds are ionic. Heavy metal oxides or salts with thiols give insoluble mercaptides, used for identifying thiols. In these, the metal atom is bound covalently to sulphur:

$$2C_2H_5SH + HgO \longrightarrow C_2H_5S{-}Hg{-}SC_2H_5 + H_2O$$

4. *Oxidation.* Very mild oxidation, for example with cupric chloride, air, sodium hypochlorite or iodine, gives disulphides. These correspond in structure, but not in reactivity to peroxides. They are easily reduced back to thiols by lithium aluminium hydride.

$$2RSH + CuCl_2 \longrightarrow RS{-}SR + 2\ CuCl + 2\ HCl$$

More vigorous oxidation of thiols, e.g. with nitric acid or permanganate ion gives sulphinic and sulphonic acids:

$$RSH \xrightarrow{[O]} R \underset{\displaystyle \overset{O}{\|}}{-S}{-}OH \longrightarrow R \underset{\displaystyle \underset{\displaystyle \overset{\|}{O}}{\overset{\overset{\displaystyle O}{\|}}{-S}}}{}{-}OH$$

a sulphinic acid   a sulphonic acid

Thioethers

The sulphur analogues of ethers (known as thioethers or sulphides) resemble the ethers physically (they are colourless liquids, insoluble in water but miscible with organic solvents), but are less inert chemically. Their names follow logically from those of ethers (e.g. $CH_3SCH_2CH_3$ can be called methyl ethyl thioether, methyl ethyl sulphide or systematically, methylthioethane).

Preparation of Thioethers

They may be prepared from the alkali salt of a thiol and an alkyl halide (see above). Symmetrical thioethers are obtained as by-products in the preparation of thiols. The thiols first formed, because of their acidity, undergo exchange with the bisulphide and then react again with alkyl halide.

$$RBr + SH^{\ominus} \longrightarrow RSH + Br^{\ominus}$$

$$RSH + SH^{\ominus} \rightleftharpoons RS^{\ominus} + H_2S$$

$$RS^{\ominus} + RBr \longrightarrow RSR + Br^{\ominus}$$

Reactions of Thioethers

1. *Formation of Sulphonium Salts*. Sulphonium salts are intermediate in stability between oxonium and ammonium salts:

trimethylsulphonium bromide

The salts are relatively stable water-soluble compounds, though on heating they are decomposed to thioether and alkyl halide. If a sulphonium halide is treated with moist silver oxide the strongly basic sulphonium hydroxide is formed:

$$(CH_3)_3S^{\oplus} \, Br^{\ominus} + AgOH \longrightarrow (CH_3)_3S^{\oplus} \, OH^{\ominus} + AgBr$$

2. *Oxidation*. Mild oxidizing conditions (e.g. dilute nitric acid, or one equivalent of hydrogen peroxide) give sulphoxides. Further oxidation with hydrogen peroxide, or with permanganate directly from the thioether, gives sulphones:

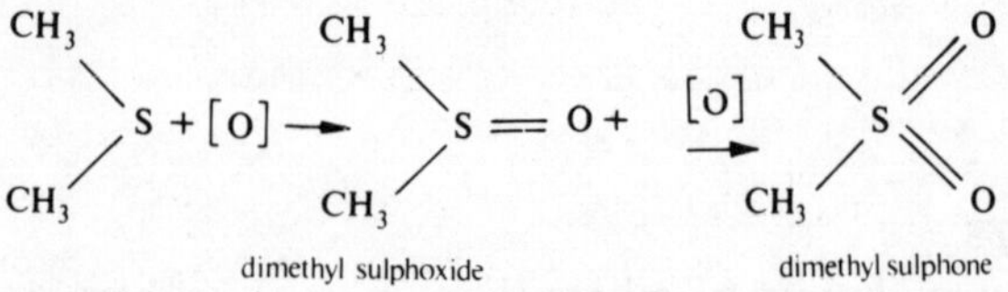

dimethyl sulphoxide   dimethyl sulphone

218

Disulphides

The preparation of disulphides has been described above. Such compounds are not commonly met except among proteins. The amino-acid cysteine (CySH) contains a thiol group and is easily oxidized to the disulphide cystine (CyS—SCy). Two protein chains each containing cysteine can be cross-linked by oxidation of the cysteine thiol groups to the disulphide. Such S—S bonds are widespread in proteins and are found, for example in silk, and the hormone insulin, which contains three such disulphide linkages.

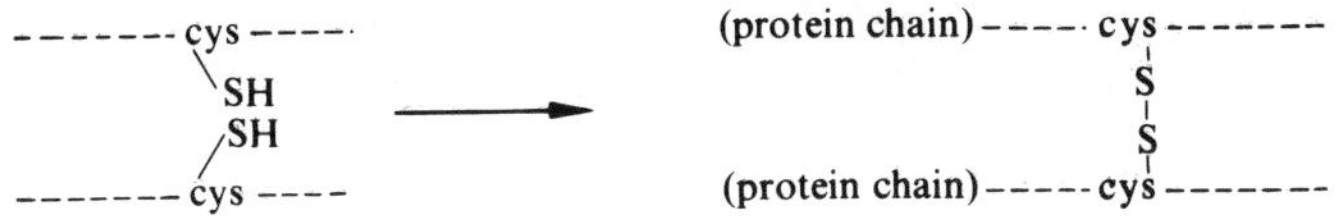

Thioaldehydes and Thioketones

As stated at the beginning of the chapter, sulphur does not readily form normal π bonds, so sulphur analogues of aldehydes and ketones are unknown or unstable in the monomeric state. For example, treatment of a ketone with hydrogen sulphide and hydrochloric acid gives a trimeric product with sulphur bonded only by sigma bonds:

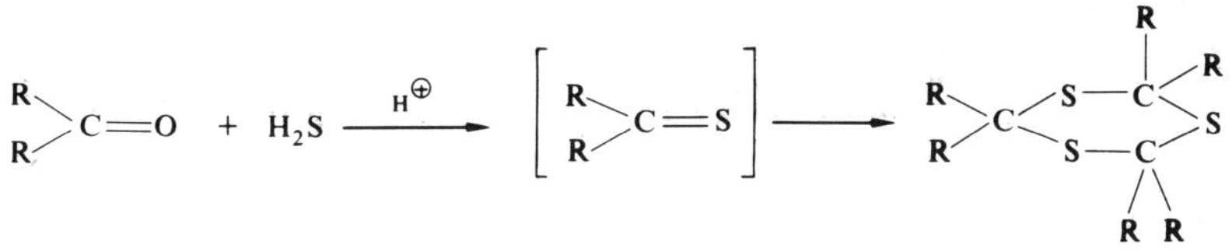

Thioacids

One or both of the oxygen atoms of a carboxylic acid can be replaced by sulphur, giving a thio-acid or dithio-acid respectively.

Thio-acids are in equilibrium between the two forms:

but because the sulphur has little tendency to form π bonds, the equilibrium is well to the right. Thio-acids can be prepared by the action of phosphorus pentasulphide on a carboxylic acid:

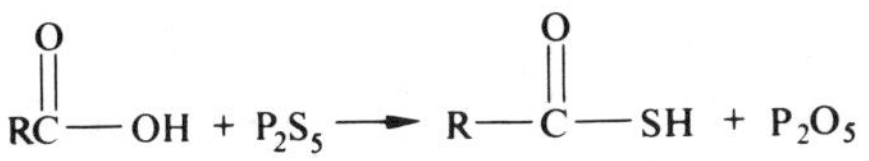

219

Dithioacids are prepared from a Grignard reagent and carbon disulphide:

$$RMgX + CS_2 \longrightarrow \overset{\displaystyle S}{\overset{\displaystyle \|}{RC}}{-}SMgX \overset{HX}{\longrightarrow} \overset{\displaystyle S}{\overset{\displaystyle \|}{RC}}SH + MgX_2$$

Both thio- and dithioacids have lower boiling points than the corresponding carboxylic acids because of the weaker hydrogen bonds formed by sulphur. Both groups of acids have very unpleasant odours. Thioacids react readily with alcohols to give normal esters, in contrast with the reaction of acids and thioalcohols:

$$\overset{\displaystyle O}{\overset{\displaystyle \|}{RC}}SH + R'OH \longrightarrow \overset{\displaystyle O}{\overset{\displaystyle \|}{RC}}OR' + H_2S$$

Thioacids react more readily because $SH^{\ominus}$ is a more stable ion than $OH^{\ominus}$, i.e. SH is a better leaving group than OH. Thioesters are made by the reaction of a thiol and acid chloride (page 217). Both thio- and dithioesters are rapidly hydrolysed by water:

$$\overset{\displaystyle O}{\overset{\displaystyle \|}{RC}}SR' + H_2O \longrightarrow \overset{\displaystyle O}{\overset{\displaystyle \|}{RC}}{-}OH + R'SH$$

Thioesters play an important part in the breakdown and synthesis of lipids and steroids in living tissues. Carboxylic acids are transferred from one enzyme reaction to another as thio-esters of the complex thiol, coenzyme A (usually written CoA—SH). For example the thioester of acetic acid with coenzyme A is the form in which acetate enters the sequence of enzyme-catalysed reactions which result in the synthesis of fatty acids, glycerides and waxes.

Higher Valency in Sulphur

Sulphur and oxygen in divalent compounds are shown with two unshared pairs of electrons. In oxygen these are not available for bonding except in the formation of oxonium salts and the special case of hydrogen bonding. In sulphur they are available for bonding with elements that contain unshared p electrons, for example oxygen.

This may be explained in terms of the atomic and molecular orbital concept by showing bond formation stepwise. Sulphur donates one of its unshared electrons to a neutral oxygen atom, this means both atoms now bear a charge and one unpaired electron (1). By combining, a σ bond is formed, but the two atoms still bear formal charges (2):

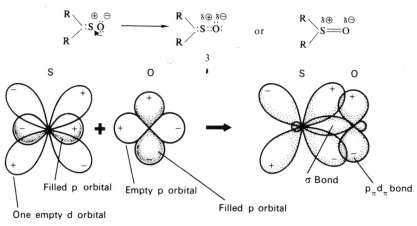

Normally the energy required to separate the charges would cancel the energy released in bond formation so that no permanent bond would result. But in the case of sulphur, and other second row elements, there are vacant $3d$ orbitals which can overlap with filled p orbitals of oxygen. A pair of non-bonding electrons from oxygen become shared with sulphur in formation of a $p_\pi d_\pi$ bond (Fig. 43) (so named because the bond has π type symmetry, but is made up of p and d orbitals). This results in donation of electronic charge to sulphur (3). Because of the greater electronegativity of oxygen, the resultant bond is polar:

Fig. 43 $p_\pi d_\pi$ *bonding between sulphur and oxygen*

By valency expansion in this way, two series of compounds are formed from thiols and thioethers by oxidation:

R—SH $\longrightarrow$ R—S—OH $\longrightarrow$ R—S—OH $\longrightarrow$ R—S—OH

sulphenic acid   sulphinic acid   sulphonic acid

R—S—R $\longrightarrow$ R—S—R $\longrightarrow$ R—S—R

sulphoxide   sulphone

Sulphoxides

Sulphoxides are made by mild oxidation of thioethers, for example with one equivalent of hydrogen peroxide. They are easily oxidized to sulphones or reduced (e.g. with zinc dust) to sulphides.

Except for dimethylsulphoxide they are crystalline solids. Dimethylsulphoxide (m.p. 8°, b.p. 189°) is a liquid, it partially decomposes during distillation at atmospheric pressure, but can be distilled unchanged at reduced pressure. It has become an important solvent in recent years because it is a good polar solvent for many inorganic and organic compounds, it is miscible with water and is non-toxic. It is obtained as a by-product in paper-making by the sulphite process. The commercial material has an unpleasant odour due to impurities.

Sulphones

Sulphones are made by oxidizing sulphides or sulphoxides with an excess of hydrogen peroxide. Sulphones are stable to further oxidation and can be reduced back to sulphides only with difficulty. Dimethylsulphone (m.p. 109°) is a water-soluble crystalline solid. Sulpholane, prepared industrially from butadiene and sulphur dioxide (page 73) is a liquid used as an industrial solvent.

$$
\underset{\substack{\text{dimethyl} \\ \text{sulphoxide}}}{\overset{\displaystyle O \atop \displaystyle \|}{CH_3SCH_3}}
\qquad
\underset{\substack{\text{dimethyl} \\ \text{sulphone}}}{\overset{\displaystyle O \atop \displaystyle \|}{\underset{\displaystyle \| \atop \displaystyle O}{CH_3SCH_3}}}
\qquad
\underset{\text{sulpholane}}{
\begin{array}{c}
CH_2 - CH_2 \\
| \quad\quad | \\
CH_2 \quad CH_2 \\
\diagdown \; S \; \diagup \\
\diagup \quad \diagdown \\
O \quad\quad O
\end{array}}
$$

Sulphonic Acids

Alkane sulphonic acids can be considered as derivatives of sulphuric acid where one OH group has been replaced by alkyl. In the same way sulphinic acids are alkyl derivatives of sulphurous acid. Sulphenic and sulphinic acids are easily further oxidized to sulphonic acids and are not commonly encountered.

Sulphonic acids can be prepared by oxidizing a thiol with potassium permanganate or hydrogen peroxide:

$$
RSH \xrightarrow{H_2O_2} \overset{\displaystyle O \atop \displaystyle \|}{\underset{\displaystyle \| \atop \displaystyle O}{RSOH}}
$$

They can also be prepared by the reaction of an alkyl halide and sulphite ion (by S_N2 mechanism):

$$CH_3CH_2Br + SO_3^{\ominus} \longrightarrow CH_3CH_2SO_3^{\ominus} + Br^{\ominus}$$

The alkylsulphonic acids, like sulphuric acid, are strong acids, soluble in water and polar organic solvents, and are hygroscopic.

With phosphorus pentachloride (but not thionyl chloride) they give sulphonyl chlorides, resembling carboxylic acid chlorides in reactivity. For example, they react with alcohols to give sulphonic esters and with amines to give sulphonamides:

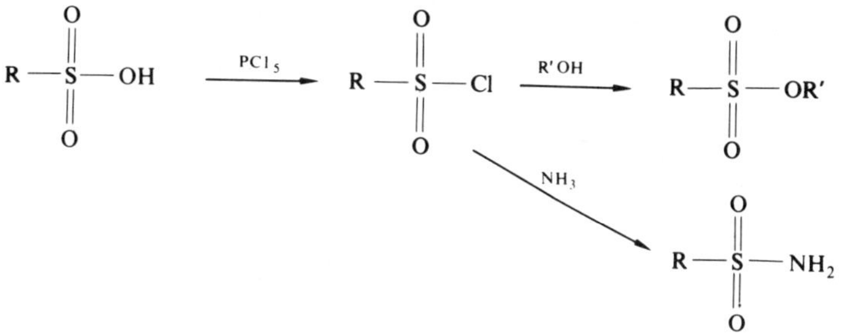

Note that sulphonic esters cannot be prepared from a sulphonic acid and alcohol.

Methanesulphonic acid (m.p. 20°, b.p. 167°/10 mm) is a colourless viscous liquid. Its calcium, barium and lead salts are soluble in water (in contrast to the sulphates). Salts of sulphonic acids with large alkyl groups are used as synthetic detergents (as are salts of alkyl esters of sulphuric acid: the two groups should not be confused):

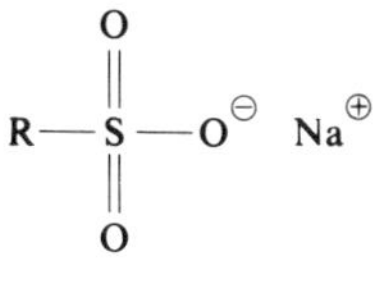

a sodium alkylsulphonate 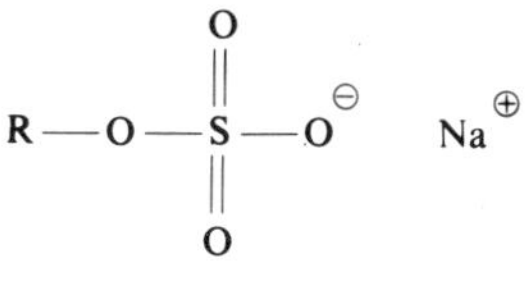   a sodium alkyl sulphate

Identification of Sulphur Compounds

Sulphur is identified in the sodium fusion (Lassaigne) test by formation of a deep red colour with sodium nitroprusside reagent. One or more of the reactions described in this chapter is then used to distinguish which class of sulphur compound is present. For example, a strongly acidic, water soluble compound will almost certainly be a sulphonic acid. A weakly acidic substance readily oxidized by permanganate would most probably be a thiol. The S—H stretching

223

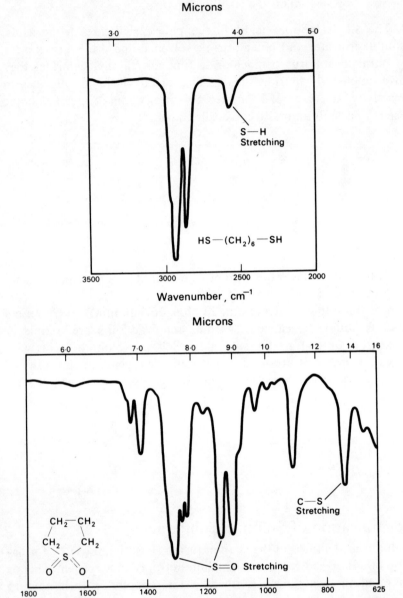

Fig. 44   *Parts of the infrared spectra of hexanedithiol and sulpholane*

224

vibration is seen as a weak infrared band at 2500 cm^{-1} (Fig. 44). Sulphur-oxygen bonds in sulphoxide, sulphones and sulphonic acid derivatives give rise to strong bands in the 1400 and 1200 cm^{-1} regions. The C—S bond absorbs near 600 cm^{-1}.

11

Alicyclic Compounds

Carbon atoms may form rings as well as straight and branched chains. The simplest form of carbon rings are found in alicyclic (*ali*phatic-*cyclic*) compounds. A great deal of their chemistry can be understood as a logical extension of the chemistry of alkanes and their functional derivatives, but alicyclic compounds possess additional properties that require special consideration. For this reason their description has been omitted up to now. In this chapter those special considerations which rings require are examined while illustrating the essential sameness of the properties of functional groups in both chains and rings.

The structures of carbon rings are essentially three-dimensional and not easy to describe on paper. The reader will find this chapter will be more easily understood if he is able to construct some rings from wire atomic models. More can be gained by examining models than can be conveyed in pages of writing or illustrations.

Nomenclature

The names of alicyclic hydrocarbons or cyclo-alkanes are obtained from the names of normal alkanes with the same number of carbon atoms, adding the prefix *cyclo-*. The Geneva system of naming is followed closely in this series and only occasionally are trivial names used. The naming of functional groups follows the system already used, thus: an alicyclic compound containing a double bond is called a cycloalkene; with two double bonds it is a cycloalkadiene, with an alcohol function, a cycloalkanol, and so on. Notice that carboxylic acids must be named differently because the functional group is attached to, and not in, the ring (see opposite).

Note that the cycloalkanes are represented by the formula C_nH_{2n}, the same as the alkenes, the cycloalkanols by C_nH_{2n-1}, the same as alcohols with one double bond, and so on. More information than the molecular formula is necessary to know to which class a

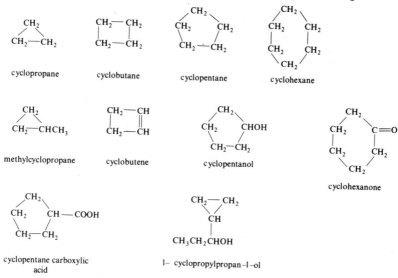

compound belongs; but a formula C_nH_{2n} must refer to *either* an alkene or a monocyclic paraffin (see page 251).

Baeyer's Strain Theory

As soon as we investigate the ways of making carbon rings of different size, we discover that these compounds vary greatly in the ease with which they are formed. To explain this, von Baeyer in 1885 pointed out that the angle between the tetrahedral valencies of carbon is 109° 28' (Fig. 45) and that some distortion is necessary to form a three, four or five membered ring.

Baeyer proposed that a *strain* is exerted on the valencies in distorting the angle between them, and the greater the strain, the greater the difficulty in forming the compound and the less stable it will be. This fits the facts quite well for small rings. Cyclopropane, where the distortion is greatest (24° 44' for each valence) is the least stable. Cyclobutane (9° 44' distortion of each bond) is more stable, while cyclopentane (only 44' of distortion) is the most easily formed and is as stable as any linear alkane.

In the molecular orbital picture of the carbon atom, distortion of the carbon bond angles to less than 104° is not possible on theoretical grounds, and the instability of small rings is due to the incomplete overlap of molecular orbitals (Fig. 46) giving weaker bonds.

Baeyer's theory implied that the cycloalkane rings are flat and therefore in cyclohexane a strain would exist in the opposite sense, since a regular hexagon has an internal angle of 120°, so the bonds

227

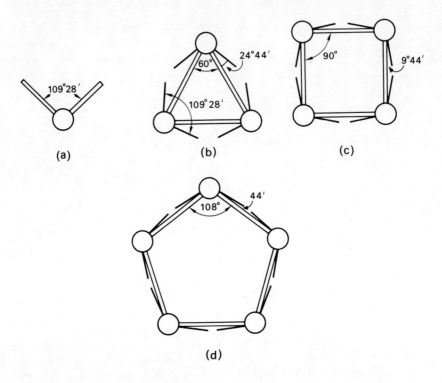

Fig. 45   The formation of small carbon rings (a) The angle between the tetrahedral valencies of carbon (b) Cyclopropane (c) Cyclobutane (d) Cyclopentane

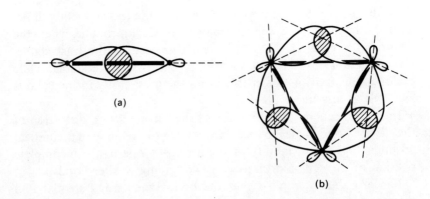

Fig. 46   Optimum overlap of sp³ hybrid orbitals in an alkane C—C bond (a), compared with less efficient overlap in cyclopropane (b)

Table 19.　Strain in cycloalkanes

	ΔH_c (kJ/mole)	$\dfrac{\Delta H_c}{\text{No. of C atoms}}$	ΔH_h (kJ/mole)	C—C bond dissociation energy (kJ/mole)
Cyclopropane	2080	693	155	238
Cyclobutane	2745	686	167	293
Cyclopentane	3320	664	68	385
Cyclohexane	3930	655	60	398
Cycloheptane	4630	661	—	389
Cyclo-octane	5310	663	—	389

would be forced apart and still larger rings would require even greater distortion. This means that cyclohexane should be less stable than cyclopentane and larger rings still less stable. It was known at that time that large rings are difficult to form; but cyclohexane is both stable and easily made. Shortly after, in 1890, Sachse pointed out that cyclohexane could form an unstrained ring by twisting into a non-planar shape. This is readily seen when a cyclohexane ring is formed with wire atomic models; the ring automatically takes on a non-planar shape (Fig. 49). Mohr in 1918 took up Sachse's half-forgotten proposals and extended them in ways which we shall consider later, but essentially the Sachse–Mohr theory accounts for the lack of strain in cyclohexane and larger rings by the non-planarity of the rings. That there is no strain can be shown by examining any of several molecular properties. We may consider the heat of combustion (ΔH_c) of the series. On burning in air all hydrocarbons produce carbon dioxide, water and heat. The energy released in burning a linear hydrocarbon is very close to 660 kJ/mole for every CH_2 group. If a compound on combustion gives more than this amount of energy, then more than the normal amount of energy must have been required to put the molecule together, i.e. the compound must possess some strain or instability. In Table 19 are given the ΔH_c values of the small ring cycloalkanes and the corresponding values per CH_2 group. Note that cyclopropane has the highest value (per CH_2) and that this falls steadily to cyclohexane and from there onward the value is as for a linear alkane. For ethylene, ΔH_c is 1410 kJ/mole or 705 kJ per CH_2 group, showing that in one sense the double bond is more 'strained' than a cyclopropane ring.

The energy released on hydrogenation (ΔH_h) is also an indicator of relative stability, and so is the bond dissociation energy, that is, the energy which must be supplied to break a single bond, in this case one of the C—C bonds of the ring. Less energy is required for cyclopropane because it is already in a strained state; the values rise to a steady 390 kJ/mole in the larger rings.

It is now known that in addition to the strain due to bending bonds, a strain caused by the repulsion between hydrogen atoms on different carbon atoms must be considered in determining the ease with which carbon rings are formed (see page 241).

Occurrence of Alicyclic Compounds

Simple cycloalkanes and cycloalkenes occur in petroleum, usually as mixtures of methyl- and dimethyl-cyclopentanes and -cyclohexanes. Pure products are difficult to obtain from petroleum, but fractions rich in cycloalkanes, called naphthenes, can be obtained from certain crude oils.

A second source of alicyclic compounds are the terpenes, found in turpentine, oil of eucalyptus and the essential oils (i.e. volatile oils obtained from the leaves, flowers and roots of many plants). The terpenes are an important series of compounds, to be discussed in a later volume. Here it is worthwhile looking at the structure of a few examples; they usually contain a six-membered ring and sometimes a three- or four-membered ring as well. The alternative formula is a simplified way of writing alicyclic rings:

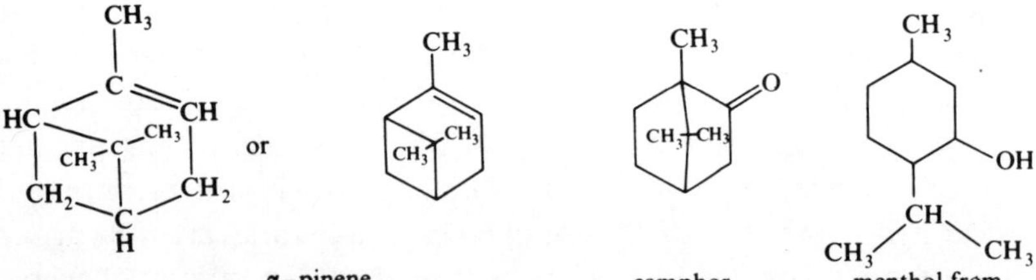

α–pinene
from turpentine

camphor

menthol from
peppermint

More complex alicyclic compounds abound among animal and plant products and many are vitally important in biological processes.

Physical Properties

Branched chain alkanes have lower boiling points than those with the same number of carbon atoms in a straight chain because branched chains are more compact and so have higher vapour pressures, but the straight chains fit better into a crystal lattice and so have higher melting points. For the same reasons alicyclic compounds tend to have higher boiling points and melting points than comparable linear compounds. Otherwise alicyclics have no characteristic appearance or properties; they rather follow in line with aliphatic compounds. Cycloalkanes and cycloalkenes range from gases to high boiling liquids. Alicyclic alcohols are generally liquids, and ketones and aldehydes are liquids with pleasant odours. The methods of preparation useful for alicyclic compounds vary with the size of the ring to be formed; these are described in the following sections.

Cyclopropane

In spite of the angle strain involved, a number of good methods for making cyclopropane rings are available. Derivatives of cyclopropane can be prepared by the addition of a carbene to an alkene. A carbene is a reactive intermediate with the properties of a strong electrophilic reagent. The simplest example is methylene CH_2: which can be produced by decomposing diazomethane or ketene in the gas phase with ultraviolet light:

$$CH_2N_2 \xrightarrow{h\nu} CH_2: + N_2$$

$$CH_2C{=}O \xrightarrow{h\nu} CH_2: + CO$$

The methylene as first produced by this reaction is a very reactive fragment and inserts itself into C—H and C—C bonds in a random way:

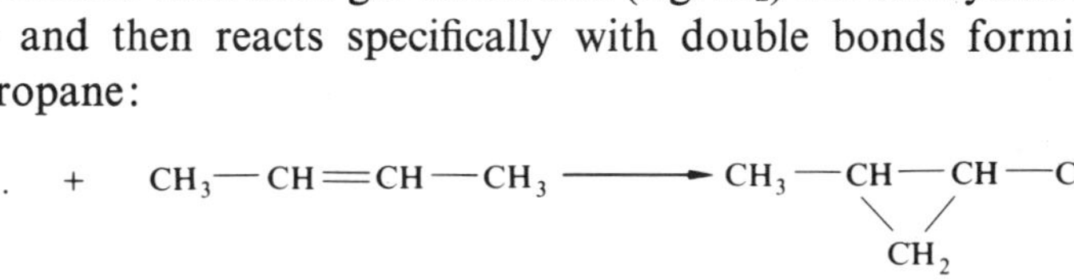

By collision with inert gas molecules (e.g. N_2) the methylene loses energy and then reacts specifically with double bonds forming a cyclopropane:

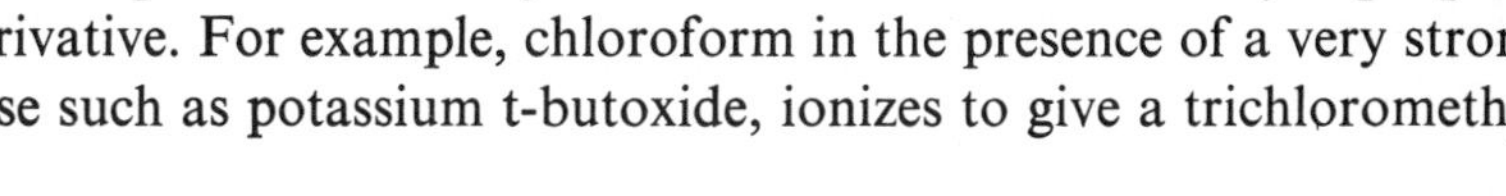

Substituted methylenes in general are called carbenes. Some simple carbenes can be produced in solution. In this case, the less reactive form is produced directly and can be converted to a cyclopropane derivative. For example, chloroform in the presence of a very strong base such as potassium t-butoxide, ionizes to give a trichloromethyl ion, which directly decomposes to a chloride ion and dichloro-carbene:

$$\text{t-BuO}^{\ominus} \quad \text{H}-\text{CCl}_3 \longrightarrow \text{t-BuOH} + \text{CCl}_3^{\ominus}$$

$$\text{CCl}_3^{\ominus} \longrightarrow \text{Cl}^{\ominus} + \text{CCl}_2\text{:}$$

If an alkene is present in the solution, a dichloro-cyclopropane is formed:

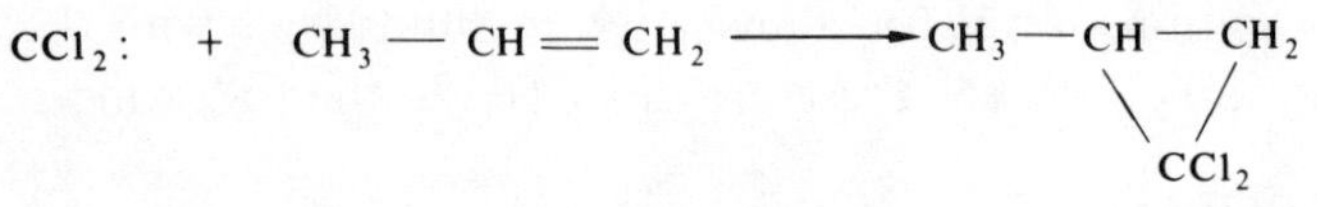

$$\text{CCl}_2\text{:} + \text{CH}_3-\text{CH}=\text{CH}_2 \longrightarrow \text{CH}_3-\text{CH}-\text{CH}_2 \quad \diagdown\diagup \quad \text{CCl}_2$$

Carbenes containing halogen are relatively more stable, and less indiscriminate in their reactions. It is thought the halogen atoms can donate some charge to the strongly electron deficient carbon atom. In molecular orbital terms, the orbitals of the unshared p electrons can overlap with the empty orbital on carbon:

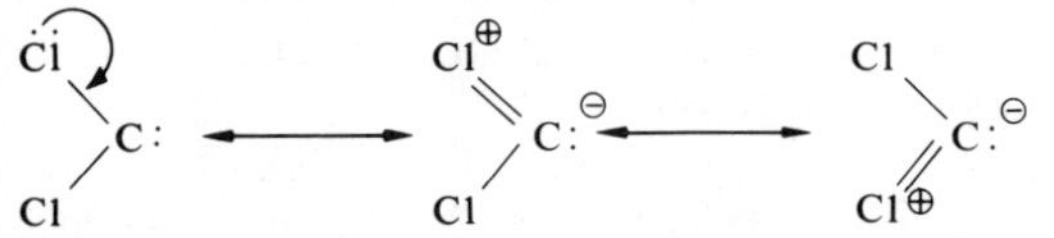

Therefore $CF_2\text{:}$, $CCl_2\text{:}$ and $CBr_2\text{:}$ can be prepared transiently in solution by a number of routes, e.g. by heating sodium trichloro-acetate above $80°$ in an inert solvent:

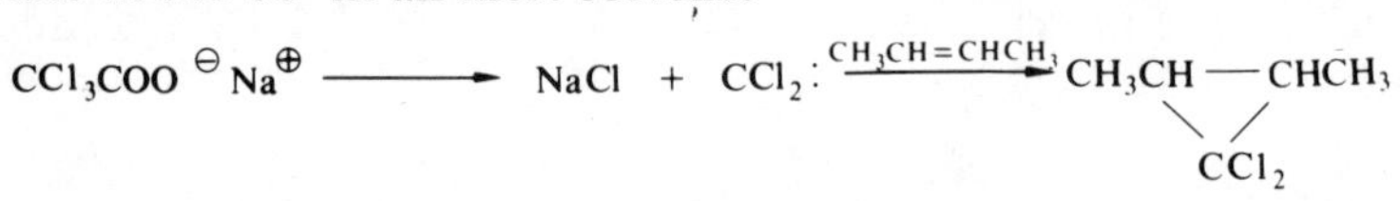

$$\text{CCl}_3\text{COO}^{\ominus}\text{Na}^{\oplus} \longrightarrow \text{NaCl} + \text{CCl}_2\text{:} \xrightarrow{\text{CH}_3\text{CH}=\text{CHCH}_3} \text{CH}_3\text{CH}-\text{CHCH}_3 \quad \diagdown\diagup \quad \text{CCl}_2$$

The last compound above (1,1-dichloro-2,3-dimethylcyclopropane) can exist in two geometrical isomers. Because the three atoms of the ring lie in a plane and substituents lie above or below the plane, cyclopropanes can exist in *cis* and *trans* forms rather like alkenes. The two cyclopropane-1,2-dicarboxylic acids below can be used to illustrate this. In the *cis* acid, the carboxyl groups are close together and together can form an anhydride, in the *trans* compound they cannot.

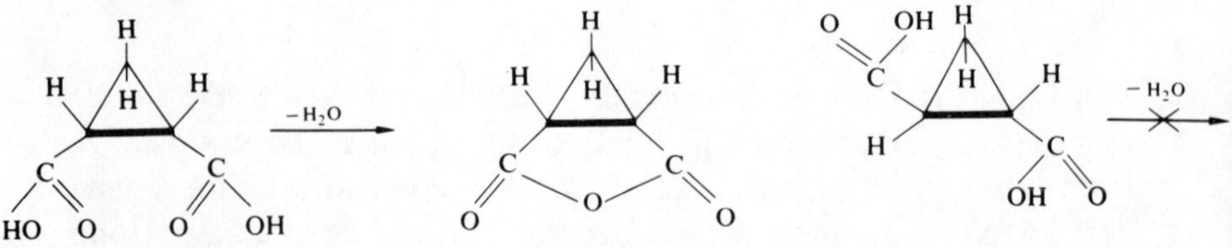

cis   *trans*

The addition of a carbene is stereospecific, that is, only one possible isomer is formed with a given alkene. The addition takes place in a *cis*-manner, either from above or below the plane of the alkene, so presumably the carbene adds to both ends of the double bond simultaneously:

cis –2–butene   a *cis*–cyclopropane

trans–2–butene   a *trans* –cyclopropane

If addition took place on first one carbon atom, then the other, we know from other reactions that there would be an opportunity for a bond to rotate during the addition

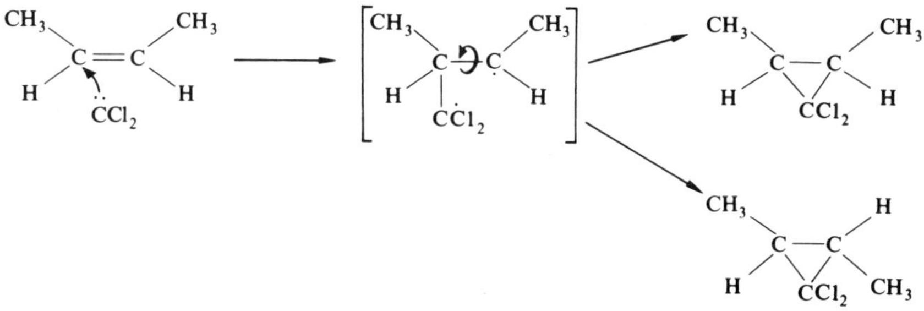

Such a mixture of products is not formed, so the second mechanism can be eliminated.

Cyclopropanes can also be made from methylene iodide, zinc and an alkene. Although the reaction is stereospecific like the addition of a carbene, and gives only one product, it does not appear to be a simple carbene reaction:

$$CH_2I_2 + Zn \longrightarrow [ICH_2ZnI]$$

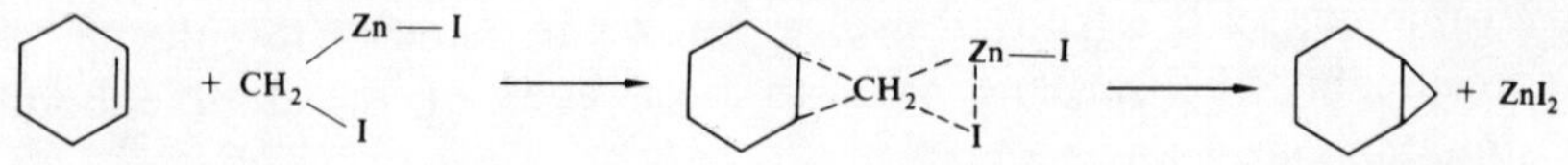

cyclohexene   norcarane

Cyclopropane itself can be made from 1,3-dibromopropane by the action of zinc (compare the formation of ethylene from its dibromide and zinc). The reaction can be regarded as an internal or intramolecular Wurtz reaction:

The method is not practical for larger rings.

Cyclopropane is a gas (b.p. $-34°$), it has good anaesthetic properties, but is dangerously explosive when mixed with air and has been replaced by halothane. Cyclopropene, a highly strained compound with a double bond in a three-membered ring, is known though it is unstable at room temperature. Cyclopropane rings are found in a number of natural products. In particular, two acids are known, related to oleic acid (page 196). Lactobacillic acid contains a cyclopropane ring and sterculic acid contains a cyclopropene ring.

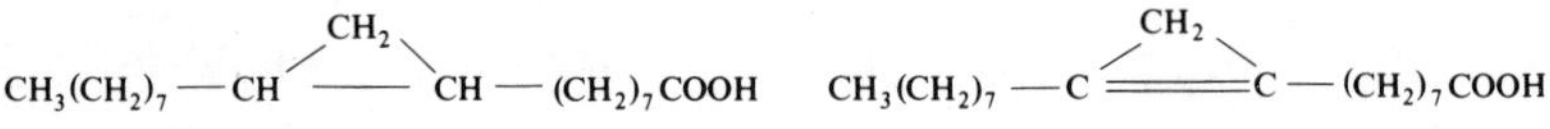

lactobacillic acid   sterculic acid

Cyclobutane

Cyclobutane rings can be prepared from a 1,4-dihalo-alkane and the sodium salt of diethyl malonate. Diethyl malonate readily forms a reactive carbanion:

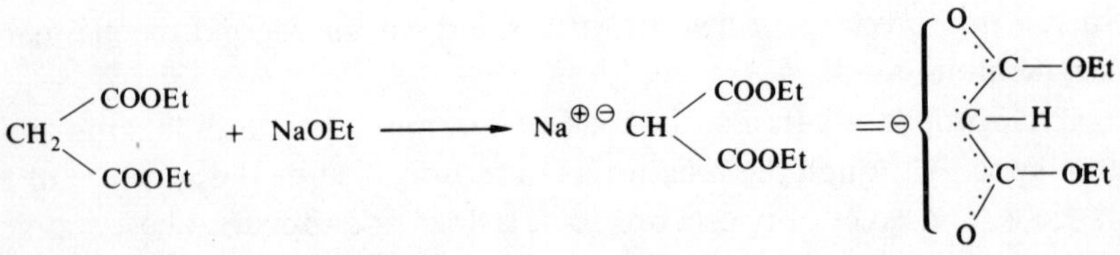

234

This ion attacks a halide by an S$_N$2 reaction:

CH$_2$—Br \quad CH$_2$⊖C—H COOEt / COOEt → CH$_2$ CH$_2$ CH$_2$Br C—H COOEt COOEt

In the presence of another equivalent of sodium ethoxide, a carbanion is again formed and further reaction occurs to give a ring compound:

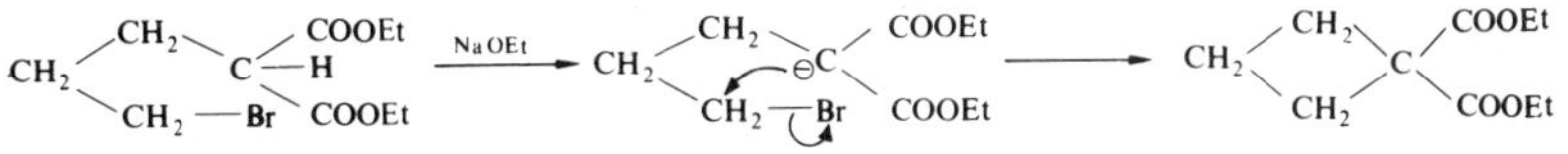

CH$_2$ CH$_2$—Br C—H COOEt COOEt →(NaOEt)→ CH$_2$ CH$_2$—Br ⊖C COOEt COOEt → CH$_2$ CH$_2$ C COOEt COOEt

The di-ester can be hydrolysed to the acid which readily decarboxylates on heating to give a cyclobutane carboxylic acid:

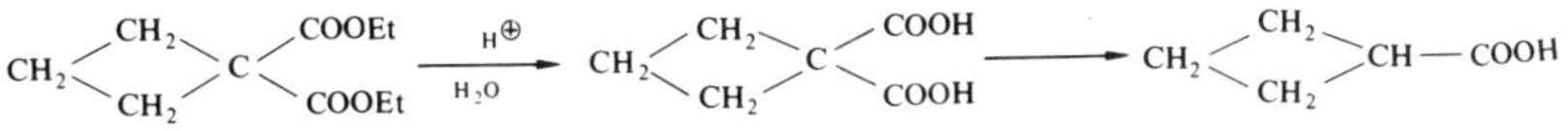

CH$_2$ CH$_2$ C COOEt COOEt →(H⊕, H$_2$O)→ CH$_2$ CH$_2$ C COOH COOH → CH$_2$ CH$_2$ CH—COOH

This reaction is of more general usefulness; three-, four- and five-membered rings can all be made using the appropriate alkyl dihalide. It is also a special case of a general reaction for making alkylated carboxylic acids.

Cyclobutane is a gas (b.p. 13°) of no commercial importance. The cyclobutane ring occurs in some natural products, such as the pinene group of terpenes (page 230). Cyclobutane derivatives can exist in *cis* and *trans* forms, but with more isomers possible than in related substituted cyclopropanes:

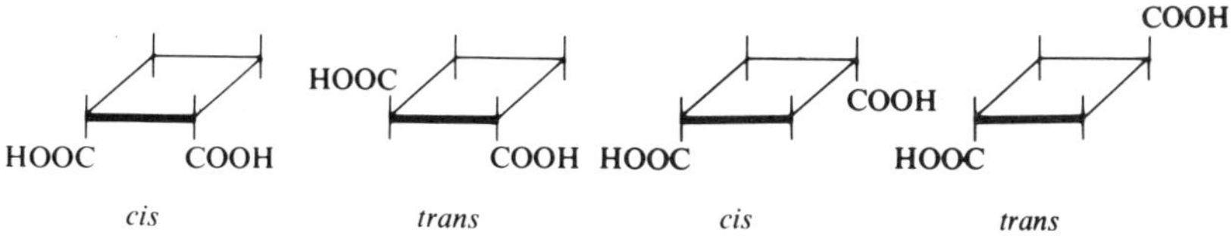

HOOC   COOH   HOOC   COOH HOOC   COOH   HOOC   COOH

cis   *trans*   *cis*   *trans*

Cyclopropane must be a planar molecule, since a plane can always be drawn through three points. Cyclobutane need not be, but any departure from planarity will increase the strain in the carbon bond angles beyond the 19·5° required to distort them inward from 109·5° to 90° in a square arrangement. Surprisingly the evidence indicates that cyclobutane is *not* flat but slightly bent (Fig. 47). The explanation is found in the mutual repulsion of groups on adjacent carbon atoms.

The C—C bond length in cyclobutane has been found to be 1·56 Å and the C—H bond 1·09 Å. The accepted radius of a hydrogen atom on the non-bonding side (the so-called van der Waals radius) is

1.2 Å. By constructing a scale diagram of part of a planar cyclo-butane molecule (Fig. 48) it is seen that the hydrogen atoms are closer together than their preferred non-bonding distance. Note also that the hydrogen atoms in this model are in the eclipsed relation-ship, as described in Chapter 3. In general, molecules such as ethane, as they rotate about the central C—C bond, spend very little time in the eclipsed position (see Fig. 7) because of this mutual repulsion. In the staggered position the groups on adjacent atoms are as far apart as possible.

Cyclobutane cannot rotate about its C—C bonds, but by distorting the ring slightly, the hydrogen atoms on adjacent carbon atoms can move a little farther apart. The slightly bent form of the cyclo-butane molecule is a compromise between strain in the carbon bond angles and distortion to avoid the crowded eclipsed positions. Though rotation is not possible, vibration causes the molecule to flip over rapidly at normal temperature between two extreme positions.

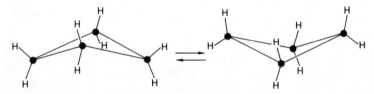

Fig. 47   Cyclobutane

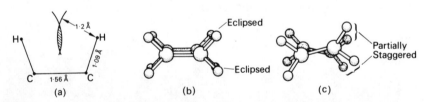

Fig. 48   Crowding in the cyclobutane molecule (a) Scale diagram showing closeness of hydrogen atoms in a planar cyclobutane (b) perspective view of planar cyclobutane showing eclipsed hydrogen atoms (c) twisted cyclobutane showing staggering of hydrogen atoms

The greater stability of cyclobutane, compared with cyclopropane is evident in the higher temperature necessary to hydrogenate cyclobutane and the fact that it does not undergo ring-opening addition with bromine.

Cyclobutene is known and behaves like an alkene. Many attempts have been made to prepare cyclobutadiene, but too much bond angle strain is apparently involved and the structure is only known in certain derivatives, e.g. tetraphenylcyclobutadiene and transition metal complexes of cyclobutadiene and tetramethylcyclobutadiene:

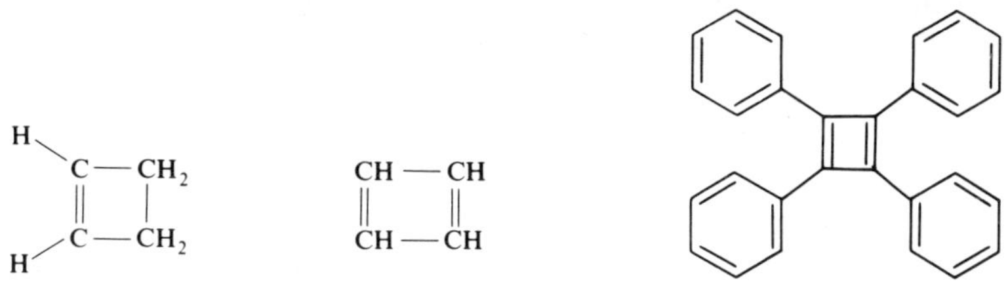

cyclobutene　　　　　cyclobutadiene　　　　　tetraphenylcyclobutadiene

Cyclopentane

There is a wider choice of methods for making cyclopentane rings than there is for cyclopropanes and cyclobutanes. First, the malonic ester method described under cyclobutane can be used.

Pyrolysis of Calcium or Barium Adipate. This is a special case of the method for making aldehydes and ketones. When the calcium or barium salt of a dicarboxylic acid is pyrolysed a cyclic ketone is formed. The method works best when making 5, 6 or 7-membered rings:

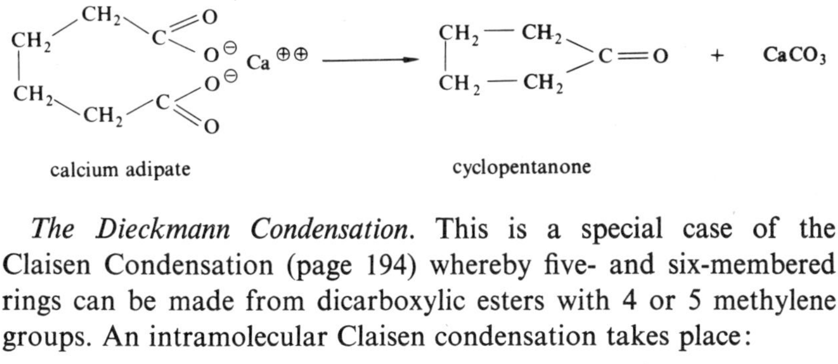

calcium adipate　　　　　　　cyclopentanone

The Dieckmann Condensation. This is a special case of the Claisen Condensation (page 194) whereby five- and six-membered rings can be made from dicarboxylic esters with 4 or 5 methylene groups. An intramolecular Claisen condensation takes place:

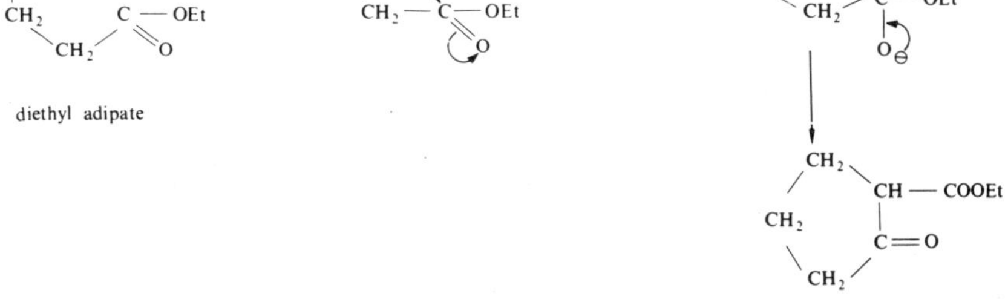

diethyl adipate

237

The product of the condensation is a β-keto-ester. This can be hydrolysed to the corresponding acid, which is unstable and decarboxylates on heating to give a ketone (compare the behaviour of malonic acid, a β-dicarboxylic acid, above):

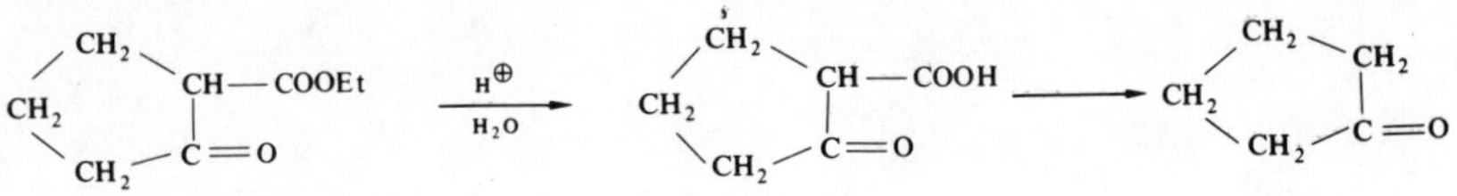

The acyloin reaction described later under large rings also gives excellent yields of cyclopentane derivatives.

Cyclopentane (b.p. 49°) occurs in certain petroleums, and derivatives containing the cyclopentane ring are common throughout nature. Methylcyclopentane is isomeric with cyclohexane and the two are interconvertible under certain catalytic conditions.

The internal angles of a regular pentagon are 108°, not very different from the normal bond angles of carbon, so there should be little angle strain in cyclopentane. In a planar cyclopentane, however, the hydrogen atoms would again be in eclipsed positions, as one considers each pair around the ring. The molecule is, in fact, slightly twisted, like cyclobutane, so that the totally eclipsed positions are avoided, which introduces a little more angle strain. The vibrations of the atoms cause the actual shape of the molecule to be changing constantly at ordinary temperature.

The cyclopentane ring therefore is only slightly strained and cannot be opened as readily as the smaller rings. It resists hydrogenation over a nickel catalyst up to 200°.

Cyclopentene can be prepared by the dehydrobromination of cyclopentyl bromide:

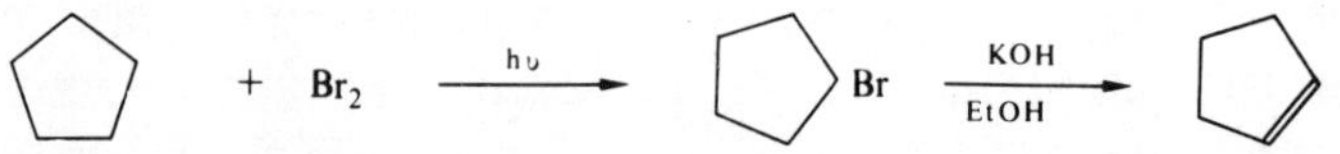

cyclopentene

Cyclopentadiene is found in the low boiling fractions of coal tar. Though more stable than cyclobutadiene it dimerizes on standing at room temperature. The dimerizing reaction is a kind of Diels–Alder reaction (see next section). By heating the dimer, the monomer can be re-formed and distilled out:

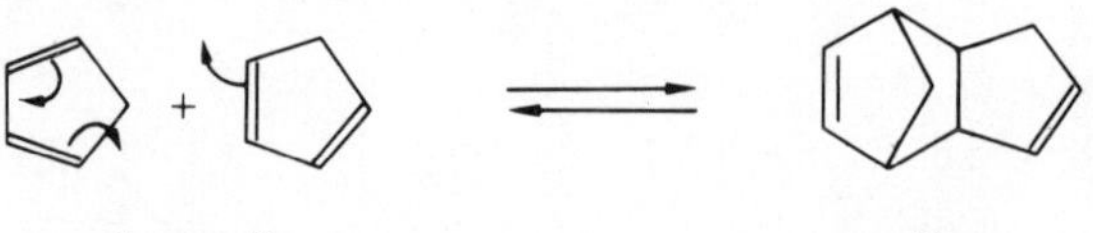

cyclopentadiene   dimer

Cyclopentadiene (as a diene) readily undergoes many Diels–Alder reactions with other dienophiles.

Cyclohexane

The methods described above for the formation of cyclopentane rings may also be used for cyclohexanes. For example, pyrolysis of the calcium or barium salt of pimelic acid will give cyclohexanone:

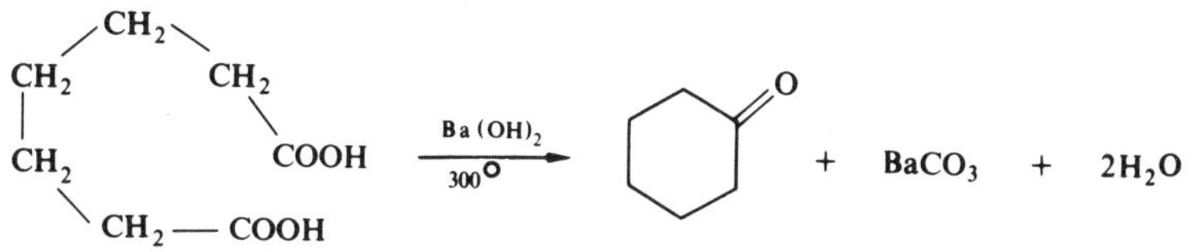

Diels–Alder Reaction. This is a very important reaction for ring formation. It involves the reaction of a conjugated diolefin (or diene) with an activated olefin (called the dienophile). The activated olefin is one where the double bond is in conjugation with a ketone, ester, acid or nitrile group, which increases the mobility of the electrons in the double bond. Usually no catalyst is required and in many cases the reaction takes place at normal temperature. An example of the Diels–Alder reaction is the condensation of butadiene (a diene) with acrolein (a dienophile) to give tetrahydrobenzaldehyde or cyclohex-3-enaldehyde:

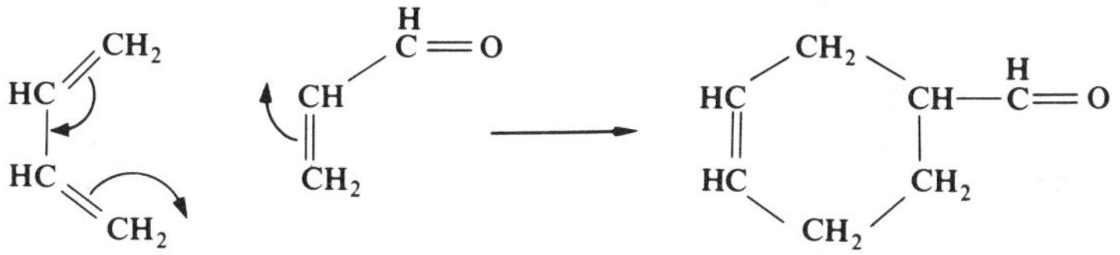

acrolein

Though we may draw a rearrangement of bonds as above, the exact mechanism of the reaction is unknown. Maleic anhydride is a very reactive dienophile, having two carbonyl groups to activate the double bond, and reacts with a large number of dienes:

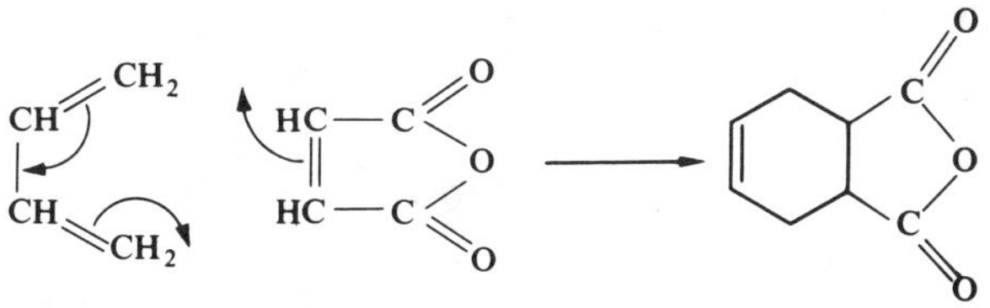

maleic anhydride   tetrahydrophthalic
                          anhydride

239

Notice that a six-membered ring is normally formed in the Diels–Alder reaction. If the diene is already a ring compound, the resulting compound has a *bridged* ring. In the example below the bridge is between carbon atoms 3 and 6:

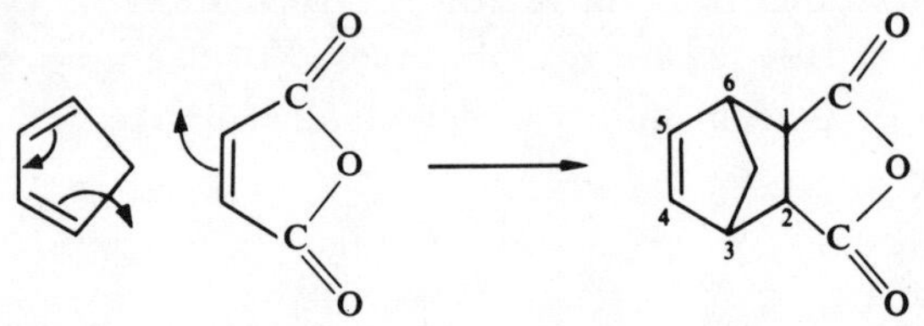

Hydrogenation of Benzene. Benzene can be catalytically hydrogenated to give cyclohexane. Other aromatic hydrocarbons and derivatives of them can also be hydrogenated; for example, phenol gives cyclohexanol. This method provides a direct route between aromatic and alicyclic compounds.

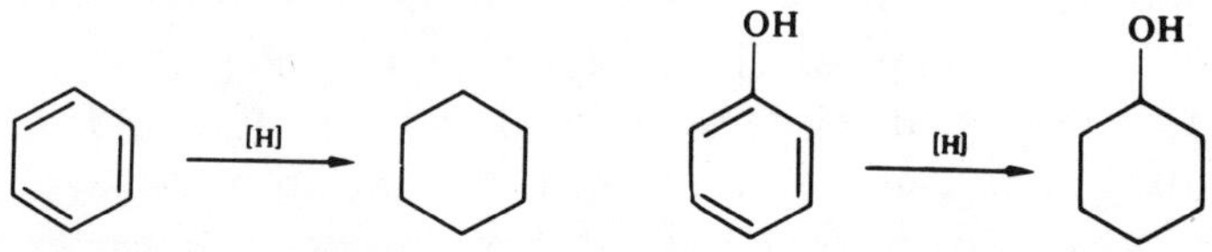

By constructing a model of cyclohexane with wire atomic models, one finds that no strain is imposed upon the tetrahedral arrangement of bonds, and that two non-planar forms are possible which can be interconverted without taking the model apart, by rotating the bonds to flip the model over from one form to the other. Sachse–Mohr theory to explain the lack of strain in cyclohexane and larger molecules predicted a non-planar ring and further, that two possible forms could exist, though only one cyclohexane was known in the laboratory. We now know that cyclohexane consists of an equilibrium mixture of the two forms (Fig. 49) which are rapidly interconverted at room temperature. The two forms are called the 'boat' and 'chair' conformations. The term conformation is used to describe molecular shapes that are readily interconverted by rotation of single carbon-carbon bonds. For example the eclipsed and staggered forms of ethane represent two conformations of ethane.

Closer examination of the two cyclohexane forms shows that in the boat conformation, the hydrogen atoms on carbons 2 and 3, and 5 and 6 (the sides of the boat) are in an eclipsed position (the bow and stern are staggered with respect to the sides); but in the chair conformation the hydrogen atoms are in the staggered position all the way round the ring (Fig. 50). This will make the chair form relatively much more stable and less crowded than the boat form. Moreover, two hydrogen atoms in the boat form (on carbons 1 and 4) are very close together, so there will be mutual repulsion (Fig. 51). Some of this crowding and the completely eclipsed position can be avoided by slightly twisting the ring to the skew-boat form. This close approach of hydrogen atoms is avoided in the chair form, and indeed the equilibrium mixture is found to contain about 99% of the chair conformation.

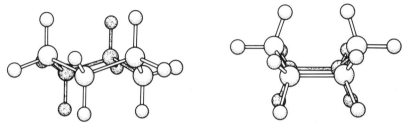

Fig. 49   Cyclohexane, the boat form (a) and the chair form (b)

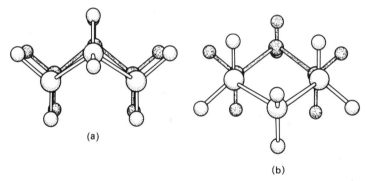

(a)

(b)

Fig. 50   Cyclohexane. End-on views of (a) boat and (b) chair forms to show eclipsed and staggered relationships of hydrogen atoms

Six of the C—H bonds of chair cyclohexane are exactly perpendicular to a plane through the mid-points of the C—C bonds (approximately, the plane of the ring) and are known as *axial* bonds (Fig. 52). The other six C—H bonds are almost within this plane and are known as *equatorial* bonds, by analogy with the poles and equator of the earth.

In cyclohexane all these hydrogen atoms are equivalent because of the rapid changes of conformation, but as soon as we replace hydrogen by a larger group it introduces another complication. A large group in the axial position on carbon atom 1 is very close to the axial hydrogen atoms on carbons 3 and 5, but with the large group in the

241

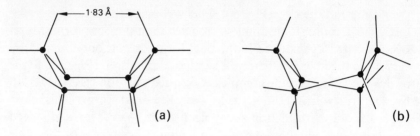

Fig. 51 Cyclohexane showing the close approach of hydrogen atoms in the boat form (a) and how this compression is partly relieved in the skew-boat form (b)

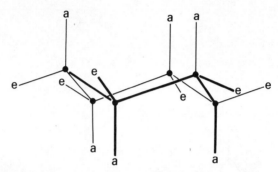

Fig. 52 Cyclohexane, chair form showing axial a and equatorial e bonds

equatorial position, this crowding is avoided. Therefore substituted cyclohexanes will tend to exist not only in the chain form but with the substituent group in the equatorial position (Fig. 53).

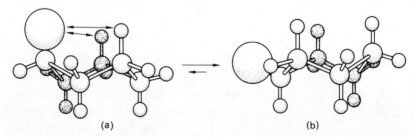

Fig. 53 A large substituent on a cyclohexane is less crowded in an equatorial position (b). Interaction with hydrogen on atoms 3 and 5 makes the axial position (a) less favourable

Since so many important chemicals from biological materials contain alicyclic rings the understanding of their properties and behaviour form an important part of chemical investigations. For example, the detailed chemistry of the steroids (which contain five alicyclic rings fused together) is understood very much in terms of conformational arguments developed from simpler rings of the type discussed above.

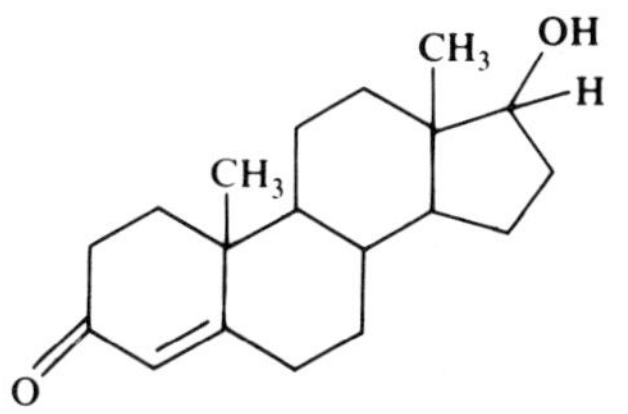

testosterone

a steroidal sex hormone

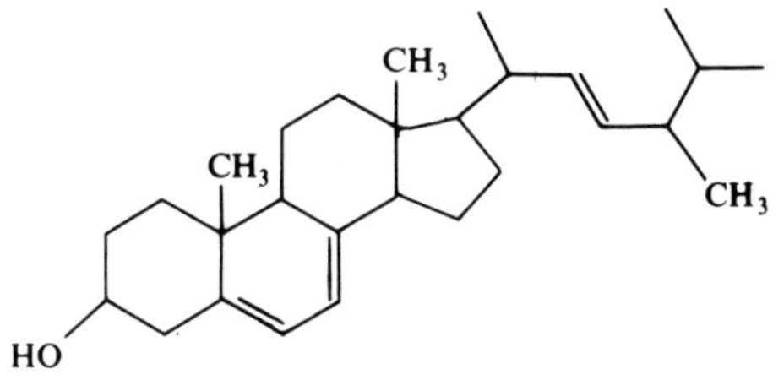

ergosterol, a steroid
of yeast, converted
to vitamin D by light

Cyclohexane occurs in petroleum, but it is obtained commercially by hydrogenating benzene or phenol. Cyclohexane and cyclohexanol are readily available from phenol. Phenol is hydrogenated to cyclohexanol, which is dehydrated to cyclohexene and then hydrogenated to cyclohexane:

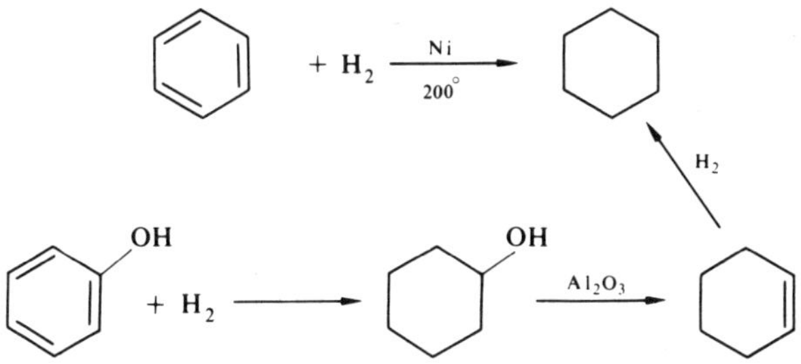

Cyclohexane can be dehydrogenated to benzene with platinum or palladium catalysts. Cyclohexane is oxidized by hot potassium permanganate or concentrated nitric acid to adipic acid. Commercially cyclohexane is oxidized by air to make adipic acid for manufacturing nylon:

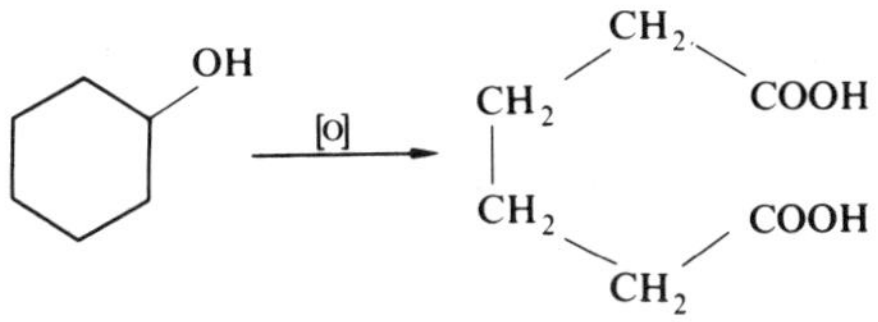

The compound decalin (or decahydronaphthalene, prepared by hydrogenating naphthalene) contains two cyclohexane rings fused together, and in this case the two forms of cyclohexane are fixed and not interconvertible, so two forms of decalin can be isolated; they

243

are called *cis* and *trans* from the arrangement of the groups around the carbon atoms common to both rings (Fig. 54).

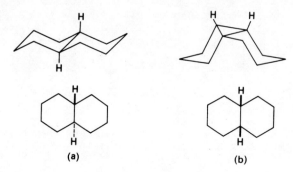

(a)   (b)

Fig. 54   The two kinds of decalin (a) trans-decalin (b.p. 186°) and (b) cis-decalin (b.p. 194°)

As a convention for writing such structures, a hydrogen above the plane of the rings is drawn with a heavy line as its bond, and a hydrogen below the plane of the rings with a dotted line.

Cycloheptane

Though it occurs in petroleum, cycloheptane is not readily available in quantity. It can be made from calcium suberate by the following sequence:

Cycloheptane derivatives are often easily isomerized to smaller rings, e.g. cycloheptane with hydrogen iodide gives methylcyclohexane and dimethylcyclopentane. The ring of cycloheptane, like all higher cycloalkanes is puckered to avoid strain.

Cyclo-octane

Cyclo-octane can be obtained from calcium azeleate by the method

244

outlined above for cycloheptane from calcium suberate, but better methods for making cyclo-octane rings have appeared. The first is the polymerization of acetylene by nickel cyanide to give cyclo-octatetraene. A more recent method is the dimerization of butadiene on heating to give cyclo-octadiene:

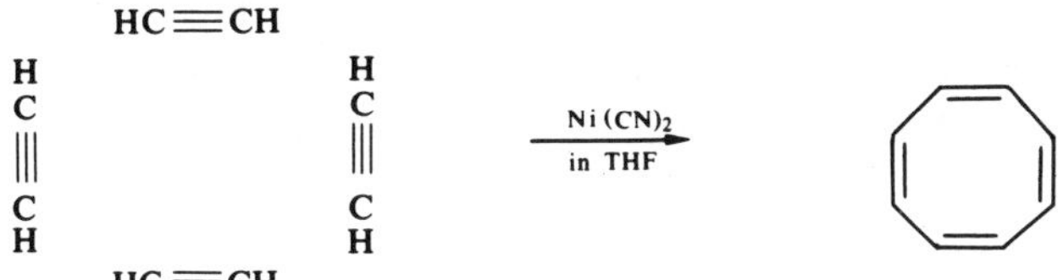

cyclo-octatetraene

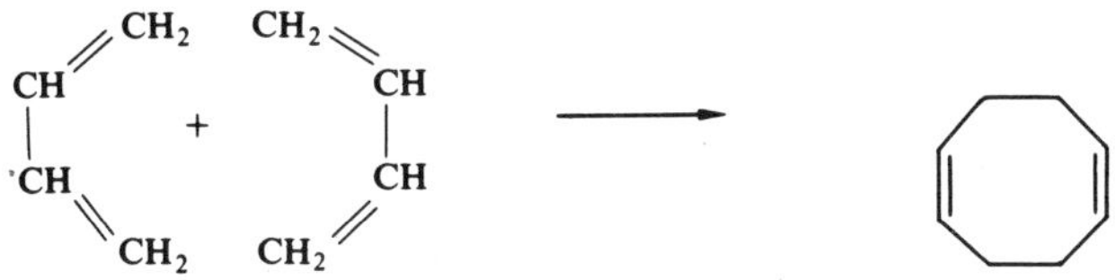

cyclo-octadiene

Cyclo-octane, the diene and tetraene all have non-planar rings. For conjugation, double bonds must be in the same plane, so the tetraene is not a conjugated olefin, but behaves as a typical unsaturated compound. Cyclo-octene is the smallest ring compound capable of containing a *trans* double bond:

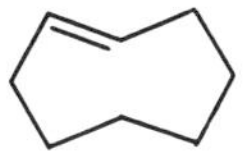

Large Rings

Baeyer's strain theory implied that large rings would be incapable of existence because of the strain in the bond angles. Later it was realized that with puckered or non-planar rings, no angle strain was involved, and so theoretically there should be no limit to the size of rings. In fact, there is a practical difficulty; the probability of the two ends of a long molecule being close enough together to react is very low, so under normal conditions, one end of a molecule is more likely to react with another molecule than with the other end of itself.

The first attempt to prepare large rings was made by Ruzicka in Switzerland in the 1920s in connection with a problem in perfumery. At one time the civet cat of Africa and the musk deer of Asia were slaughtered to obtain small quantities of civet and musk, two substances important in the production of perfumes. From these substances were obtained two pure ketones, civetone ($C_{17}H_{30}O$) and muscone ($C_{16}H_{30}O$). Ruzicka, by degradation studies, showed that civetone has the structure below. Later, when he had succeeded in synthesizing some of these large-ring ketones, he was able to select from among several possibilities, the structure of muscone. He developed a method of pyrolysing the thorium salts of dicarboxylic acids to give a useful yield of cyclic ketones. In this way he succeeded in producing rings of up to 34 carbon atoms. He also reduced the ketones, by the two methods outlined below, to give the corresponding cycloalkanes:

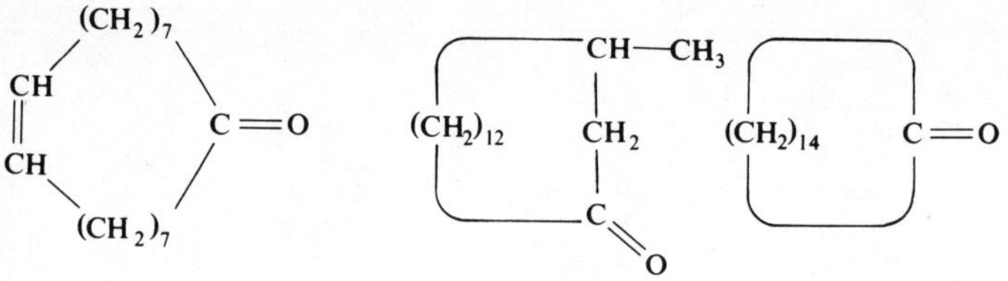

civetone

muscone

exaltone

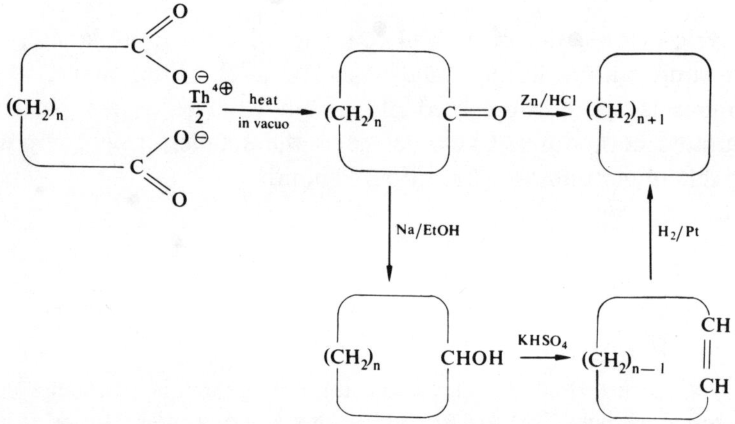

$$(CH_2)_n \overset{C=O,\ O^{\ominus}}{\underset{C=O,\ O^{\ominus}}{\bigg|}} \cdot \frac{Th^{4\oplus}}{2} \xrightarrow[\text{in vacuo}]{\text{heat}} (CH_2)_n\ C{=}O \xrightarrow{Zn/HCl} (CH_2)_{n+1}$$

$(CH_2)_n\ C{=}O \xrightarrow{Na/EtOH} (CH_2)_n\ CHOH \xrightarrow{KHSO_4} (CH_2)_{n-1} \overset{CH}{\underset{CH}{\|}} \xrightarrow{H_2/Pt} (CH_2)_{n+1}$

All the cyclic ketones up to about C_{20} have characteristic odours, but it was found that the C_{15} to C_{17} simple cyclic ketones have a strong musk or civet-like odour, and today, cyclopentadecanone (known commercially as exaltone) is produced industrially as a substitute for musk and civet.

A later method for making large rings is the cyclization of a dinitrile with a strong base. The reaction is carried out in solution, and by working with very dilute solutions, the tendency for long chains to form is diminished and high yields of cyclic ketones are obtained. Some diketone of twice the molecular weight is also formed. Lithium derivatives of secondary amines are used as base because they are soluble in the reaction mixture. The condensation is of the Claisen type, giving first an imido-nitrile which is hydrolysed to a keto-nitrile and then a β-ketoacid which is decarboxylated to the ketone by heating:

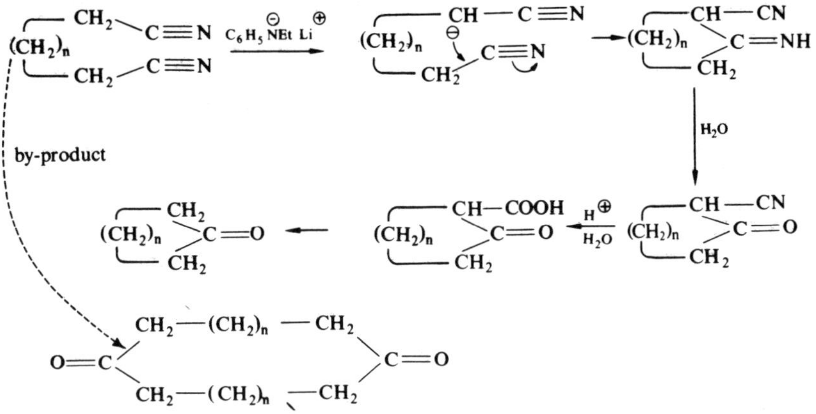

The method that gives best yields for large rings, and is very useful for C_5 and C_6 rings too, is the acyloin synthesis. A di-ester is treated with molten sodium under an inert high boiling solvent, such as xylene (dimethylbenzene). Sodium transfers an electron to the ester group and the resulting radicals dimerize and liberate sodium ethoxide to give an α-diketone which is further reduced by sodium to the acyloin:

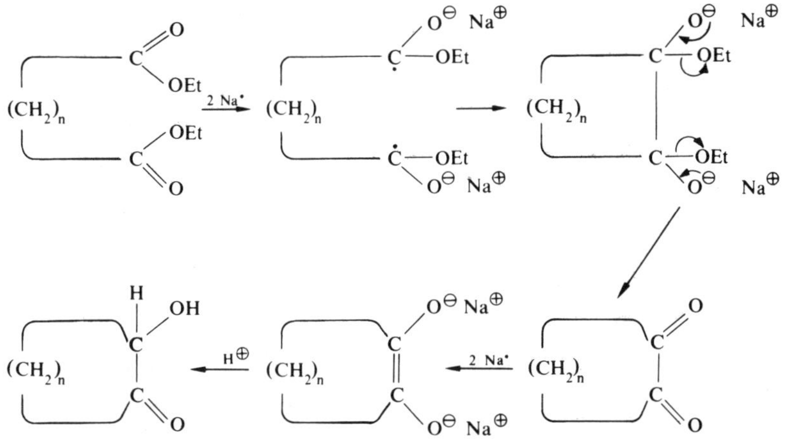

The keto-alcohol (known as an acyloin) is released on acidifying the reaction mixture; in this way ring closure and reduction of the di-ester occur together. Over 90% yield of C_{18}—C_{20} rings can be

247

obtained in this way. The reaction occurs on the surface of the molten sodium, and it seems very probable that the high yields result because the two ester groups at the end of the chains are brought together on the metal surface.

A number of large rings can be made by catalytic reactions from simple alkenes on an industrial scale. For example, 1,5,9-cyclo-dodecatriene can be made by the trimerization of butadiene with Ziegler catalysts. At present two isomers can be made as shown:

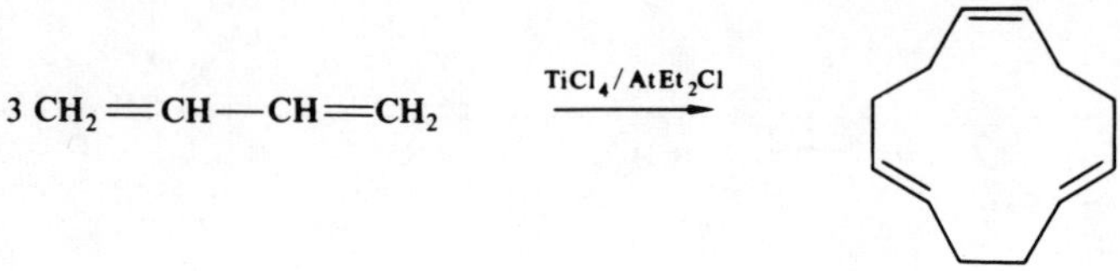

$$3 \ CH_2=CH-CH=CH_2 \quad \xrightarrow{TiCl_4/AtEt_2Cl}$$

trans, trans, cis - cyclododecatriene

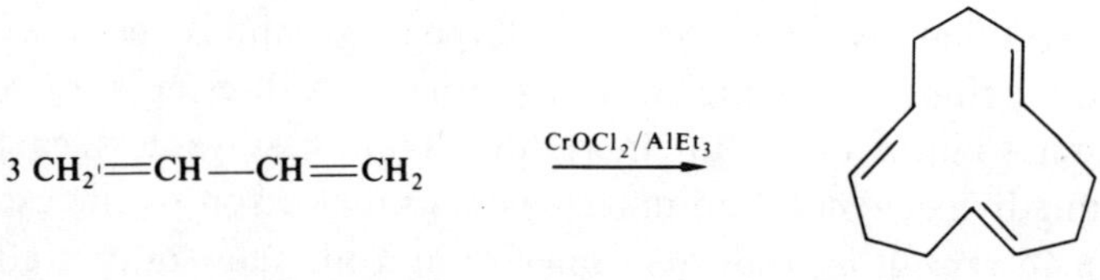

$$3 \ CH_2'=CH-CH=CH_2 \quad \xrightarrow{CrOCl_2/AlEt_3}$$

trans, trans, trans - cyclododecatriene

The same catalysts can be used to produce a C_{10} ring from a mixture of butadiene and ethylene:

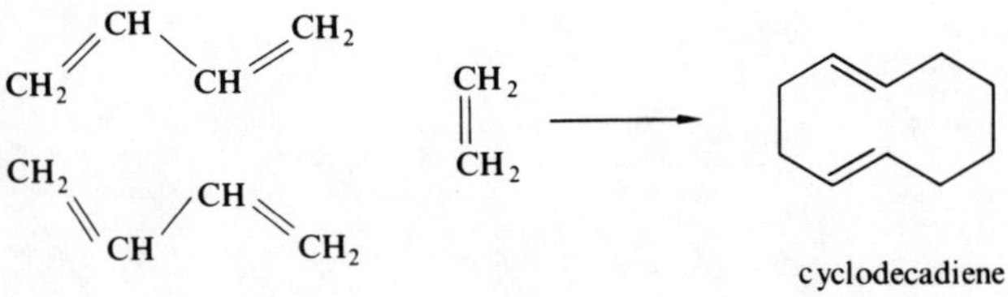

cyclodecadiene

When more is understood about how these catalysts work it is possible that even larger rings will be made in this simple way.

Even with the latest methods, yields of large rings in the C_8—C_{11} range are often inferior, because although angle strain is relieved entirely by puckering of the ring, repulsion between hydrogen atoms on opposite sides of the ring is significant. With rings above C_{11}, strainless conformations exist in which this interaction across the ring is negligible.

Reactions

There are only two characteristic reactions of alicyclic compounds; breaking the ring and rearrangement from one ring size to another. All other reactions are special cases of the reactions already studied for alkanes, alkenes, alcohols, etc., and the same transformations can be carried out on the ring compounds.

Ring opening is easiest for small rings, having a high degree of strain. The ease of fission is used as a test to distinguish between these rings.

1. *Hydrogenation.* Cyclopropane is hydrogenated to propane at 80–120° when using a nickel catalyst:

$$\triangle \quad \xrightarrow[80°]{H_2, Ni} \quad CH_3CH_2CH_3$$

Cyclobutane with the same catalyst requires a temperature of 180° to give butane; cyclopentane and cyclohexane only react above 250°. For comparison, ethylene gives ethane at 40° with the same catalyst.

2. *Hydrogen Halides.* Cyclopropane reacts with hydrogen bromide like an olefin giving propyl bromide:

$$\triangle \quad + \quad HBr \quad \longrightarrow \quad CH_3CH_2CH_2Br$$

Cyclobutane and higher members are unaffected. Hydrogen iodide opens the ring of cyclopropane and cyclobutane, giving the n-iodides:

$$\square \quad + \quad HI \quad \longrightarrow \quad CH_3CH_2CH_2CH_2I$$

3. *Bromine.* Cyclopropane reacts with bromine to give 1,3-dibromopropane. Larger rings are not affected:

$$\triangle \quad + \quad Br_2 \quad \longrightarrow \quad BrCH_2CH_2CH_2Br$$

These three ring opening reactions used as qualitative tests would tend to place cyclopropane with the olefins, but cyclopropane is *not* oxidized by potassium permanganate or ozone. The permanganate test is usually sufficient to distinguish a cyclopropane from an olefin.

4. *Oxidative Ring Opening.* For compounds other than cyclo-alkanes, the ring can be opened in a number of ways, for example, a cycloalkene can be oxidized by permanganate to a dicarboxylic acid; a cyclic ketone or alcohol is oxidized by aqueous nitric acid also to

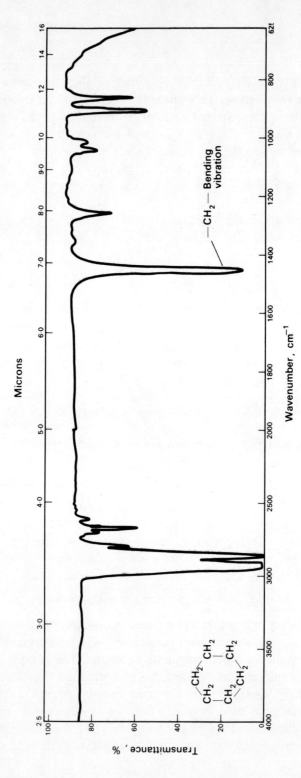

Fig. 55 Infrared spectrum of cyclohexane

a dicarboxylic acid, when the conditions are carefully controlled:

$$\text{(cyclopentanone)} =O \xrightarrow{\text{HNO}_3} \quad CH_2 \begin{array}{c} CH_2 - COOH \\ \\ CH_2 - COOH \end{array}$$

5. *Ring Contraction and Expansion.* A number of reactions are known where the size of the ring is altered. Treating cyclohexane with an aqueous solution of aluminium chloride leads to an equilibrium mixture containing a large amount of methylcyclopentane. Treating cyclopropylmethylamine with nitrous acid gives some cyclobutyl alcohol. Such changes can usually be traced to some secondary effect, such as the rearrangement of a primary carbonium ion to a secondary or tertiary ion (see page 65):

$$\triangle - CH_2NH_2 \xrightarrow{\text{HNO}_2} \left[\triangle \overset{\oplus}{CH_2} \right]$$

$$\square - OH \xleftarrow{\text{H}_2O} \left[\square^{\oplus} \right]$$

Identification

There are no specific means of identifying cycloalkanes as a class. If the molecular formula is known to be of the type C_nH_{2n} and the substance cannot be hydrogenated at room temperature, the presence of a ring may be inferred, though many highly hindered alkenes are not hydrogenated under these conditions. Similarly a compound of formula C_nH_{2n-2} which is not unsaturated must contain two rings. Cyclopropane rings can be distinguished from double bonds by treatment with cold dilute permanganate as already described; cyclopropanes are inert, double bonds react, but both groups add bromine in the absence of light.

Since there are no CH_3 groups in simple cycloalkanes the infrared absorption at 1380 cm^{-1} due to bending of a CH_3 group which can often be recognized in alkanes is absent (cf. Fig. 55 and Fig. 3). The C—H stretching absorption of cyclopropanes is at 3020–3050 cm^{-1}, between that of alkenes and alkanes. In small ring compounds, the frequency of absorption of a functional group such as $C=O$ is often strongly dependent on the size of the ring to which it is attached. Nuclear magnetic resonance spectrometry (N.M.R.) provides the best method of recognizing C—H bonds on small rings.

Mass Spectrometry

Mass spectrometry has become an important tool for the study of molecular structure in recent years, with the availability of mass spectrometers which can be used with the high molecular weight substances, found in organic chemistry.

The substance to be examined is introduced as a solid, liquid or gas, into the apparatus which is evacuated to a very low pressure (about 10^{-7} mm mercury) and there the liquid and solid substances are volatilized, with heat, if necessary. The vapour molecules are sucked into the path of a beam of electrons, which ionize the molecules by knocking out another electron, leaving the molecule charged and with an unpaired electron:

$$M + e \rightarrow M^+ + 2e$$

This molecular ion decomposes into a number of fragments of smaller size, which are sorted by mass (actually by the ratio of mass to charge m/e, but for most ions produced the charge is $1e$,) and focused by a magnetic field. The number of ions of different sorts are measured by a sensitive electrometer and a graph of ion abundance against mass units is plotted by a recorder (Fig. 56). This pattern of molecular fragments produced in varying abundance, dependent upon their relative stability is called the cracking pattern of the compound, and in a given instrument, and keeping certain conditions, such as the energy of the bombarding electrons constant, the cracking pattern is characteristic of the compound and can in many cases tell us a great deal about the structure of the compound.

Molecular Weight

The molecular ion, the ion produced by removal of one electron from the intact molecule is the ion of highest mass, but it is not always detected. Sometimes the compound may decompose before volatilizing or the molecular ion may be so unstable that all such ions decompose further before reaching the electrometer, but in the case of about 90% of compounds examined the molecular ion is seen.

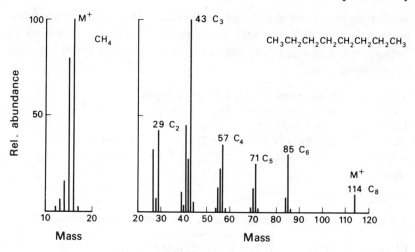

Fig. 56   Mass spectra of methane and octane

Some other information is necessary therefore to decide if the highest mass observed is the molecular ion. Once confirmed, the information is very useful, and it can distinguish between formulae such as $C_{20}H_{40}O_2$ and $C_{21}H_{42}O_2$ which analysis for carbon and hydrogen composition cannot do.

Normally we are dealing with average molecular weights, because many elements consist of two or more isotopes. The molecular weight of hydrogen chloride is 36·461 because chlorine consists of two isotopes ^{35}Cl and ^{37}Cl and these are present naturally in the ratio of 75% and 25% respectively. Hydrogen contains a small amount of deuterium. The mass spectrometer will pick out the ions produced from $^1H^{35}Cl$ and $^1H^{37}Cl$ in the ratio 75:25; there will also be very small amounts of $^2H^{37}Cl$.

The atomic masses are based on a scale of $^{12}C = 12·0000$. Some other common masses are given in Table 20. The most accurate mass spectrometers can measure masses to 1 part in 33,000 so that very accurate values of molecular weight are obtained. The use of accurately determined molecular weights can tell us the molecular formula. For example, ethylene, nitrogen and carbon monoxide have molecular weights of 28. But using the accurate masses in Table 20, ethylene has a molecular weight of 28·0313, nitrogen 28·0061, and carbon monoxide 27·9949. A high resolution mass spectrometer can easily distinguish between these and tell us the exact molecular formula of our compound. Tables of accurate molecular weights and corresponding formulae have been prepared with the aid of computers to ease the burden of calculating the correct formula from the accurately measured mass. The relative abundance of isotopes (such as is found in chlorine and bromine, see Table 21) can help to indicate what elements are present even when one cannot measure the accurate mass.

Table 20.   Some accurate masses of isotopes

^1H	1·007824	^{32}S	31·972072
^{12}C	12·000000	^{35}Cl	34·968853
^{14}N	14·003074	^{79}Br	78·91839
^{16}O	15·994914	^{127}I	126·90466

Table 21.   Percent abundance of isotopes of common elements

Element	Mass	%	Element	Mass	%
H	1	99·985	S	32	95·00
	2	0·015		33	0·76
C	12	98·89		34	4·22
	13	1·11	Cl	35	75·53
N	14	99·63		37	24·47
	15	0·37	Br	79	50·54
O	16	99·76		81	49·49
	17	0·037	I	127	100·00
	18	0·204			

Note too that all compounds consisting of C and H, or C, H and O, must have even molecular weights and if they contain C, H, O and N, the weight to the nearest integer must be odd if there are an odd number of nitrogen atoms (e.g. CH_5N, mol. wt. 31; $C_2H_8N_2$, mol. wt. 60).

Molecular Structure

The study of the cracking patterns, the way molecules decompose when ionized by electron impact in the mass spectrometer, has given rise to a new kind of chemistry in recent years, which is sometimes referred to as that of electron impact reactions. Some of these reactions closely resemble reactions we understand in conventional chemistry, and some of them introduce entirely new kinds of bond breaking and rearrangement. The purpose in studying these reactions is to interpret molecular structure in unknown compounds. Since a complete mass spectrum can be obtained with a fraction of a milligram, skilful interpretation can yield much information about molecular weight, element composition and molecular structure from this tiny sample of pure material.

Exactly how the structures of the fragment ions are known is beyond the scope of this account. It must suffice to say that accurate mass determination of the ions will give the 'molecular' formula of a fragment, and much information on structure is gained from the study of *meta-stable ions* and by replacing specific hydrogen atoms by deuterium (^2H) atoms (the latter is particularly useful in studying rearrangement reactions).

1. *Hydrocarbons.* Consider an electron removed from a methane molecule. The ionized molecule is normally written inside brackets when we do not know from where the electron has been removed, but in this simple case of equivalent bonds we can show it as removed from one of them:

$$CH_4 + e^- \longrightarrow CH_4^{+} \quad \text{or} \quad H\!:\!\overset{\text{H}}{\underset{\text{H}}{\text{C}}}\!:\!H + 2e^-$$

This can now decompose in at least two ways. By using an arrow with a single barb, we can indicate how one electron moves and keep the double-barbed arrow of organic mechanism for cases where two electrons move together:

$$H\!-\!\overset{\text{H}}{\underset{\text{H}}{\text{C}}} + H \longrightarrow CH_3\cdot \ + \ H^+$$

$$H\!-\!\overset{\text{H}}{\underset{\text{H}}{\text{C}}} + H \longrightarrow CH_3^{+} \ + \ H^{\cdot}$$

Both of these fragmentations can occur, but one will be more probable than the other, and will give rise to a larger ion current at the electrometer. The radicals formed ($CH_3\cdot$ and $H\cdot$) have no net charge, are not detected by the electrometer and are pumped out of the instrument by the vacuum system. Some of the methyl carbonium ions which have an even number of electrons, may decompose further.

$$H\!-\!\overset{+}{\underset{\text{H}}{\text{C}}}\!-\!H \longrightarrow CH_2^{+} \ + \ H\cdot$$

$$\text{or} \quad H\!-\!\overset{+}{\underset{\text{H}}{\text{C}}}\!-\!H \longrightarrow CH_2\!: \ + \ H^+$$

255

In addition to the molecular ion of CH_4 (mass 16) one might expect to see ions of mass 15, 14, 13 (corresponding to CH_3, CH_2 and CH) (Fig. 56) in varying abundance depending upon the probability of a fragmentation occurring and the stability of the fragments formed.

A straight hydrocarbon chain may break anywhere when ionized by electron impact:

$$[R_3C \frown CR_3]^{+} \longrightarrow R_3C^+ + \cdot CR_3$$

Therefore ions corresponding to the whole molecule, and a series of smaller fragments formed by the loss of $CH_3\cdot$, $C_2H_5\cdot$, $C_3H_7\cdot$, etc., should be detected. In this fragmentation the charge tends to be left with the smaller portion of the molecule, so that these ions are more abundant (see Fig. 56):

$$C_2H_5 \overset{.}{:} C_4H_9 \longrightarrow \underset{29}{C_2H_5^+} + C_4H_9\cdot \qquad \text{more probable}$$
$$\searrow \underset{57}{C_2H_5\cdot + C_4H_9^+} \qquad \text{less probable}$$

Each major peak on the chart is associated with close neighbours from the loss of an additional $H\cdot$ or H_2. When a chain is branched, fragmentation will occur preferentially at the branching point, and since the order of stability of carbonium ions is tertiary > secondary > primary, the branched fragment will give rise to an abundant ion: also the largest chain will tend to be lost as a radical:

$$CH_3CH_2 \overset{\overset{CH_3}{|}}{\underset{\overset{|}{H}}{\dashv}} C \dashv (CH_2)_3CH_3 \left\langle \begin{array}{l} \overset{\overset{CH_3}{|}}{CH_3CH_2\overset{+}{CH}} + CH_3(CH_2)_3\cdot \\ \underset{57}{} \\ CH_3CH_2\cdot + \overset{\overset{CH_3}{|}}{{}^+CH(CH_2)_3CH_3} \\ \underset{85}{} \end{array} \right.$$

The stronger peaks therefore indicate the position of the chain branching (Fig. 57).

All of these reactions involve the decomposition of a single molecule. At the very low pressure inside the spectrometer, the possibility of collision, or of reaction between two ions or fragments is very low.

2. *Alkenes.* When a double bond is present, fragmentation occurs chiefly at the β—C—C bond; this gives rise to allyl radicals, which have enhanced stability because the charge is spread over two atoms:

256

$$RCH = CH - CH_2 \overset{2}{\underset{3}{\not{|}}} R' \rightarrow RCH = CH - CH_2^+ \leftrightarrow \overset{+}{R}CH - CH = CH_2 + R'.$$

These ions will be much more prominent than the ions produced by breaking of the alkyl chain, though both kinds of fragmentation occur. Note that an allyl ion e.g. C_3H_5 is two mass units less than an alkyl ion, e.g. C_3H_7, so abundant ions from alkenes and branched alkyl fragments will not be confused. Unfortunately double bonds sometimes isomerize along the chain very readily, so that though the spectrum may be recognized as that of an alkene, the *position* of the double bond may not be reliably inferred.

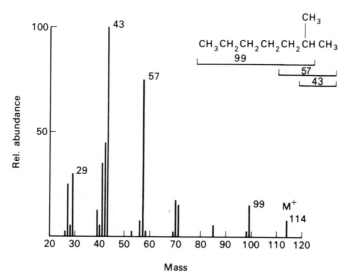

Fig. 57   Mass spectrum of 2-methylheptane

3. *Alkyl Halides.* The halogen atom tends to be readily lost, chiefly as HX from fluorides and chlorides and chiefly as X from bromides and iodides. Because both Cl (35 and 37) and Br (79 and 81) have isotopes two mass units apart, their presence in a compound is easily detected by examining the molecular ion (Fig. 58). Two other unfamiliar reactions also occur; flourine and chlorine form $CH_2 = X^+$ ions:

$$R-CH_2-X^+ \rightarrow R\cdot + CH_2 = X^+ \qquad \text{33 for F, 49 and 51 for Cl}$$

However, where there is a long alkyl chain, a five-membered ring is formed:

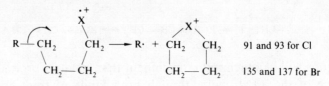

$$\text{91 and 93 for Cl}$$

$$\text{135 and 137 for Br}$$

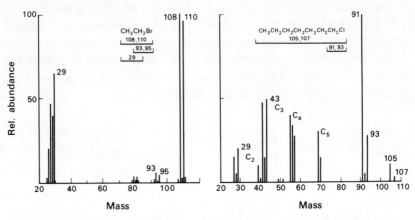

Fig. 58   Mass spectra of ethyl bromide and heptyl chloride

4. *Alcohols.* There are two characteristic reactions of alcohols involving cleavage near the functional group, as well as the normal pattern of breakdown of the alkyl chain. The first, called α-cleavage, gives rise to an ion of mass 31; this relatively stable ion can be compared with oxonium ions in solution which contain a triply bonded oxygen atom carrying a positive charge:

$$RCH_2 \!\!-\!\! CH_2 \!\!-\!\! \overset{+\cdot}{O}H \longrightarrow RCH_2\cdot + CH_2 \!\!=\!\! \overset{+}{O}H$$

$$31$$

In secondary and tertiary alcohols, once again the largest R group is expelled preferentially:

$$R\!\!-\!\!\overset{\overset{\displaystyle R''}{|}}{\underset{\underset{\displaystyle R'}{|}}{C}}\!\!-\!\!\overset{+\cdot}{O}H \longrightarrow R\!\!-\!\!\overset{\underset{\underset{\displaystyle R'}{|}}{C}}{=}\!\!\overset{+}{O}H + R'' \qquad R''\text{ largest}$$

$$M\!-\!R''$$

In the second reaction, a molecule of water is eliminated to give an alkene (or possibly a cyclopropane), and this will then give the further breakdown pattern of an alkene:

258

$$R-\underset{\underset{H}{|}}{\overset{\overset{CH_2}{\diagup}\quad\diagdown}{CH}}\;\;\overset{+\cdot}{O}-H \longrightarrow R-\underset{\underset{H}{|}}{\overset{\overset{CH_2}{\diagup}\quad\diagdown}{CH}}\;\;\overset{+}{O}-H \longrightarrow [RCH\!=\!CH_2]^{+} + H_2O$$

M — 18

This reaction takes place very readily, so that molecular ions of alcohols are very weak, and in some cases, particularly tertiary alcohols, are not seen at all. The ion of highest mass is then the molecular ion less H_2O (M—18).

5. *Ethers*. Ethers undergo α-cleavage like alcohols:

$$R\!-\!CH_2\!-\!\overset{+\cdot}{O}\!-\!R' \longrightarrow R\cdot + CH_2\!=\!\overset{+}{O}\!-\!R'$$

this ion can undergo further elimination:

$$CH_2\!=\!\overset{+}{O}\!-\!\overset{|}{\underset{|}{C}}R_2 \longrightarrow CH_2\!=\!\overset{+}{O}H + \underset{CR_2}{\overset{CH_2}{\|}}$$

Cleavage can take place at the oxygen atom, but the alkoxy group is almost entirely lost as a radical.

$$R\!-\!\overset{+\cdot}{O}\!-\!R' \longrightarrow RO\cdot + R'^{+}$$

6. *Amines*. Because the atomic weight of nitrogen is even and the valency is odd, a simple amine will have an odd-numbered molecular weight; all compounds of C, H and O (only) have even-numbered molecular weights. The molecular ions of amines are often very weak, like alcohols. Amines undergo α-cleavage like ethers followed by further rearrangement and elimination of an alkene:

$$R\!-\!CH_2\!-\!\underset{\underset{H}{|}}{\overset{+\cdot}{N}}\!-\!CR_2CR_2H \longrightarrow R\cdot + CH_2\!=\!\underset{\underset{H}{|}}{\overset{+}{N}}\!-\!CR_2 \longrightarrow CH_2\!=\!\overset{+}{N}H_2 + \underset{CR_2}{\overset{CR_2}{\|}}$$

30

7. *Aldehydes and Ketones*. Simple aldehydes and ketones undergo α-cleavage:

$$\underset{H}{\overset{R}{\diagdown}}C\!=\!\overset{+}{O}\!\cdot \quad or \quad \underset{H}{\overset{R}{\diagdown}}C\!=\!\overset{+}{O}\!\cdot \longrightarrow RC\!\equiv\!O^{+} + H\cdot \; or \; HC\!\equiv\!O^{+} + R\cdot$$

29

$$\underset{R'}{\overset{R}{\diagdown}}C\!=\!\overset{+}{O}\!\cdot \longrightarrow RC\!\equiv\!O^{+} + R'C\!\equiv\!O^{+} \; etc.$$

259

Simple cleavage at the functional group also occurs as in ethers to give an alkyl ion and an acyl radical (Fig. 59):

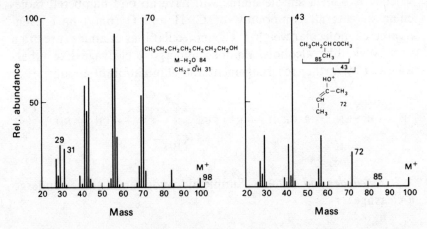

Another important rearrangement causes β-cleavage in aldehydes and ketones if there is a γ-hydrogen atom available. This reaction is general for a wide range of carbonyl compounds, including esters and amides, and is known as the McLafferty rearrangement. It resembles the kind of unimolecular rearrangement met in solution chemistry:

This rearrangement gives ions of even mass, whereas the other fragments have odd masses.

In molecules of higher molecular weight, these reactions are not as prominent and the fragmentation pattern of a hydrocarbon becomes more evident.

Fig. 59   *Mass spectra of 1-heptanol and 3-methylpentan-2-one*

8. *Acetals and Ketals.* Aldehydes and ketones are sometimes converted to ethylene acetals and ketals (page 195) because in fragmenting, these produce prominent ions which are stabilized by spreading the charge on to two oxygen atoms:

260

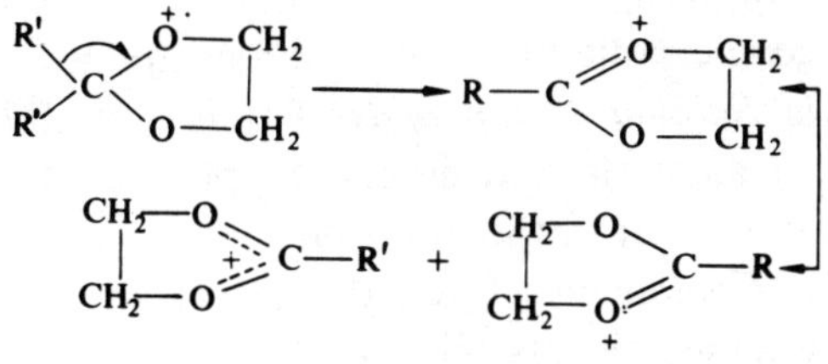

$$R' \diagdown \overset{\underset{R''}{|}}{C} \diagup \overset{+\cdot}{O}-CH_2 \atop O-CH_2 \quad \longrightarrow \quad R-C \diagup \overset{+}{O}-CH_2 \atop O-CH_2$$

$$\begin{array}{c} CH_2-O \\ CH_2-O \end{array} \overset{+\cdot}{C}-R' + \begin{array}{c} CH_2-O \\ CH_2-\overset{+}{O} \end{array} C-R$$

$$M-R \qquad\qquad M-R'$$

9. *Acid Derivatives.* Acids are usually converted to their more volatile methyl esters for mass spectrometry. Esters in general undergo the same kinds of cleavage and rearrangement as ketones:

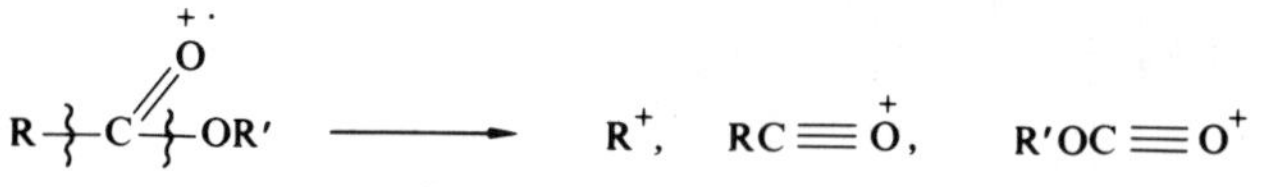

$$R \!\!\left.\!\!-\!\!\overset{\overset{+\cdot}{O}}{\overset{\|}{C}}\!\!-\!\!\right|\!\!OR' \quad \longrightarrow \quad R^+, \quad RC \!\equiv\! \overset{+}{O}, \quad R'O\overset{+}{C} \!\equiv\! O^+$$

The alcohol portion OR′ is preferentially lost as a radical. When β-cleavage is possible, it tends to occur on the acid side first, and then if the alcohol has a sufficiently long chain it can occur again:

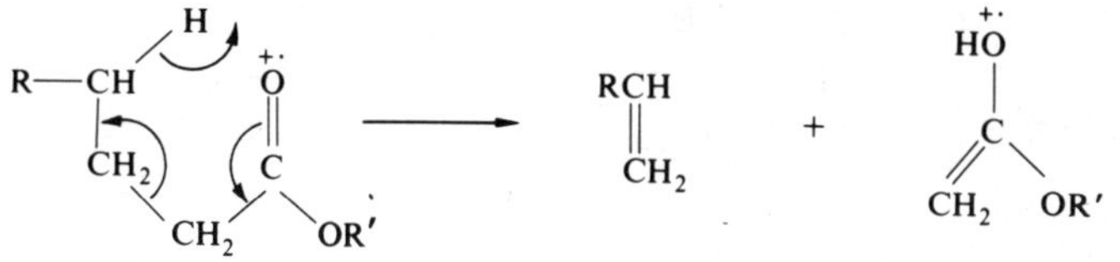

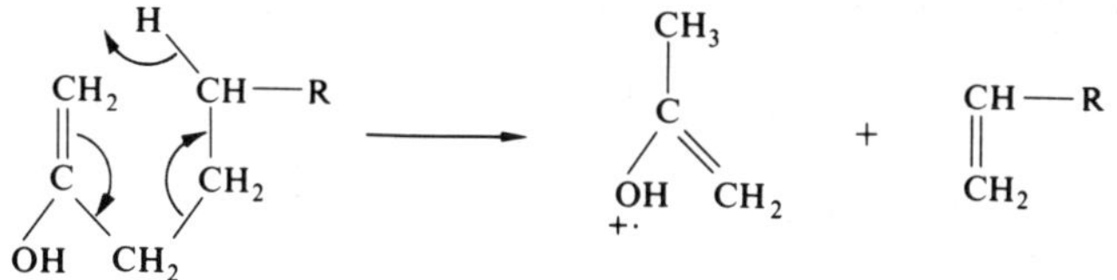

$$R-CH \diagup \!\! \overset{H}{\diagdown} \quad \overset{+\cdot}{O} \quad \longrightarrow \quad \begin{array}{c} RCH \\ \| \\ CH_2 \end{array} + \quad \overset{\overset{+\cdot}{HO}}{\underset{CH_2 \quad OR'}{C}}$$

$$\begin{array}{cc} CH_2 & \overset{H}{\diagdown} CH-R \\ \| & (CH_2 \\ \underset{+\cdot}{\overset{C}{\underset{OH \quad CH_2}{}}} \end{array} \quad \longrightarrow \quad \begin{array}{c} CH_3 \\ | \\ C \\ \diagup \diagdown \\ \underset{+\cdot}{OH} \quad CH_2 \end{array} + \begin{array}{c} CH-R \\ \| \\ CH_2 \end{array}$$

$$57$$

Acetates characteristically lose a molecule of acetic acid:

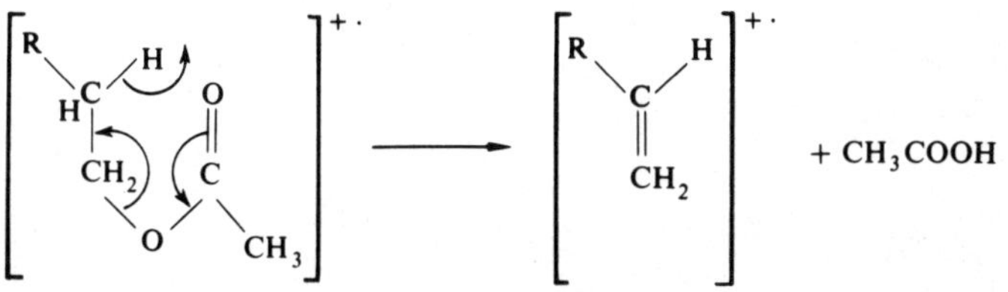

$$\left[\begin{array}{c} R \diagdown \overset{H}{\diagup} \\ HC \diagdown C \overset{O}{\diagup} \\ | \quad (C \\ CH_2 \quad \diagdown \\ O \quad CH_3 \end{array} \right]^{+\cdot} \quad \longrightarrow \quad \left[\begin{array}{c} R \diagdown \overset{H}{\diagup} \\ C \\ \| \\ CH_2 \end{array} \right]^{+\cdot} + CH_3COOH$$

$$M-60$$

The spectra of simple amides resemble those of esters and substituted amides undergo the same kinds of fission as amines.

10. *Aliphatic Sulphur Compounds.* Thiols and thioethers follow similar rules to alcohols and ethers. Sulphur contains 4% of ^{34}S and this will be evident in the molecular ion.

11. *Alicyclic Compounds.* Cycloalkanes give stronger molecular ions than do alkanes because cleavage can give an ion of the same mass, which may then decompose further in a variety of ways depending upon the functional group:

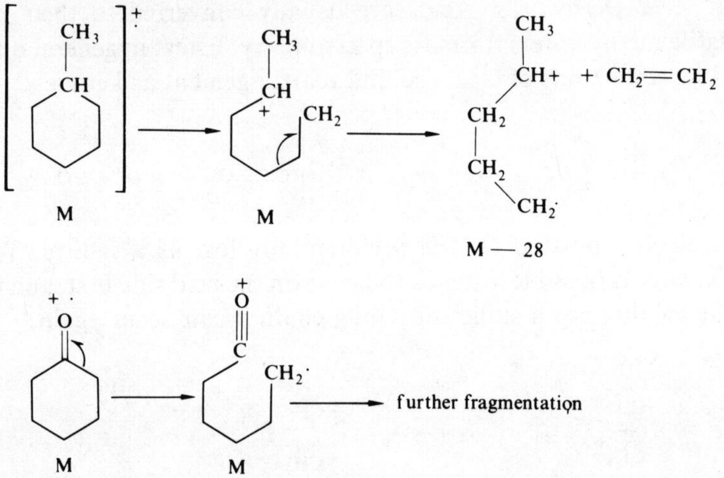

Cyclohexenes undergo an interesting rearrangement and fragmentation that resembles a reversed Diels–Alder reaction:

Further Reading

The following selection of readings is appended either to expand or delve more deeply into some special topics and thereby introduce the reader to a wider source of reading, including *Education in Chemistry* published by the Chemical Society and the *Journal of Chemical Education* published by the American Chemical Society.

Introduction

The Development of Modern Chemistry, A. J. Idhe, Harper, 1964, Chapters 7 and 8, Organic Chemistry.

'Wöhler and the Vital Force', L. Hartman, *J. Chem. Educ.* 1957, **34**, 141.

The Origins of the Theory of Chemical Structure, G. V. Bykov, *J. Chem. Educ.* 1962, **39**, 220.

Chapter One

An Introduction to Chemical Nomenclature, R. S. Cahn, Butterworth, 3rd Edition 1968.

Organic Chemistry, L. F. Fieser and M. Fieser, Reinhold, 3rd Edition 1956, Chapter 5, Petroleum.

'Hydrocarbons in Petroleum', F. D. Rossini, *J. Chem. Educ.* 1960, **37**, 554.

Spectroscopy in Chemistry, R. C. Whitfield, Longmans, 1969, Chapter 4, Infrared Spectra.

Chapter Two

Two Lectures on an Outline of an Electrochemical (Electronic) Theory of the Course of Organic Reactions, R. Robinson, Institute of Chemistry of Great Britain and Ireland, 1932. [This pamphlet is now difficult to obtain.]

'Intramolecular Effects on the Course of Chemical Change', R. Robinson, *Endeavour*, 1954, **13**, 173.

Qualitative Organic Analysis, W. Kemp, McGraw-Hill, 1970, Chapter 1, Chemical Examination of an Unknown Compound.

'Economic Aspects of Chlorine Production', D. Cooper, *Educ. in Chem.* 1970, 186.

Chapter Three

'The Markovnikov Rule', G. Jones, *J. Chem. Educ.* 1961, **39,** 297.
The Geometry of Molecules, C. C. Price, McGraw-Hill, 1971, Chapter 4, Properties of Polymers.
Chemical Bonding, A. L. Companion, McGraw-Hill, 1964, Chapters 1 to 4.
'New Selective Reducing Agents', H. C. Brown, *J. Chem. Educ.* 1961, **38,** 173.

Chapter Four

'New Trends in Petroleum Chemicals', G. Jones, *Educ. in Chem.* 1965, **2,** 107.
'Reaction Mechanisms and Resonance Structures', D. H. Williams, *Educ. in Chem.* 1971, 214.

Chapter Five

Organic Reaction Mechanisms, R. Breslow, Benjamin, 1965, Chapter 3, Nucleophilic Aliphatic Substitution.
'Hydrogen Bonding and the Physical Properties of Substances', L. N. Ferguson, *J. Chem. Educ.* 1956, **33,** 267.
Valence, C. A. Coulson, Oxford, 2nd Edition 1961, pp. 344–356. The Hydrogen Bond.
The Chemical Bond, L. Pauling, Oxford, 1967, Chapter 12, The Hydrogen Bond.

Chapter Seven

A Shorter Sidgwick's Organic Chemistry of Nitrogen, I. T. Millar and H. D. Springall, Oxford, 1969, Chapter 1, General Features and Chapter 3, Aliphatic Amines.
Concerning Amines, D. Ginsburg, Pergamon, 1967.

Chapter Eight

The Chemistry of Carbonyl Compounds, C. D. Gutsche, Prentice-Hall, 1967, Chapter 1, Structure and Properties.
'Enolization; An Electronic Interpretation', S. Zuffanti, *J. Chem. Educ.* 1945, **22,** 230.
'Chemistry of the Grignard Reagent', J. H. J. Peet, *Educ. in Chem.* 1968, **5,** 109.

Chapter Nine

Basic Principles of Organic Chemistry, J. D. Roberts and M. C.

Caserio, Benjamin, 1965, Chapter 16, Carboxylic Acids and Derivatives.

A Shorter Sidgwick's Organic Chemistry of Nitrogen, I. T. Millar and H. D. Springall, Oxford, 1969, Chapter 6, Amides.

'Modern Detergents', M. Bell, *Educ. in Chem.* 1968, **5,** 75.

Chapter Ten

Basic Principles of Organic Chemistry, J. D. Roberts and M. C. Caserio, Benjamin, 1965, Chapter 21, Organo-sulphur Compounds.

Organic Chemistry, N. L. Allinger, M. P. Cava *et al.*, Worth, 1971, Chapter 26, Organo-sulphur Chemistry.

Chapter Eleven

Alicyclic Chemistry, G. H. Whitham, Macdonald, 1963.

The Chemistry of the Steroids, W. Klyne, Methuen, 1957.

Organic Chemistry, N. L. Allinger, M. P. Cava *et al.*, Worth, 1971, pp. 884–902, Making and Breaking of Carbocyclic Rings.

Appendix

Spectroscopy in Chemistry, R. C. Whitfield, Longmans, 1969, Chapter 5, Mass Spectroscopy.

Modern Methods of Chemical Analysis, R. L. Pecsok and L. D. Shields, Wiley, 1968, Chapter 13, Mass Spectrometry.

Spectroscopic Methods in Organic Chemistry, D. H. Williams and I. Fleming, McGraw-Hill, 2nd Edition 1973, Chapter 5, Mass Spectra.

Index

Page numbers in bold indicate a main reference; tables, illustrations and formulae are shown in italic

267

carbon monoxide *cont.*
  reaction of, 178
carbonic acid, salts and reaction, 179
carbonium ions, 17, 29, 31, 32
  amines and, 134
  from alcohols, 48–9
  from amines, 139
  in alkene additions, 29, 59, 61, 65
  ionization to, 52
  primary, 32, 48, 59
    rearrangement of, 139, 251
  rearrangement of, 65
  relative stability, 29
  secondary, 48, 139
  S_N1 and E1 reactions, 52
  solvation of, 51–2
  substitution and elimination from, 35
  tertiary, 48, 65, 139
    in esterification, 182
carbon rings, alicyclic compounds, 226
carbon tetrabromide, *39*
carbon tetrachloride, **42–3**
  physical data, *39*
  preparation of, 15
  as solvent for infrared spectra, 123
carbon tetrafluoride, 42
carbon tetra-iodide, *39*
carbonyl compounds, 152
  addition reactions, **154–9**
  mass spectra of, 259–60
  tautomerism in, 160
  unsaturated, 153
carbonyl group, 145, **151–2**, 164
  addition reactions, 154–9
  addition versus displacement at, 189
  in acids, 172, 185
  in acid halides, 199
  in amides, 204
  in Diels-Alder reaction, 239
  in esters, 194
  infrared absorption, 168–9, *170–1*, 209–14
  nucleophilic attack at, **154–5**, 189–90, 193
  orbital arrangement in, *151*
  planarity of, *151*
  polarization of π-electrons, *151–2*, 172
  reduction to CH_2, 154
carboxylate ions, **172–4**, 189
  as leaving group, 189–90
  compared with enolate ion, 163
  molecular orbital arrangement, *173*
  soaps and, 197, *198*

carboxylate salts, 179
  decarboxylation of, 33, 184–5
  esters from, 190
  Hunsdiecker reaction, 33, 185
carboxyl free radical, 185
carboxyl group, 12, *172*
  identification of, 179, 209
  ionized, infrared absorption of, 209
  trigonal, 181
carboxylic acids, 110, 145, **172–214**
  acidic strength, 173–4, *175*
  acyl derivative formation, 179–80
  branched, table of, *177*
  calcium salts, pyrolysis of, 147–8, 185, 237, 239, 244
  decarboxylation of, 11–12, 184–5
  derivatives, *see* acid halides, amides, anhydrides, esters, nitriles
  esterification of, **180–2**, 190
    mechanism of, 180–2
  halogenation of, 185
  hindered, esterification, 182
  hydrogen bonding in, 175
  identification of, 209
  in petroleum, 19
  iodoform reaction, 161
  K_a values, 173–4, *175*
  nomenclature, **175–6**, 226
  physical properties, **191**
  preparation of, **176–9**
    from amides, 206
    from esters, 192
    from Grignard reagents, 159, 177
    from methyl ketones, 161
    from nitriles, 208
  reactions, **179–86**
  reduction of, 108, 184
  salts, 179
  table of, *175*
  test for, 179, 209
  thioacids and, 219–20
carboxylic acid chlorides, 200–2, *see also* acid chlorides
carboxylic esters, *see* esters
carnaubyl alcohol, 120
carnauba wax, 120
catalysis:
  by acids, 157, 164
  in Claisen condensation, 150, 194
  in esterification, 114, 181, 182
catalysts, 23
  amine preparation and, 136
  carbonylation, 178

catalysts *cont.*
  classification of, 23–4
  copper acetylide, 94
  cyclization, 73, 95, 298
  dehalogenation, 11, 37
  dehydrating, 47, 72, 112, 130
  dehydrogenation, 72,112,119,147,243
  epoxidation, 58
  hydration, 117, 119
  hydroformylation, 109
  hydrogenation, 22, 55, 90, 107, 193–4
    surface adsorption, *56*
  in margarine hydrogenation, 197
  in Oxo process, 109
  in petroleum refining, 21–4
  Lindlar's, 90
  polymerization, 68, 69, 77
  Raney nickel, 107, 153
  transesterification, 190
  transition metal complexes, 73, 109
  Ziegler, 67, 68, 77, 248
catalytic 'cracking', *see* cracking
catalytic oxidation, 116
catalytic dehalogenation, 11
cationoid reagents, 34
cellophane, 70
cellulose, 70, 115, 121
  ethanol from, 117
cellulose acetate, 188
chain-initiating reaction, 15
chain-lengthening reactions, 136
chain-propagating reaction, 15
chain reaction, 15, 30, 62–3, 68
chain-shortening reactions, 161
'chair' conformation, 240, *241*, 242
chloral, *115*, 161
chloride ion as leaving group, 189–90
chlorine, 14
  alkene addition, 59–60
  atomic masses, 254
  effect of light on, 14
  electronegativity, 28, 189
  free radicals of, 14–15, 62
    unselective attack by, 27
  isotopes in mass spectra, 253, 257, *258*
chloro-alkanes, *see* alkyl chlorides
2-chlorobutadiene, *see* chloroprene
chloroethane, *25, 39*
chloroform, *15, 39,* **42**, 161
  decomposition of, 42
  dichlorocarbene from 231–2
  in haloform reaction, 161
  orthoformic esters from, 148, 195

rubber solvent, 76
  stabilization with ethanol, 42
chloromethane, *15, 39*, 42, 43
chloropentanes, 120
chloroprene, 77, 96
chloropropane, *16*, 17
chloropropyl radicals, 62
chromic acid, *see also* chromium
    trioxide
  oxidation with, 111, 152
    mechanism of, *111*
chromium trioxide, 188
civetone, **246**
Claisen condensation, 150, **194**, 237, 247
Clemmensen reduction, 154
Clostridium acetobutylicum, 119
coal, hydrogenation of, 24
coal gas, 9
coal tar, 238
coenzyme A, 220
conformation, *53*, 54, *241–2*, 244
  defined, 240
co-polymer, 77
copper acetylide, as catalyst, 94
Couper, Archibald Scott, 2
Couper formulae, 4
covalency, 2
  limitation to, 3
cracking, 17, 47, 74, *see also* pyrolysis
  catalytic, 21
  petroleum, 21, 55
  process, 22
cracking patterns, 254
cross-links, 76
crotonaldehyde, *162*
cuprous acetylide, 94
cyanhydrins, 156, 208
  aldehydes and ketones and, 155–6
  catalyst effect on formation, 155
cyanides, *see* nitriles
  formation of, 36
  hydrolysis, 176–7
  ion, as nucleophile, 35
cyclic ketones, 231, 237, 246–7, 249
cycloalkadiene, 248
cycloalkanes, 226, 229, 231
  higher, 244
  identification, **251**
  infrared spectrum, *250*
  in petroleum, 19, 230
  large rings, 246
  mass spectra, 261
  ring opening reactions, 249

oxidation *cont.*
  thiols, 217, 221–2
oxidative ring opening, 249
oximes, *156*
oxonium ions, **31–2**, 47–9, 128, 258
oxonium salts, 129, 131
'Oxo' process, **109–10**, 120, 150
oxyacetylene flame, 90
oxygen:
  atomic mass, *254*
  bonding to sulphur, 220–1
  comparison with sulphur, 215
  electronegativity of, 28, 101, 151–2, 221
  isotopes, 254
  peroxide formation, 30, 59, 152
  unshared electron pairs, 102, 129
ozone, 57–8, 249
ozonolysis, **57–8**, 148–9
  of rubber, 75–6

paint, 92, 196, 206
  solvents, 120, 191
palmitic acid, *175*, 176, 196, 197
paraffin hydrocarbons, 13–14, 19, *see also* alkanes
paraffin oil, 21
paraformaldehyde, *165*, 166
paraldehyde, *167*, 168
pelargonic acid, *175*
pentacosane, *7*
pentadecane, *7*
pentaerythritol, *167*
  tetranitrate, 167
pentane, *6*, 13
  isomers of, *7*
pentanoic acid, *175*
  derivatives, *208*
pentanol(s), *106*, 120–1, *see also* amyl alcohols
n-pentanol, 120
pent-1-ene, *55*
pent-2-ene:
  trans-isomer, spectrum, *85*
peracetic acid, *58*
perfluoro-alkanes, 33, 42, 43
perfumery, 120, 191, 245
permanganate test, 84, 97, 249
peroxides, 29–30, 152, 217
  epoxidation and, 59
  in ethers, 130, 131
  test for, 130

Perspex, *see* polymethyl methacrylate
PETN, 167
petrochemical processes, 23
petroleum, 9, 10, **19–24**
  acetylene from, 89
  alcohols from, 109
  alicyclic compounds and, 230
  alkenes in, 47
  butadiene from, 72
  catalytic cracking, 17, 21–2, 23
  cyclo-alkanes in, 229–30, 238, 243
  detergents from, 199
  ethanol from, 118
  isoprene from, 74
  olefins from, 47, 55, 109
  synthetic, **24**
  thiols in, 11, 216
petroleum chemical industry, 23
petroleum 'ether', 21
phenol, 165, 240, 243
phenylhydrazine, 156, 169
phenylhydrazone, 156
phorone, 164
phosgene, 42, 43
phosphonium salts, *55*, 165
phosphorane, Wittig reaction and, 54, 165
phosphorus halides, 32, 39–40, 115, 159–60
  acid halide preparation, 183, 200, 223
  formation of, 185
phosphorus pentasulphide, 219
phosphorus pentoxide, 183, 203, 206
phosphorus, red, 32, 43, 185
phosphoryl chloride, 114
pimelic acid, 239
pinene, *230*
pivalic acid, *164*, *177*, 182
pivalic aldehyde, *164*
plasticizers, 69, 191
platinum as catalyst, 23, 55, *56*
Plexiglass, *see* polymethyl methacrylate
poly-enes, 70–7
polyethylene, 23, **67–69**, 70
polyfluoro-compounds, 42, 43
polyhalo-alkanes, **39–43**
polyhydroxy alcohols, 121
polyisobutylene, 67, 69, 77
polyisoprene, 75–6, 77
  cis isomer, 75
  synthetic, 77
  trans isomer, 76

Aliphatic and Alicyclic Compounds

sodium metal *cont.*
  carbon tetrachloride and, 43
  butadiene polymerization, 73, 77
  drying of ethers, 127
  in acyloin reaction, 194–5, 247
  reduction of esters, 107, 194
  reduction of nitriles, 136, 208
  Wurtz reaction, 11, 36
sodium nitroprusside reagent, 223
sodium soaps, 197
sodium trichloroacetate, 232
sodium xanthates, *112–13*
solvation, **51**, *52*
spectroscopy, 18
spirits of wine, 125
starch, fermentation of, 117, 119
stearic acid, *175*, 176, *196*
  soaps and, 197
sterculic acid, *234*
stereochemistry, **53**
  cyclobutane, *236*
  cyclohexane, 238
  cyclopentane, 240–2
  fused rings, *244*
  Newman projections, *53*
  'saw-horse' projections, *53*
stereospecific, 231
steric hindrance, 155, 182, 193
  in aldol reaction, 163
  in derivative formation, 169
steroids, 220, 242
structural isomers, 5–6, 7, 10
strychnine, 139
styrene, *69*, 77
substitution reactions, *see* nucleophilic
      substitution
succinimide, *63*
sucrose, 117
sugar ethers, 126
sugars, ethanol from, 117
sulphenic acids, *221*, 222
sulphides, 217, 222
  in petroleum, 19, 23
sulphinic acids, *217*, *221*, 222
sulpholane, *73*, 222
  infrared spectrum, *224*
sulpholene, *73*
sulphonamides, *223*
sulphones, *218*, *221*, 222, 225
sulphonic acids, *217*, *221*, **222**–3, 225
  chlorides, 223
  esters, 223
  salts, 179, 199, 223

sulphonium salts, 218
sulphoxides, 218, *221*, **222**, 225
sulphur:
  alkyl derivatives of, **215–25**
    identification of, 223–5
    infrared spectra, *224*
    mass spectra, 261
  atomic mass, *254*
  comparison with oxygen, 215
  catalyst 'poison', 149
  higher valency in, 220–1
  isotopes, 254, 261
  petroleum, 19, 22, 23
  $p_\pi d_\pi$ bonding, *221*
  vulcanized rubber, 76, 77
sulphuric acid:
  acetylene and, 91, 97, 149
  alcohol dehydration, 47–50, 112, 182
  alkene addition, 63, 65, 105, 119, 120
  as acid catalyst, 24, 31–2, 58, 128, 181
  esters of, 105, 114, 125
  ether preparation, 125, 126–7, 130
  nitriles and, 205, 208
  polymerization of aldehydes, 166, 167,
      168
  test for alkenes and ethers, 84, 131
sulphur-oxygen bonds, 225
synthetic detergents, 199, 223
  manufacture, 106, 194, 206
synthesis gas, 24, 116, 186

tautomers, 160
tautomerism, 160, 162
terminal alkene, 50
terpenes, 74, 135, 230, 235
testosterone, *243*
tetra-alkylammonium hydroxides, 138
tetra-alkylammonium salts, 38, 133–4,
      135, 136
tetrabromomethane, *39*
tetrabromopropane, *75*
tetrachloroethane, *91*
tetrachloromethane, *see* carbon
      tetrachloride
tetracosane, *7*
tetracosanol, 120
tetradecane, *7*
tetradecanoic acid, *175*
tetraethyl lead, 37, 43
tetrafluoroethylene, *43*
tetrahydrobenzaldehyde, *239*
tetrahydrofuran, *131*
tetrahydrophthalic anhydride, *239*